W0090705

Springer Series in Synergetics   Editor: Hermann Haken

Synergetics, an interdisciplinary field of research, is concerned with the cooperation of individual parts of a system that produces macroscopic spatial, temporal or functional structures. It deals with deterministic as well as stochastic processes.

M. Markus S.C. Müller
G. Nicolis (Eds.)

From Chemical
to Biological Organization

With 202 Figures

Springer-Verlag Berlin Heidelberg New York
London Paris Tokyo

Dr. Mario Markus
Dr. Stefan C. Müller

Max-Planck-Institut für Ernährungsphysiologie, Rheinlanddamm 201,
D-4600 Dortmund, Fed. Rep. of Germany

Professor Dr. Grégoire Nicolis

Faculté des Sciences, Université Libre de Bruxelles,
Campus Plaine, C. P. 231, Boulevard du Triomphe, B-1050 Bruxelles, Belgium

Series Editor:

Professor Dr. Dr. h. c. mult. Hermann Haken
Institut für Theoretische Physik und Synergetik der Universität Stuttgart, Pfaffenwaldring 57/IV,
D-7000 Stuttgart 80, Fed. Rep. of Germany and
Center for Complex Systems, Florida Atlantic University,
Boca Raton, FL 33431, USA

ISBN-13: 978-3-642-73690-2   e-ISBN-13: 978-3-642-73688-9
DOI: 10.1007/978-3-642-73688-9

This work is subject to copyright. All rights are reserved, whether the whole or part of the material is concerned, specifically the rights of translation, reprinting, reuse of illustrations, recitation, broadcasting, reproduction on microfilms or in other ways, and storage in data banks. Duplication of this publication or parts thereof is only permitted under the provisions of the German Copyright Law of September 9, 1965, in its version of June 24, 1985, and a copyright fee must always be paid. Violations fall under the prosecution act of the German Copyright Law.

© Springer-Verlag Berlin Heidelberg 1988
Softcover reprint of the hardcover 1st edition 1988

The use of registered names, trademarks, etc. in this publication does not imply, even in the absence of a specific statement, that such names are exempt from the relevant protective laws and regulations and therefore free for general use.

2153/3150-543210

Preface

Open nonlinear systems are capable of self-organization in time and space. This realization constitutes a major breakthrough of modern science, and is currently at the origin of explosive developments in chemistry, physics and biology.

Observations and numerical computations of nonlinear systems surprise us by their inexhaustible and sometimes nonintuitive variety of structures with different shapes and functions. However, in addition to this variety one finds upon closer inspection that nonlinear phenomena share universal aspects: (1) kinetic and thermodynamic principles; (2) global behaviour, for instance the routes to chaos; and (3) novel experimental approaches, for example methods to determine dimensions of attractors reconstructed from time series.

The universal aspects shared by a variety of nonlinear phenomena in time and space make it possible to bridge the gap between inanimate and living matter at various levels of complexity, both in theory and experiment. It is our hope that the present book, which contains a wide range of examples from very diverse areas of scientific research, provides convincing evidence of the emergence of nonlinearity as a new scientific paradigm.

Benno Hess, head of the Max-Planck-Institut für Ernährungsphysiologie, and truly a pioneer in the investigation of nonlinear phenomena in chemistry and biology, celebrated his 65th birthday on February 22, 1987. On this occasion, Ilya Prigogine presented a colloquium in Dortmund on the foundations of irreversibility and sent us a manuscript (in collaboration with E. Kestemont and M. Mareschal) which is included in the present volume. This anniversary was followed by a meeting which gathered 150 scientists from 16 nations at the Max-Planck-Institut in Dortmund, March 16–19, 1987. The meeting was devoted to regulation in bioenergetics and to nonlinear phenomena in the biosciences. The most impressive feature of this event was that chemists, biologists, ecologists, physicists, mathematicians and medical doctors found themselves speaking the same language – and talking with each other. One could realize that nonlinear phenomena are currently abolishing the scientific Tower of Babel. Inspired by this exciting atmosphere, we made plans to edit this book and invited several participants to send us contributions from their area of research. Subsequently, we asked other scientists to write articles that would further strengthen this feeling of universality within the phenomenal variety of self-organizing systems in chemistry and biology. With this volume, we also intend to express our gratitude to Professor Hess for the stimulation that he has given to so many of us. We are grateful to scientists of highest distinction who took the time to prepare a contribution, honouring him with their work.

Spontaneous emergence of order does not apply to international conferences. Without Theo Plesser's efficient organization the conference would never have been realized in such a successful manner. We wish to thank him for his deep involvement. We also acknowledge the financial support of the Instituts Internationaux de Physique et de Chimie, fondés par E. Solvay, Brussels, for the preparation of this book. Special thanks are due to Ms. Helga Wagner for her assistance before, during and after the conference, as well as to Ms. Bettina Plettenberg, Ms. Angelika Rohde, Ms. Christine Riemer and Ms. Ina Wilms for typing most of the manuscripts. Finally, we thank Dr. H. Lotsch, Ms. D. Hollis and Ms. J. Meyer of Springer-Verlag for their patience and publishing efficiency.

Dortmund, Brussels
February, 1988

M. Markus
S.C. Müller
G. Nicolis

Contents

Part III   Biochemical Organization

Part IV   Cellular and Intercellular Organization

Part V   From Complex Cellular Networks to the Brain

Part VI   Ecological, Epidemiological and Economical Organization

Part I

General Concepts

From Chemical to Biological Organization: A Snapshot

E. Katchalski-Katzir

Department of Biophysics, The Weizmann Institute of Science, Rehovot 76100, Israel

This lecture was delivered at a Dinner held on March 19, 1987, in honor of Professor Benno Hess on his 65th birthday:

I am delighted to be here this evening, Benno, to join you in celebrating your 65th birthday. From your many students, collaborators and friends here and abroad I bring you greetings and good wishes for continued good health, happiness and success in your work. At the Symposium held in your honor, we devoted many hours to stimulating discussion on the scientific topics closest to your heart. This is the way that we as scientists can best pay tribute to a colleague for whom we have the highest regard, both as a scientist and as a human being.

When I was still at an early stage of my own career I had already perceived that a true scientist is one who achieves three main objectives in his lifetime: firstly, he should carry out original research that contributes significantly to our understanding of nature and natural phenomena; secondly, he should act as an inspiration and mentor to his students, while training them to follow in the best traditions of scientific investigation and scholarship; and thirdly, he should use his influence as an outstanding scientific personality to help build up a center of learning and research, where the finest scientific minds will come together and address themselves to problems and activities of common concern.

You, Benno, have achieved all three of these objectives with distinction. Your original work, both theoretical and experimental, on the mechanisms involved in the control of energy transducing proteins and on nonlinear phenomena in the biosciences deals with some of the most basic characteristics of living organisms. You have acquired a following of talented students, who are continuing to extend your own findings. As Director of the Max-Planck-Institut für Ernährungsphysiologie for over 20 years you have succeeded in creating a center of excellence here in Dortmund that attracts young scholars from all over the world.

I have followed with admiration and keen interest your research work on bacteriorhodopsin. I am fascinated by your synoptic views on its photochemical reaction cycle, your findings on light-driven protonation changes of internal aspartic acids of bacteriorhodopsin, and your observation that electric field-induced conformational changes occur in bacteriorhodopsin when embedded in purple membrane films. Your work in these fields sheds new light on some of the most basic concepts involved in bioenergetics. Moreover, you are to be congratulated for the meticulous work being done in your laboratory on proteins involved in energy transduction _in vivo_.

Bioenergetics is only one of your areas of interest, however. Another is nonlinear phenomena in the biosciences, to which your attention was directed as a result of developments in irreversible thermodynamcis and in the understanding of nonlinear phenomena in chemistry and physics. Your remarkable analysis of the factors and conditions under which nonlinearity prevails in living organisms led to your demonstration that under conditions far from equilibrium living systems can achieve multiple steady states, oscillatory states and chaotic states. Your fine work on reaction oscillations in glycolysis and the peroxidase reaction will no doubt become a classic in the field.

Also of great interest are your recent studies on order and chaos in biochemistry. You have shown that oscillating dynamics play a prominent role in biological clock functions, in inter- and intracellular signal transmission and in cellular differentiation. Furthermore, you have pointed out that oscillations having no recognizable frequency (chaotic oscillations) have been observed in enzymic systems; and that these chaotic oscillations may be useful in enhancing biological viability and may even be related to pathological phenomena. In short, you have opened up a new field of theoretical and experimental importance in biology, a field in which order and chaos will be studied and compared simultaneously in various biological processes.

The Hess family has no shortage of artistic talent; this may well account for Benno's fascination with the core structure of spiral-shaped travelling waves of chemical activity appearing in a thin excitable layer of the Belousov-Zhabotinskii reaction, in which the oxidation and decarboxylation of malonic acid by bromate ions is catalyzed by ferroin. Benno was so taken by the exquisite patterns obtained as a result of these complex nonlinear chemical reactions that he had some of them reproduced for publication in an artistic calender.

As his friends and colleagues are no doubt aware, Benno's interests are by no means restricted to science but extend also to social problems in general and particularly to questions related to medical bioethics. In a lecture entitled "Man and Animal" delivered in 1985, he stressed that the value of science lies not only in its practical nature, but that one of its highest aims is to broaden our understanding of man and society. Science is thus an integral part of human culture. Indeed, Benno himself serves both science and society, in his various capacities as Vice-President of the Max-Planck-Gesellschaft for the Advancement of Science and as a member of the German Academy of Science Leopoldina, the Heidelberg Academy of Science, and the Instituts Internationaux de Physique et de Chimie Solvay, Brussels.

Dortmund, this green and pleasant island in the heart of the Ruhr district, is a city that in 1945 lay in ruins. Through the devoted efforts of liberal-minded men and women the city has been rebuilt and its cultural and scientific life reborn. Benno is one of those who have played a central role in promoting the manifold activities which enrich the quality of life.

At the Symposium organized in honor of our esteemed colleague, we as members of the international scientific community are again reminded that science represents a universal language. Scientists the world over share the same aims - to contribute through the search for truth to the wellbeing of all mankind. In this sense one can truly say that we are one family.

As an Israeli scientist and a former President of the State of Israel, I bring you warm greetings from your friends and well-wishers in my country. Scientists in Israel look back with pleasure to your visit in November 1986 when you delivered the Aharon Katzir-Katchalski lectures at the Weizmann Institute. You took as your general theme "Biological Foundations of Human Behavior", and it was a moving experience for me to realise that you share the hope of my late brother, Aharon, "that the understanding of living organisms will provide man with a way of better understanding himself, and that such understanding will suggest relationships between human biology and ethical behavior." On that occasion, while following your stimulating discussions of complex dynamics in chemistry and biology on the mechanism and control of ordered and chaotic states, I recalled the saying in the Babylonian Talmud: "A scholar takes precedence over a king of Israel; for if a scholar dies no one can replace him, while if a king dies, all Israel is eligible for the kingship."

We in Israel have special reasons for regarding you, my dear friend, with affection and esteem. For you are remembered, among a small band of German scientists, along with Gentner, Lynen, Straub, Eigen and others, who have helped to reconstruct the bridges of scientific contact between post-war Germany and Israel. Through the devoted labors of this distinguished group, the lines of communication have been built up and strengthened over the years, leading to the establishment of highly valued professional and personal contacts on both sides.

The distinguished physicist Isidore Isaac Rabi, on looking back over the work of a lifetime, had this to say: "Physics brought me closer to God. That feeling stayed with me throughout my years in science. Whenever one of my students came to me with a scientific project, I asked only one question: 'Will it bring you nearer to God?' They always understood what I meant."

You, my dear Benno, through your explorations and revelations of some of the most elemental phenomena of nature, can truly be said to have brought us nearer to God. May you be blessed with a long and healthy life, in which you will continue to stand as an example to us all.

Pattern Formation: Thermodynamics or Kinetics?

H. Haken

Institut für Theoretische Physik und Synergetik, Universität Stuttgart,
Pfaffenwaldring 57, D-7000 Stuttgart 80, Fed. Rep. of Germany

1. Introduction

Undoubtedly, contemporary science is becoming more and more hectic and more and more diverse. Thus, the 65th birthday of a scientist is a good occasion to relax and to try to regain a wider scope of our views. Among the problems dealt with by Benno Hess were the glycolytic cycle, chemical oscillations in slime mold, and pattern formation in the Belousov-Zhabotinski reaction. Besides their intrinsic interest, these problems are of utmost importance for our understanding of biological systems and for morphogenesis. But there is a still wider scope, namely these experimental and theoretical studies are of importance for our understanding of complex systems and how to treat them adequately by theory. The spatial patterns and oscillations studied in these experiments occur spontaneously, i.e. they are not forced upon the system by specific controls from the outside. Rather we are dealing here with phenomena of self-organization.

In the following I wish to discuss what general methods or concepts we have nowadays to cope with these problems. At first sight it might appear that scientists agree upon the general methods to be used here, but as a recent controversy (cf. [1]) to which I will come back later in this article demonstrates quite clearly, there is still disagreement even about quite fundamental aspects. I do hope that my article will help to clarify the present status of the theoretical approaches to complex systems. It seems to be in order to start with a quite general statement. Namely, in chemistry or physics we must not forget that these sciences are after all inductive, and that any theory must be checked against experiments. This is the only way we can gain, at least to some extent, security about our view of the world. We have two methodologies at hand, namely the macroscopic or phenomenological approach and the microscopic approach which attempts at deriving macroscopically observed phenomena from basic equations which in the case of chemistry must eventually be the basic physical laws established by quantum mechanics and electrodynamics.

I quite intentionally exclude here thermodynamics as being basic for reasons which will become clear, I hope, in the course of my article. The most salient macroscopic approaches in chemistry, physical chemistry, and physics are certainly thermodynamics on the one hand and the kinetic theory on the other hand. According to the ideal of western science, we wish to treat macroscopic phenomena by decomposing systems into their individual parts, and then we attempt at deriving the laws of thermodynamics or kinetic theory from first principles where perhaps the latter seems to be still closer to the microscopic level. The question we wish to discuss is whether thermodynamics, kinetic theory, and statistical theory are equivalent, or whether one is superior over the other. Or in other words, we shall discuss the question how fundamental are the laws of thermodynamics in its present state. As we shall see below, time and again it will be very important to treat simple model systems against which the various concepts and methods can be checked.

But now let us start with basic ideas of thermodynamics. To be quite clear and explicit, I shall record a number of features of the thermodynamic approach in order to elucidate on the one hand its power, but on the other hand also its limitations. Thermodynamics being a macroscopic approach, we first have to introduce adequate macroscopic variables which describe a system. Such variables may be volume, pressure, temperature, concentrations of molecules etc.. A basic concept is that of thermal equilibrium. When we consider an isolated or closed system, we know that it moves towards a unique state, namely that of thermal equilibrium. When the system is disturbed in that state it will return to it. Clearly the concept of an isolated or closed system is an idealization. Whenever we observe a system we interact with it by light, or pressure, or in other ways so that it is brought away from thermal equilibrium. But in general we assume in contradistinction to the microscopic events treated by quantum mechanics that such perturbations are unimportant.

The concept central to thermodynamics is, of course, entropy. What I stress from the very outset is that we have to strictly stick to operational definitions. According to the fathers of the concept of entropy, it is defined by

$$dS = \frac{dQ_{rev}}{T} , \qquad (1.1)$$

where dQ_{rev} is the reversibly added or substracted heat, whereas T is the absolute temperature. According to the original definition we are dealing at all stages with systems in thermal equilibrium which is, in particular, retained by guiding all processes infinitely slowly. We note that introducing temperature T leads us to delicate questions. Of course, any system in thermal equilibrium has such a uniquely defined temperature. But for instance living objects such as humans have also a specific temperature, though they are not in thermal equilibrium. But an arbitrary system does not need to possess a unique temperature as we shall see below. Equilibrium states can be also connected by irreversible processes where in an isolated or closed system

$$dS \geq 0 \qquad (1.2)$$

holds. The next definition which then bridges the gap between microscopic and macroscopic theory is due to Boltzmann with the famous formula

$$S = k \ln W , \qquad (1.3)$$

where W is the number of microscopic realizations of a given macrostate. Then, it is stated that a system is in thermal equilibrium once S has become a maximum. According to Boltzmann's understanding who actually primarily treated gases, maximum entropy is equivalent with maximal disorder, or in other words with maximal microscopic chaos. On the other hand on the macroscopic scale the system is homogeneous. Or in other words, any spatial inhomogeneity has a smaller probability than the homogeneous state. Boltzmann's principle was carried further by Gibbs who, in particular, stressed the role of constraints.

Finally, we have to deal here with the concept of Shannon entropy or Shannon information [2] which is defined by

$$S = - \sum_j p_j \ln p_j . \qquad (1.4)$$

Here, the j's characterize specific states over which the system can be distributed and p the corresponding probabilities. Jaynes [3] has formulated a quite general entropy principle by the idea of making unbiased guesses on incompletely known systems. According to this idea (1.4) must be maximized under given constraints. In thermodynamics the given constraints are rather obvious, namely they are based on the conservation laws of energy, particle numbers etc.. As it appears to me these constraints are generally adopted and agreed upon. I stress this point in particular because below we shall be confronted with a different situation. As we know from statistical physics, a number of theorems

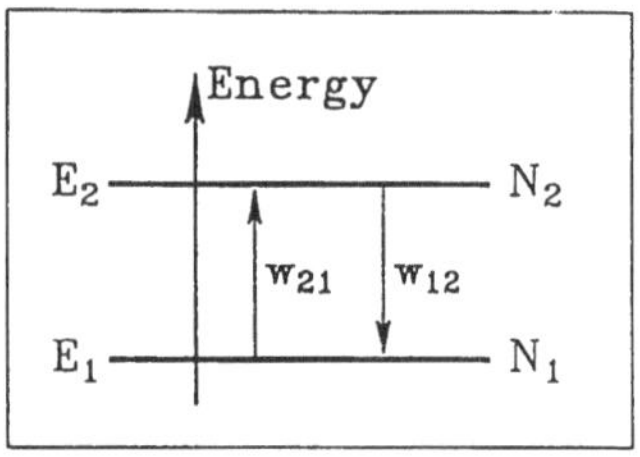

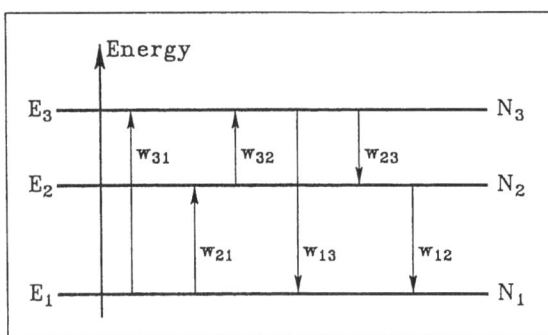

Fig. 1: Scheme of transitions in a 2-level atom

hold for systems in thermal equilibrium. For instance the equipartition theorem according to which each degree of freedom carries $1/2$ kT as energy.

Then the principle of detailed balance holds. I explain it in the following manner: Let us consider the very simple example of a single atom which is coupled to a heat bath. If this atom has only two levels (cf. fig. 1) then we shall have transitions from the lower state to the upper state and vice versa. Let us consider these rates in detail and let us denote the transition probabilities from state 1 to 2 by w_{21} and for the transition from 2 to 1 by w_{12}. Then depending on the actual occupation numbers N_1 and N_2, detailed balance requires the relation

$$w_{21} N_1 = w_{12} N_2 . \qquad (1.5)$$

But because the electron of a two-level atom can be in only one of the two states and the sum of N_1 and N_2 is equal to unity, (1.5) is a triviality. Things change, however, when we consider a three-level atom (cf. fig. 2). Here detailed balance requires that the transitions between each pair of levels occur with the same rate from one level to the other as well as vice versa. Thus we have quite generally

$$w_{ki} N_i = w_{ik} N_k. \qquad (1.6)$$

In conclusion we may note that by means of Jaynes' principle in connection with the adequate constraints one may recover a number of well-known results of thermodynamics, e.g. the relation between internal energy, free energy and entropy, where the absolute temperature appears as an inverse Lagrange parameter. The crucial point is, of course, in how far these results apply to systems away from thermal equilibrium.

2. A reminder of irreversible thermodynamics

In its classical form irreversible thermodynamics deals with transport processes, e.g. Fick's law, and relaxation processes, e.g. spin relaxation. In this area a number of theorems had been derived, e.g. the

Fig. 2: Scheme of transitions
in a 3-level atom

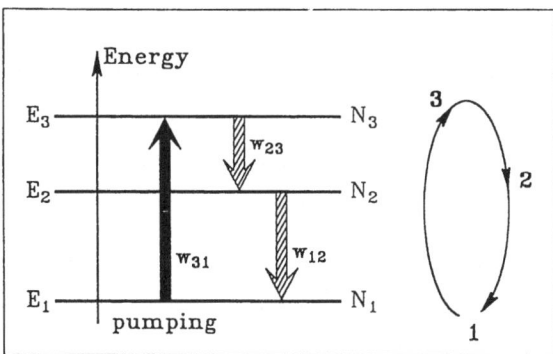

Fig. 3: Scheme of transitions in a 3-level atom without detailed balance

dissipation-fluctuation theorem. It is assumed that the principle of detailed balance still holds by means of which one may derive the Onsager reciprocity relations. Finally, Prigogine's minimum entropy production principle holds here. Again, entropy plays a central role in this field. There are relationships for the production and transport of entropy. Above all entropy is considered as a potential, the derivative of which with respect to the variables of the system can be considered as a driving force to thermal equilibrium. It may be noted that transport and relaxation can be treated also by means of a microscopic theory using concepts of quantum or classical statistics. Here we may mention as examples heat conduction, electric conduction, or relaxation phenomena, in particular in spin physics. These theories allow, in particular, an at least in principle microscopic calculation of relaxation times.

3. Far from equilibrium

The basic approaches dealing with systems far from thermal equilibrium are synergetics on the one hand [4] and the theory of dissipative structures on the other hand [5]. Both theories have a similar goal, though their starting points were rather different. Synergetics started from physics, in particular laser physics, while the theory of dissipative structures started from chemical and bio-chemical systems. But I believe there is now a considerable convergence of ideas. At any rate we are dealing here with open systems i.e. with systems which are maintained far from equilibrium by a flux of energy and/or matter through them. The first question we wish to discuss is, can one define entropy S for such systems? And if there are discrepancies between different results on the definition of S, how to decide which one is correct.

I wish to give the reader the impression that systems far from equilibrium present a rather strange world, at least for those who have been acquainted with the laws of thermodynamics. First of all we immediately run into difficulties when we wish to define temperature. Let us consider a still rather simple example, namely hot carriers in a crystal. Then the temperature of the carriers, i.e. the electrons, is much higher than that of the atoms of the systems. Quite evidently the equipartition theorem is lost. Or let us consider another example, namely the occupation numbers of a 3-level atom which is now coupled to reservoirs causing transitions between different levels. One can easily construct pumping processes by which one can obtain prescribed occupation numbers of the levels 1,2, and 3, and we immediately recognize that they are no more connected by any type of Boltzmann distribution function, i.e. we can no more define any unique temperature.

Let us consider figure (3) showing again a three-level atom pumped from the outside. Then here detailed balance is lost and one may have a circular process as indicated on the r.h.s. of that figure. This example shows quite clearly that systems far from thermal equilibrium need not show detailed balance. Rather detailed balance is an exceptional case. Furthermore we mention that inspite of many efforts, there is no straightforward extension of the dissipation-fluctuation theorem valid for systems away from thermal equilibrium. Finally we realize that the classical definition of entropy by thermodynamics according to (1.1) is no more possible. First of all temperature T is, in general, no more defined. But what is still more important, in (1.1) we have to assume that the processes go on infinitely slowly. But let us consider structures which are quite adequately called dissipative structures. Here dissipation tries to bring the system back to thermal equilibrium. But this dissipation is fought against by pumping energy or matter into the system, i.e. a competition is going on between processes in which energy is fed into the system and those in which energy is dissipated. That means in other words that now time constants (or rate constants) play a decisive role.

These structures can be maintained only by processes going on in finite times. This is quite clearly exhibited by the approach of synergetics which we shall indicate below. But to anticipate its main result, in systems driven away from equilibrium there is an establishment of different collective modes which compete with each other. Some of them can use the input of energy and/or matter better than others, and those modes determine the macroscopic behavior of the system. In other words, energy is now concentrated in a few modes so that quite evidently the equipartition theorem is invalidated. These modes are like surface waves on a deep sea. Quite clearly their contribution to entropy is entirely negligible compared to that of the deep sea. But on the other hand it is these modes which determine the spatial or temporal patterns.

Let us now consider an often quoted relation

$$dS = dS_i + dS_e \ , \qquad\qquad (3.1)$$

where dS is the total entropy change, whereas dS_i is the change of internal entropy and dS_e the change by the coupling of the system to the external world. It was assumed by some authors that (3.1) decreases when an ordered state is reached in a non-equilibrium phase transition. But there are two difficulties. The first one is that according to what I have said before, a thermodynamic definition of S or dS is impossible. Therefore we must resort to a microscopic definition. But then the question arises what the proper constraints, for instance in Jaynes' principle, are in order to derive S, S_i, S_e. In order to arrive at a decision let us consider the microscopic theory. Such a theory has been developed for instance in laser physics where one starts from the fundamental quantum mechanical equations [6]. Then after a number of transformations one establishes a Fokker-Planck equation which, in turn, allows us to calculate the distribution function p_i. These distribution functions have been checked in great detail in marvellous experiments by a number of experimental laser physicists and excellent agreement was found (see,eg.[7]). Therefore we are sure that the distribution functions are correct. But by means of these distribution functions we may form the expression for the entropy S according to (1.4). In other words we may calculate the entropy of laser light or, as we shall see below, rather the information of laser light. The question now arises, can we derive S according to (1.4) by means of Jaynes' principle? The answer is yes, provided we use constraints which are entirely different from those of equilibrium thermodynamics [8].

Namely, the observables are now correlation functions of the intensity and intensity fluctuations. We could show that these correlation functions are capable of reproducing the distribution functions according to Jaynes' principle for a wide class of experiments, namely those on non-equilibrium phase transitions. In these phenomena patterns are formed by self-organization in

10

systems far from thermal equilibrium. Several authors including myself have found that [9], [10]

$$dS > 0 \qquad\qquad (3.2)$$

holds, when a system is driven beyond its instability point so that it acquires a structure. This result is quite in contradiction to the expectation expressed by dS < 0 which at first sight seems to be intuitively appealing. The reason has been elucidated by Klimontovich [10]. In fact he shows that while (3.2) holds, also the total energy is increasing. In other words we are not comparing two states of equal energy as it would be so in a closed system, but we are dealing with systems having quite different energies. When we renormalize the system so that we can define the same energy then dS < 0 holds again [10]. I think this example shows quite clearly how careful we must be with the concept of entropy in system far from equilibrium and this is one of the reasons I have proposed to call S as defined by (1.4) in non-equilibrium systems _information_. In fact it can be shown that for instance laser light can store the information as expressed by S. At this moment I have to discuss a possible criticism which has been raised time and again against Jaynes' principle. Namely, this criticism precisely concerns the question of the adequate constraints. It is claimed that the choice of constraints introduces some kind of subjectivity into the theory.

This is a point which requires some careful discussion which I will sketch here only briefly, however. This discussion has, on the one hand, a philosophical aspect, but on the other hand a very practical one. Let us recall Popper's idea [11] that science proceeds by trial and error and that it cannot prove theorems in the natural sciences but rather can only falsify them, and then proceed to theorems which seem to be more adequate. In other words, science is continuously undergoing a learning process, and this is precisely the situation in which we are in the present moment what the formulation of a macroscopic theory for non-equilibrium systems is concerned. We all agree that in systems in thermal equilibrium the adequate constraints are provided by the conservation laws. Such a general consensus has not yet been reached in systems far from thermal equilibrium. Here I have shown that in fact the constraints on correlation functions mentioned above are adequate for the class of non-equilibrium phase transitions leading to spatial patterns [8]. Therefore, the question of finding adequate constraints is twofold. First of all we have to find a consensus on which class of phenomenon we are talking, and secondly we have to find a consensus what the adequate constraints are. Quite evidently in this way subjectivity is eliminated by this consensus, and the only way to break this consensus is to falsify the statement. But in the present case falsification just means that in such a case one is no more dealing with the class of non-equilibrium phase transitions, but rather with a different class of phenomena. Or in other words, I think the proper extension of prior concepts of thermodynamics to non-equilibrium systems consists in finding and defining more and more classes of common behavior. So eventually, thermodynamics is becoming a special case concerning thermal equilibrium.

Or to put the statement differently, we now recognize that a number of relations of classical thermodynamics are no more applicable to systems away from thermal equilibrium but must be replaced by new relationships, part of which can be derived by Jaynes' principle under the adequate constraints.

So let us eventually discuss the role of kinetic equations which in a way have a middle position between a purely microscopic and a purely macroscopic thermodynamic approach. When one wishes to derive kinetic equations, at least in principle, one must proceed in the following steps. One starts from the laws of quantum mechanics from which at least in principle we may derive the binding forces of molecules, or scattering cross-sections etc.. Then, for instance by applying the golden rule of quantum mechanics, in a way irreversibility is introduced. In this author's opinion the question of a proper derivation of irreversible equations from reversible ones is somewhat overstressed. Let us discuss this point by means of the example of the recurrence time. Even if a system has a finite recurrence time, in practice it is so large that it will never occur in any practical experiment.

Therefore the question of recurrence times may be of interest for discussions on cosmology which, in my opinion, are still somewhat idle because we still have too little information on the properties of the universe and its basic laws. So I take here a rather pragmatic point of view, namely that for any practical purposes it is quite sufficient to apply any of the well-developed methods to derive irreversible processes. Then taking into account appropriate dephasings we may form averages which then are the basis of kinetic equations. As was exemplified by the atomic system but can be equally well demonstrated for chemical systems, in general detailed balance is lost, in particular when the systems are open, and detailed balance can be claimed only to hold for systems in/or close to thermal equilibrium. A very important point is the dependence of the kinetic equations on external parameters, namely the influx (or non-influx) of energy and/or matter. There are quite different regimes of behavior, for instance behavior close to or in thermal equilibrium, or behavior far away. What I have said with respect to detailed balance and the parameter dependence, just reiterates a statement which was made by a number of authors many times, but quite clearly in a recent discussion of Leefver et al. [1] in rebuttal of papers by Cray et al. [12]. Actually this rebuttal was one of the reasons that I felt encouraged to write this contribution in order to help to clarify a number of points which seem to be still controversial, but which on the other hand make the field of non-equilibrium phenomena so exciting.

In conclusion I wish to sketch the approaches which were developed mainly in the context of laser theory, but which are now being rather generally adopted. Here we have two approaches, namely the microscopic (or mesoscopic) approach and the macroscopic approach. In both cases a system is described by a set of adequate variables which are lumped together to a state vector, $\underline{q}$. In the first case we have either equations which are derived from a truly microscopic theory or from kinetic equations which are in general of the form

$$\underline{\dot{q}} = \underline{N}(\underline{q}, \alpha) + \underline{F}(t). \qquad (3.3)$$

The non-linear function $\underline{N}$ may contain reaction terms as well as diffusion or other processes, while $\underline{F}(t)$ represents fluctuating forces which are of importance when the system undergoes a non-equilibrium phase transition. $\underline{N}$ depends on one or a set of control parameters, α. When a control parameter is changed, a previous stable solution $\underline{q}_0$ may become unstable. The stability analysis then reveals a set of modes with increasing amplitudes which serve as order parameters and modes which are still damped. According to the slaving principle of synergetics [13], the damped modes may be eliminated so that one then eventually deals with a closed set of order parameter equations. Thus the dynamics of a system far from thermal equilibrium is governed by few variables, namely the order parameters, at least close to transition points. This then makes it possible not only to classify various kinds of spatial or temporal behavior, but also to calculate evolving patterns in physics, chemistry and biological model systems. Going over to a Fokker-Planck equation one may derive the distribution functions in the steady state, at least, in a number of important cases such as the non-equilibrium phase transitions mentioned above. In this way one may calculate the information (or entropy). As it turns out, close to non-equilibrium phase transitions, information can be decomposed into that of the order parameters which may change dramatically and those of the slaved modes which change very smoothly [8].

Let us comment on the macroscopic approach. For the class of non-equilibrium phase transitions we may also calculate S by means of Jaynes' principle using adequate constraints. While the approach by kinetic equations is rather general, I think a good deal of work must still be done to make the maximum information entropy principle available for still broader classes, for instance those referring to deterministic chaos. All in all I hope to have demonstrated that systems far from equilibrium are a fascinating topic of research and that we are still far away from having here a complete theory of the same range as thermodynamics happened to be for systems in thermal equilibrium. But I think in the field of synergetics and related fields a considerable progress has been made over the past decade or so.

References

1. R. Lefever, G. Nicolis, P. Borckmans: The Brusselator: it does oscillate all the same, preprint 1987

2. C.E. Shannon, W. Weaver: The Mathematical Theory of Communication (Univ. of Illinois Press, Urbana 1949)

3. E.T. Jaynes: Phys. Rev. 106, 4, 620 (1957), Phys. Rev. 108, 171 (1957)

4. H. Haken: Synergetics. An Introduction (Springer, Heidelberg, New York 1977, 3rd ed. 1983)

5. G. Nicolis, I. Prigogine: Self-Organization in Nonequilibrium Systems (Wiley, New York 1977)

6. H. Haken: Laser Theory (Springer, Heidelberg, New York, 2nd corr. printing 1984)

7. see e.g. F.T. Arecchi in: Quantum Optics, Proceedings of the 1967 Varenna School (Academic Press, New York 1969)

8. H. Haken: Z.Phys. B 63, 487 (1986)

9. H. Haken: Z.Phys. B 62, 255 (1986)

10. Yu.L. Klimontovich in: Lasers and Synergetics, ed. by R. Graham and A. Wunderlin, (Springer, Heidelberg, New York 1987)

11. K. Popper: Conjectures and Refutations: The Growth of Scientific Knowledge (Routledge Kegan Paul, London; Basic Books Inc., New York 1963)

12. B.F. Gray, Morley-Buchanau in: J.Chem.Soc., Faraday Trans. 2, 81, 77 (1985)

13. H. Haken: Advanced Synergetics (Springer, Heidelberg, New York 1983)

Spatiotemporal Organization in Biological and Chemical Systems: Historical Review

J.J. Tyson[1] *and M.L. Kagan*[2]

[1]Department of Biology, Virginia Tech, Blacksburg, VA 24061, USA
[2]Department of Organic Chemistry, Hebrew University, Jerusalem, Israel

1. Introduction

The ability to tell time and location is characteristic of living organisms. We are all aware of our daily cycle of sleep and wakefulness, connected to underlying circadian rhythms of body temperature and hormone levels. Equally familiar to us are the relentless beating of our heart and the slower rhythm of breathing. Proper function of many of our organs depends not only on temporal periodicity but also on spatially organized behavior. For example, during one heart beat, a wave of muscular contraction must pass over the surface of the ventricle at the proper speed and in the right direction so as to pump blood efficiently through the aorta. Similarly, proper digestion relies on the correct tempo and direction of peristaltic waves in the intestine. Surely the most dramatic example of clocks and maps in biological systems is provided by the developing embryo, as the *bauplan* of a complicated multicellular organism unfolds in a spectacular display of cellular differentiation and morphogenetic movements strictly coordinated in space and time. To account for the essential, intrinsic spatiotemporal organization of biological organisms in terms of the basic principles of biophysical chemistry, a new approach to biology is necessary.

In the last century biology was exclusively an empirical science of organisms, concerned primarily with the description and classification of the great diversity of living things. Anatomy, physiology, development, genetics, evolution, phylogeny ... these were the triumphs of nineteenth century biology. In the first half of this century the biochemical basis of life was pursued with great intensity and success. Fermentation, oxidation, biosynthesis, enzymes, substrates, energy transduction ... these molecular details gave us deep insight into the mechanics of a living cell. In 1953, however, attention shifted to macromolecular structure and function with the discovery of the DNA double helix and the clear challenge posed by the information coding and decoding problem. Fascination with structure dominates modern biology, and rightly so, considering the success of this approach and the mind-bending opportunities provided by the possibility of engineering the genomes of ourselves and other organisms. But the structure/function paradigm of modern molecular biology is basically a static view of life. Time and space, rhythm and pattern, clocks and maps are missing from this conception of the machinery of life. The study of the *dynamics* of biological systems provides the new approach necessary to understand life's rhythms and patterns.

2. Historical Setting

Just as the static structural view of life is based ultimately on the principles of chemical bonding, so the dynamical view is rooted in the emerging understanding of temporal oscillations and spatial symmetry breaking in chemical reaction systems. The study of spatiotemporal organization in chemical systems has had a stormy history which delayed the development of dynamical theories in biology.

As early as 1828 FECHNER [1] reported on fluctuations in the potential of an iron electrode in a weak acid solution. Nearly 100 years later HEDGES and MYERS [2] reviewed the field of nonlinear electrochemical oscillations and pattern formation, and remarked, "It may be that biological periodicities have their origin in a physical property of matter..." In 1872 another remarkable heterogeneous phenomenon, the beating mercury heart, in which a drop of mercury exhibits pulsating geometrical distortions, was reported by LIPPMANN [3]. Electrochemical oscillations, surface-chemistry effects, and membrane phenomena have been studied regularly and thoroughly since then [4].

Stationary spatial patterns of periodic precipitation were first reported by LIESEGANG [5] in 1896. This spontaneous breaking of spatial symmetry caught the imaginations of many, and hundreds of papers have been published on the subject of Liesegang rings [6].

In combustion chemistry, temporal oscillations and traveling waves of combustion have long been known. Lord RAYLEIGH [7] in 1928 investigated the pulsating glow of phosphorus. Oscillations were found to last up to a week, and waves of light could be observed to travel along glass pipes for distances of several centimeters. Since the 1930s much fundamental work has been done in the field of thermokinetic oscillations and combustion waves [8,8a].

Though temporal and spatial periodicities in heterogeneous systems seem to have been readily accepted, such was not the case for sporadic reports of similar phenomena in homogeneous liquid-phase chemical reactions. In 1906 LUTHER [9] described experiments on the propagation of chemical waves in solutions of various oxidizing and reducing agents and correctly attributed the motion of these waves to the autocatalytic production of a chemical intermediate and its subsequent diffusion to regions ahead of the wave. Luther speculatively linked such traveling chemical waves to nerve impulse propagation, an analogy that is wrong in detail but right in principle. Despite Luther's enthusiasm for the subject in 1906, it was not pursued by himself or anyone else for 60 years [10].

One of the first claims for a homogeneous oscillating reaction was made by MORGAN [11] in 1916, for the periodic formation of carbon monoxide by the action of sulphuric acid on formic acid. The accepted explanation for the periodic frothing of the solution is supersaturation, although Morgan also reported that when sulphuric acid is replaced by nitric acid there appear to be blue oscillations within the solution as well as periodic bubble formation. Possibly because of the dangers of measuring pressure changes of carbon monoxide gas or because of the perceived impossibility of oscillations in homogeneous liquid phase, the Morgan reaction received little attention.

Ironically, the discovery by BRAY [12] of a seemingly authentic homogeneous periodic chemical reaction proved to be the end of the study of such phenomena in solution chemistry for nearly forty years. Bray found that when two reactions, involving hydrogen peroxide as an oxidizing agent in one and as a reducing agent in the other, are coupled, temporal periodicitiy is seen in the release of oxygen and in iodine coloration. Further work was published with LIEBHAFSKY [13] who, strangely enough, left the field to return nearly 40 years later picking up exactly where he left off. This "dropping-of-the-subject" was indicative of the general disbelief with which the result was received. RICE and REIFF [14] claimed that the observed periodicity was due to impure reagents and dust particles in the solution, implying that the entire phenomenon was an artifact of poor chemistry. Oscillations in homogeneous chemical reactions were thought to be proscribed by the second law of thermodynamics.

Theoretical studies on the possibility of periodicity in homogeneous chemical systems were undertaken by LOTKA [15] in the early twentieth century. By coupling two autocatalytic reactions, he demonstrated theoretically that reaction kinetic equations can have solutions corresponding to sustained oscillations in chemical concentrations. But Lotka did not follow up this line of research. Rather, he joined the Metropolitan Life Insurance Company and pursued his interest in demography. In the 1940s and 50s, BONHOEFFER, FRANCK, and others developed theoretical tools for understanding oscillations, primarily in heterogeneous reaction systems [4]. They emphasized the concepts of autocatalytic, positive and negative feedback, excitability, and limit cycles.

The next major theoretical advance was made by TURING [16], who showed that reaction-diffusion equations can admit solutions that are spatially periodic. That is, the spatially homogeneous solution can lose stability and develop into a stationary spatial pattern. Using the first digital computer (that he helped design), Turing simulated his model, producing solutions that he likened to zebra stripes, leopard spots, leaf whorls, etc. Turing's work on the chemical basis of morphogenesis is remarkable for its uniqueness, not seeming to come from anywhere and not seeming to be taken anywhere.

The last chapter of this story of rejection, disbelief and false starts was written in Russia in the 1950s, when BELOUSOV discovered temporal oscillations in the oxidation of citric acid by acidic bromate in the presence of cerium ions. His careful studies of this reaction were repeatedly refused publication, and so he gave up his study of the reaction that would eventually carry his name [17,18].

In the 1960's the climate of opinion began to warm to the idea of spontaneous spatiotemporal organization in homogeneous liquid-phase chemical reactions, and to the idea that such organization underlies certain biological phenomena. This change of opinion came about as well-documented experimental examples became known and as a theoretical framework for the field became established. On the experimental side, some important advances were:

1. the observation of damped oscillations in the concentrations of various intermediates in the dark reactions of photosynthesis [19], and of cyclic changes in volume and respiration rate in mitochondria [20];

2. the discovery of glycolytic oscillations in yeast cell cultures [21-23] and the demonstration that allosteric properties of the enzyme phosphofructokinase were primarily responsible for generating the oscillations [24,25];

3. the discovery of damped oscillations in the horseradish peroxidase reaction [26];

4. the discovery of periodic enzyme synthesis during the cell cycle [27,28];

5. ZHABOTINSKII's [29] studies of temporal oscillations in the bromate-malonic acid reaction, derived from Belousov's original work on the bromate-citric acid reaction;

6. the demonstration by WINFREE [30] that phase resetting of circadian rhythms bears all the characteristics of a spontaneous biochemical or biophysical oscillator.

On the theoretical side, advances needed to be made in our understanding of both the thermodynamic and kinetic aspects of spatiotemporal organization. Thermodynamic questions were addressed primarily by PRIGOGINE and GLANSDORFF [31], who showed that, in open chemical reaction systems operating far from equilibrium, the branch of steady states which develops from the equilibrium state can lose thermodynamic stability. Beyond the stability limit, temporal and/or spatial structures may develop. Prigogine calls these "dissipative structures" to emphasize that self-organized, macroscopically ordered patterns in time and space are supported, in thermodynamic terms, by the dissipation of free energy in open systems far from equilibrium.

Useful as the nonequilibrium thermodynamic theory was in dispelling the unfounded prejudice of physical chemists against oscillatory processes in homogeneous solution chemistry, it was not particularly constructive in elucidating
the mechanism and detailed properties of actual oscillatory systems. Such practical questions are better addressed by a kinetic approach based on the differential equations describing local chemical reactions and diffusion. Naturally,
therefore, there arose simple kinetic models, in the spirit of Lotka, Bonhoeffer
and Turing, to explain the new experimental discoveries. CHERNAVSKII [32] put
forward a detailed kinetic theory of the photosynthetic oscillations, HIGGINS
[33] and SEL'KOV [34] developed models for the glycolytic oscillator, GOODWIN
[35] proposed a model for periodic enzyme synthesis, DEGN and MAYER [36] applied
Lotka's ideas to the horseradish peroxidase oscillations, LEFEVER and NICOLIS
[37] investigated the "Brusselator" as a general kinetic model of spatiotemporal
organization.

In 1968 an international conference took place in Prague, Czechoslovakia, with
a follow-up meeting in Finland the next year. At these meetings, the proceedings
of which were eventually published in 1973 under the title Biological and Biochemical Oscillators, there occurred the first general, formal exchange of ideas
between East and West, between chemist and biologist, between experimentalist and
theorist. The Prague meeting proved something of a turning point in our story.
The cross-fertilization of ideas, the encouragement of like-minded colleagues,
and the challenge of new problem sparked a flame of interest in the dynamics of
chemical and biological systems that has spread and grown to this day.

In a sense the history - or, rather, the prehistory - of the field ends in
1968. The struggle to win acceptance, recognition, respectability, and a steady
supply of new researchers was over. Since 1968 we have seen the field consolidate
its early achievements and expand in new directions. To round out our story we
mention a few advances in the field since 1968:

1. the discovery of periodic traveling waves in the BZ reaction by ZHABOTINSKII
   [38] and WINFREE [39];

2. the realization of a close relationship between chemical waves in the BZ
   reaction, aggregation waves in the slime mould Dictyostelium discoideum, and
   waves of neuromuscular activity in cardiac tissue;

3. the elucidation of the mechanism of the BZ reaction by FIELD, KÖRÖS and NOYES
   [40]; and the development of the "Oregonator" model by FIELD and NOYES [41];

4. the prediction by RÖSSLER [42] of deterministic chemical chaos and its subsequent discovery by many groups;

5. the more-or-less rational appproach to discovering new inorganic chemical
   oscillators, as pioneered by EPSTEIN and his coworkers [43];

6. the elaboration of Turing-models for pattern formation in developing or regenerating organs by GIERER, MEINHARDT, MURRAY and others [44];

7. the revival of interest in the oscillating reactions discovered early on by
   MORGAN and BRAY [45-47].

3. The Biology-Chemistry-Mathematics Connection

To study the dynamics of biological systems we must necessarily adopt mathematical methods. The natural languages to describe spatial and temporal organization
are ordinary and partial differential equations, functional and integral differential equations, difference equations, cellular automata. The natural tools are
linear stability analysis, bifurcation theory, regular and singular perturbation
theory, and numerical analysis. These concepts, being unfamiliar to most biolo-

gists, present a barrier to communication and understanding. As theorists we should strive to make this barrier as innocuous as possible, but we should also remember the important role played by theory and mathematical reasoning in other areas of biology: for example, mathematical population genetics in the study of evolution, equilibrium thermodynamics in the study of bioenergetics, and x-ray diffraction theory in the determination of molecular structure.

The connection between biological and chemical systems is three-fold. In the first place, purely chemical systems, notably the Belousov-Zhabotinskii reaction, provide direct experimental confirmation of the reality of temporal oscillations and spatial patterns - notions that are central to our ideas of biological organization in time and space. In the second place, clearly the mechanisms of biological organization will be rooted in biochemical and biophysical interactions, so ultimately biodynamics will be reducible to chemicophysical dynamics. We can make this reduction in a few cases, such as glycolytic oscillations in yeast, cyclic-AMP signalling in slime molds, the regulation of gene expression in bacteria, and the propagation of action potentials in excitable membranes. However, we know very little about the actual mechanisms of most biological regulatory systems. What are the molecular mechanisms of circadian oscillations, of cell cycle progression, of neurological information processing, of embryological development, of limb regeneration? In these cases we can only guess at mechanisms. So the third connection between biology and chemistry is by analogy. Studies of purely chemical reaction systems serve to sharpen our intuition and technical skill concerning spatiotemporal organization in dynamical systems.

The Belousov-Zhabotinskii reaction has played an important role in this regard. Though the detailed mechanism of the BZ reaction is surely unrelated to any of the biological systems mentioned, there are deep structural analogies that transcend mechanistic details. For instance, excitability, oscillations, and chemical wave propagation observed in the BZ reaction bear striking resemblances to cyclic-AMP signalling and aggregation in Dictyostelium discoideum and to signal propagation in neurological and cardiac tissue. These resemblances are firmly based on the qualitative theory of dynamical systems, and on this basis the BZ reaction can be exploited as a reliable analog of spatiotemporal organization in the biological systems. The advantages of working with the purely chemical system are two-fold. First, it is easier to perform controlled, accurate experiments on the chemical system. Second, because the mechanism of the reaction is now well understood, we can build realistic models of the system's behavior and expect to achieve quantitative agreement between theory and experiment. Quantitative tests of our models are essential in proving the practical relevance of our theoretical notions of limit cycles, cusp catastrophes, strange attractors, traveling waves; but they are hardly ever possible in biological systems where the mechanisms are much less understood than in the BZ reaction.

4. Dynamical Diseases

Each advance in our understanding of biological systems has led to dramatic improvements in human health care. For instance, the discovery of the microbial world led ultimately to the germ theory of disease and to the adoption of sterile techniques in medicine. The delineation of biochemical pathways led to the development of antibiotics and to a scientific understanding of nutrition. The unravelling of the genetic code and of information flow in cells provided both a molecular picture of genetic diseases and also a means to attack such maladies by genetic engineering.

In like manner, the dynamical approach to biological systems sheds light on another class of diseases, the so-called "dynamical diseases" [48]. A dynamical disease is an abnormal, pathological pattern of temporal or spatial organization in a physiological control system. A control circuit that should hold the physiological system at a steady state breaks out into disruptive oscillations; a circuit that should generate regular oscillations damps out to a steady state or degenerates into chaotic oscillations; a regular spatial pattern is lost or becomes disorganized. For instance, jet lag and various sleep disorders are malfunctions of our circadian temporal organization. Sudden infant death syndrome may involve the abrupt spontaneous extinction of the breathing rhythm. Cardiac arrhythmias and ventricular fibrillation are maladies in the temporal and spatial organization of the pacemakers and muscular activity of the heart. Certain congenital defects may result from the disruption of normal spatiotemporal organization in the developing embryo. Epileptic seizures and hallucinations may reflect aberrant spatiotemporal organization in the brain. The unrestrained proliferation of some cancer cells may be attributable to malfunctions of the mechanisms controlling cell growth and division.

The recognition, understanding, and scientific treatment of dynamical diseases could prove to be an important advance in human health, as important perhaps as the germ theory of disease and the development of antibiotics. We should not be shy about the practical medical implications of the dynamical view of biosystems.

5. Conclusion

By reviewing the history and some recent developments in the field of biodynamics we have tried to make a case for a new paradigm in biology, a dynamical approach which emphasizes the organization of biological systems in space and time. The dynamical approach complements the essentially static view of life provided by modern macromolecular biology. Eventually the dynamical approach will be firmly rooted in molecular details of biological control mechanisms, but even then it will serve a necessary role in relating mechanistic details to observable behavior. At present, however, the mechanistic details are unknown for many interesting biological problems. In such a climate, the dynamical approach can play an important role in assessing possible mechanisms and in providing general, model-independent insight into how dynamical systems must behave.

Three prototypes have motivated much of the development of the theoretical and experimental foundations of spatiotemporal organization in biological and chemical systems: glycolytic oscillations in yeast cells; oscillations, bistability, chaos and traveling waves in the Belousov-Zhabotinskii reaction; and oscillations, signal-relaying, and aggregation in Dictyostelium discoideum. Careful study of the original literature on these three systems would provide a good introduction to the field for any newcomer. It is appropriate on this occasion, the celebration of Benno Hess's 65th birthday, to point out that Benno was intimately involved from the start in the study of all three prototypes, and he has played a leading role in many important studies of all three problems.

When the "official" history of this field is written, probably much ink will be spilled over paradigm shifts, revolution vs. evolution, "dissipative structures" or "synergetics", etc. In closing, we would like to point out an analogy that may be helpful in understanding the historical development of this branch of science. In dynamical systems theory, "excitability" is an important concept. An excitable system has a locally stable steady state, meaning that small perturbations away from the steady state are rapidly damped out. But sufficiently large perturbations, ones that cross a certain threshold, elicit an abrupt transition away from the steady state toward some excited state. Nerve axons show this sort of excitability. So do cardiac muscle, Belousov-Zhabotinskii reagent, and Dictyostelium amoebae. Science also is an excitable medium. Small perturbations, small deviations from accepted modes of thought, are rapidly damped out and the steady state (normative science) is maintained. Luther, Lotka, Morgan, Bray, Turing were small, isolated perturbations whose effects rapidly disappeared. Only in the 1960s did a sufficiently large and coherent disturbance occur to carry the system across threshold and off on a wild excursion. What the new steady state will look like is still unclear.

References

1. M.G.Th. Fechner: Schweigger's J. Chem. Phys. $\underline{53}$, 129 (1928)
2. E.S. Hedges, J.E. Myers: The Problem of Physico-Chemical Periodicity (E. Arnold, London 1926)
3. G. Lippmann: Ann. Phys. Chem. $\underline{149}$, 4 (1972)
4. U.F. Franck: In Biological and Biochemical Oscillators, ed. by B. Chance, E.K. Pye, A. Ghosh, B. Hess (Academic Press, New York London 1973) p.7
5. R.E. Liesegang: Z. Phys. Chem. $\underline{23}$, 365 (1896)
6. K.H. Stern: Chem. Rev. $\underline{54}$, 79 (1954)
7. Rayleigh: Proc. Roy. Soc. $\underline{99A}$, 372 (1921)
8. J.J. Griffiths: In Oscillations and Traveling Waves in Chemical Systems, ed. by R.J. Field, M. Burger (Wiley-Interscience, New York 1985) p.529
8a. B. Lewis, G. von Elbe: Combustion, Flames and Explosions of Gases (Academic Press, New York, London 1987). Earlier editions were published in 1938, 1951, and 1961
9. R. Luther: Z. Elektrochemie $\underline{12}$, 596 (1906)
10. K. Showalter, J.J. Tyson: J. Chem. Educ. (in press)
11. J.S. Morgan: J. Chem. Soc. Trans. $\underline{109}$, 274 (1916)
12. W.C. Bray: J. Am. Chem. Soc. $\underline{43}$, 1262 (1921)
13. W.C. Bray, H.A. Liebhafsky: J. Am. Chem. Soc. $\underline{53}$, 38 (1931)
14. F.O. Rice, M. Reiff: J. Phys. Chem. $\underline{31}$, 1352 (1927)
15. A.J. Lotka: J. Am. Chem. Soc. $\underline{42}$, 1595 (1920)
16. A.M. Turing: Phil. Trans. R. Soc. Lond. $\underline{B237}$, 37 (1952)
17. B.P. Belousov: In Oscillations and Traveling Waves in Chemical Systems, ed. by R.J. Field, M. Burger (Wiley-Interscience, New York 1985) p.605
18. A.T. Winfree: J. Chem. Educ. $\underline{61}$, 661 (1986)
19. A.T. Wilson, M. Calvin: J. Am. Chem. Soc. $\underline{77}$, 5948 (1955)
20. N.E. Mustafa, K. Utsumi, L. Packer: Biochem. Biophys. Res. Commun. $\underline{24}$, 381 (1966)
21. L.N.M. Duysens, J. Amesz: Biochim. Biophys. Acta $\underline{24}$, 19 (1957)
22. B. Chance, R.W. Estabrook, A. Ghosh: Proc. Nat. Acad. Sci. USA $\underline{51}$, 1244 (1964)
23. B. Hess, K. Brand, K. Pye: Biochem. Biophys. Res. Commun. $\underline{23}$, 102 (1966)
24. A. Ghosh, B. Chance: Biochem. Biophys. Res. Commun. $\underline{16}$, 174 (1964)
25. J.J. Higgins: Proc. Nat. Acad. Sci. USA $\underline{51}$, 989 (1964)
26. I. Yamazaki, K. Yokota, R. Nakajima: Biochem. Biophys. Res. Commun. $\underline{21}$, 582 (1965)
27. J. Gorman, P. Tauro, M. LaBerge, J. Halvorson: Biochem. Biophys. Res. Commun. $\underline{15}$, 43 (1964)
28. M. Masters, P.L. Kuempel, A.B. Pardee: Biochem. Biophys. Res. Commun. $\underline{15}$, 38 (1964)
29. A.M. Zhabotinskii: Biophysics $\underline{9}$, 329 (1964)
30. A.T. Winfree: In Biological and Biochemical Oscillators, ed. by B.Chance, E.K. Pye, A. Ghosh, B. Hess (Academic Press, New York, London 1973) p. 461, 479
31. P. Glansdorff, I. Prigogine: Thermodynamic Theory of Structure, Stability and Fluctuations (Wiley, New York 1971)
32. N.M. Chernavskaya, D.S. Chernavskii: Sov. Phys. Usp. $\underline{4}$, 850 (1961)
33. J.J. Higgins: Ind. Eng. Chem. $\underline{59}$, 19 (1967)
34. E.E. Selkov: Eur. J. Biochem. $\underline{4}$, 79 (1968)
35. B.C. Goodwin: Adv. Enz. Regul. $\underline{3}$, 425 (1965)
36. H. Degn, D. Mayer: Biochim. Biophys. Acta $\underline{180}$, 291 (1969)
37. R. Lefever: J. Chem. Phys. $\underline{49}$, 4977 (1968); J. Theor. Biol. $\underline{30}$, 267 (1971)
38. A.N. Zaikin, A.M. Zhabotinskii: Nature $\underline{225}$, 535 (1970)
39. A.T. Winfree: Science $\underline{175}$, 634 (1972)
40. R.J. Field, E. Körös, R.M. Noyes: J. Am. Chem. Soc. $\underline{94}$, 8649 (1972)
41. R.J. Field, R.M.Noyes: J. Chem. Phys. $\underline{60}$, 1877 (1974)
42. O.E. Rössler: Z. Naturforsch. $\underline{31a}$, 259 (1976)
43. I.R. Epstein, K. Kustin, P. DeKepper, M. Orban: Sci. Amer. $\underline{284}$(3), 112 (1983)
44. Theories of Biological Pattern Formation (The Royal Society, London 1981)
45. H. Degn: Acta Chem. Scand. $\underline{21}$, 1057 (1967)

46. S.D. Furrow: In <u>Oscillations and Traveling Waves in Chemical Systems</u>, ed. by R.J. Field, M. Burger (Wiley-Interscience, New York 1985) p. 171
47. P. Bowers, R.M. Noyes: In <u>Oscillations and Traveling Waves in Chemical Systems</u>, ed. by R.J. Field, M. Burger (Wiley-Interscience, New York 1985) p.473
48. L. Glass, M.C. Mackey: Ann. N.Y. Acad. Sci. <u>316</u>, 214 (1979)

Velocity Correlations and Irreversibility:
A Molecular Dynamics Approach

I. Prigogine*, E. Kestemont, and M. Mareschal

Faculté des Sciences, Université Libre de Bruxelles,
Campus Plaine, CP. 231, Bd. du Triomphe, B-1050 Brussels, Belgium

1. INTRODUCTION

It is a great privilege to contribute to this volume dedicated to Benno Hess. The impact of his work goes far beyond biochemistry. It has stressed the importance of the role of time and irreversibility in the description of biological and chemical systems. This is now well accepted on the phenomenological level. However the relation of irreversibility with the underlying microscopic dynamics remains a subject of some controversy. Let us remember the classical point of view expressed concisely by Smoluchowski /1/ : "If we continued our observation for an immeasurably long time, all processes would appear to be reversible". With Smoluchowski, Chandrasekhar /2/ concludes that "a process appears irreversible (or reversible) according to whether the initial state is characterized by a long (short) average time of recurrence compared to the time during which the system is under observation". Irreversibility would then appear as an artifact due to the time scale of observation!

Recent developments have put this conception into doubt. We know that a simple microscopic model like a system of hard spheres is characterized by a Lyapounov exponent, which implies a finite temporal horizon. As a result, the concept of a trajectory for long time predictions is limited, and the existence of a Poincaré's recurrence time becomes irrelevant for times much longer than a Lyapounov time /3,4/. These new ideas make it interesting to use molecular dynamics simulations to clarify the relation of irreversibility and classical mechanics.

2. SIMULATIONS

Molecular dynamics simulation is a technique which makes use of the huge capacity of modern computers in order to solve numerically Newton's equations of motion. The systems considered generally consist of a few hundred model particles, usually with periodic boundary conditions. Since the first numerical experiments, reported by Alder and Wainwright /5/ on hard-core systems, relatively few studies of the approach to equilibrium have been carried out. An interesting example of a molecular dynamics calculation has been given by Orban and Bellemans /6/ to describe the evolution of the Boltzmann H-function towards its equilibrium value.

This function is defined by

$$H(t) = \int dr dv f(r,v,t) \cdot \ln f(r,v,t)$$

and is related to the entropy for sufficiently low density systems. Here $f(r,v,t)dr dv$ is the number of particles which at time t have positions in a region dr around r, and velocities in a region dv around v. For a dilute fluid, the Boltzmann equation describes the evolution of $f(r,v,t)$ and predicts that, for long times, it will reach an equilibrium value, with a maxwellian velocity distribution.

(*) Also at the Center for Statistical Mechanics and Thermodynamics, the University of Texas at Austin

In a finite system, one has to replace integration by summation and discretize f(r,v,t) in velocity and configuration space by use of finite regions, Δv and Δr instead of dr, dv. Observing the discrete distribution function at a serie of times during the evolution of the phase space point x(t) along its trajectory, starting from a given initial condition x(o), one obtains a measured value for the H function.

The system studied by Orban and Bellemans was made of an assembly of 100 hard disks at a low density. The initial condition, x(o), consisted in random (but non-overlapping) positions and velocities of equal magnitude but randomly oriented. The evolution in time of the H function is shown on Figure 1.

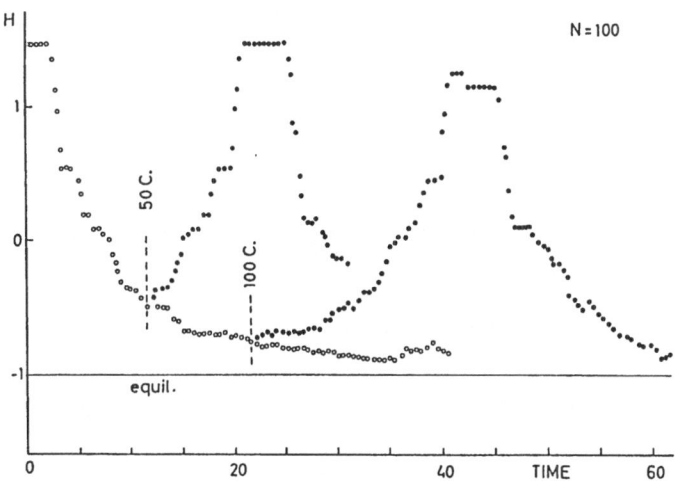

Figure 1 : Evolution of the H function in the simulation of Orban and Bellemans. The effect of reversing the velocities after one and two relaxation times is displayed.

They also investigated the behaviour of H(t) upon reversing particles velocities at t/τ equal to 1 or 2, where τ is the relaxation time, that is the average time between 2 collisions. From figure 1 it is clear that the H function reaches its equilibrium value after nearly 2 collisions per particle. Reversing velocities at time τ, one observes that H(t) goes back to its initial value after a time 2τ. Reversing the velocities at a later time, numerical errors in the trajectory computation accumulate and the recovery of the initial value of H(t) becomes less accurate.

This calculation illustrates very well the so-called Loschmidt paradox; inverting the velocities of the particles of a system going towards equilibrium, one is able to generate configurations whose evolution is characterized by a (temporary) divergence of the H function from its equilibrium value. This has been analyzed some time ago by our group /7,8,9/, both in particular models and in the framework of our general approach. The argument focused on the role of the correlations which exist initially in the system. Depending on these correlations, the evolution may be kinetic (with a decrease of H(t) towards its equilibrium value) or antikinetic (with a corresponding increase). What came out of this analysis was that the absence of initial correlations results in a kinetic evolution, persisting for a time of, at least, a recurrence time, which is, as we already stressed, much larger than any time interval for which the notion of trajectory remains meaningful. Starting from an uncorrelated state, collisional processes create intricate correlations between positions and velocities and, as time proceeds, these correlations involve an always increasing number of particles. The flow of correlations is then related to the irreversible behaviour

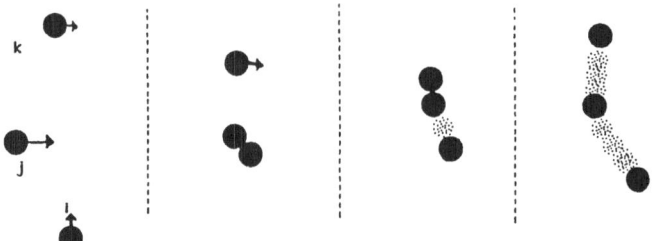

<u>Figure 2</u> : Successive binary collisions between uncorrelated particles lead to a
flow of correlations.

observed. This flow of correlation is illustrated in Figure 2 : initially particles i,j and k are uncorrelated. A first collision between i and j results in
the creation of a pair correlation. A subsequent collision between j and k
contributes at the same time to the decay of the pair correlation and the creation of a 3-particle correlation. When any of these particles collide with a
fourth particle, that collision will contribute to create a four-particle correlation, and so on. Once equilibrium is reached, the flow of correlations reaches
a steady state, with the creation of correlations between uncorrelated particles
and, their decay due to the interaction with the surrounding particles.

We have carried out some new numerical experiments in order to characterize
the correlations built in the system during collisions. However, instead of
starting from a non-equilibrium configuration, we have performed our measurements
on an equilibrum system of hard disks. Of course, the correlations between the
velocities vanish when averaged over an equilibrium ensemble. However, one can
focus one's attention to particular pairs of particles (for instance, particles
which just collided) and measure how their velocities are related.
This is shown on Figure 3. The system to which this figure refers is an
assembly of 400 hard disks with periodic boundary conditions. We plot the value
of the scalar product, $\langle v_i(t).v_j(t)\rangle$, of the velocities of particles i and j as a
function of time after their collision; the average is performed over all colli-
sons, more than a million, taking place in the system during the experiment.
As can be seen from Figure 3, there is a strong angular correlation between
the velocities of the colliding particles. On the average, at collision, velocities are oriented in opposite directions. As time proceeds, that correlation
will decay due to the interaction of i and j with surrounding particles. For
short times this corresponds to collisions of i (or j) with a third particle k,
which will tend to make v_i and v_j more parallel, creating also correlations
between particles i,j and k. For longer times, the decay of the correlation
between i and j will involve more and more surrounding particles and can be described by a hydrodynamical slow process of exchange of momentum. Note that for
a dilute system (n=0,01) the decay is purely exponential as can be expected from
kinetic theory.
The existence of a correlation between the velocities of i and j is not in
contradiction with the fact that at equilibrium one does not expect such correlations : equilibrium implies only that the ergodic long time limit of the correlations vanishes. Of course the velocity correlations between a particle and its
neighbours, without specifying that a collision has, or not, taken place, should
vanish. We have indeed verified that the corresponding scalar products of the
velocity of any given particle with the velocities of its neighbours average to
a negligible quantity.
The remarkable feature of the numerical experiment, in this example, is that
it allows us to focus on particular functions measured at particular times. We
are able to follow in time the decay of velocity correlations built during the
collision process, although this decay is not related to any macroscopic transport process in the system.

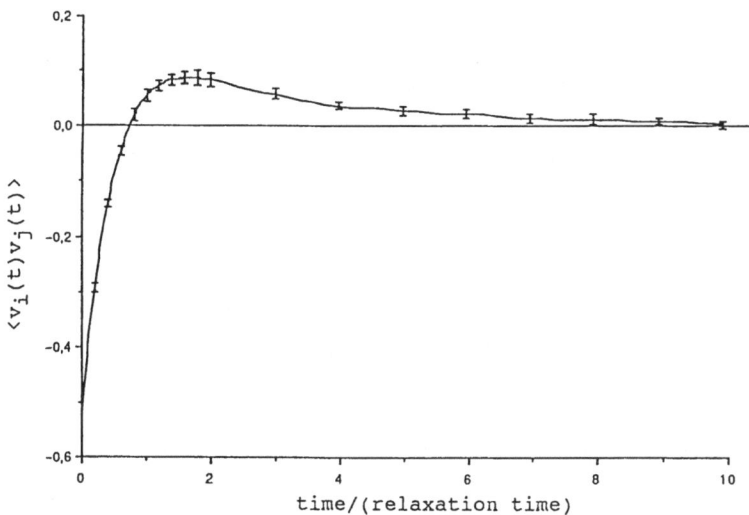

<u>Figure 3</u> : Decay of the velocity correlations after collision. The scalar product $\langle v_i(t)v_j(t)\rangle$, is plotted as a function of the time after the collision and the average is performed over all colliding pairs in the run.

3. CONCLUSIONS

The persistence of equilibrium, as can be seen from the numerical simulations, requires a continuous decay of the correlations that are built in by the collisions. This continuous decay is a manifestation of an arrow of time which permits to maintain "molecular chaos".

We come in this way to a most unexpected conclusion. The arrow of time exists even at equilibrium. Of course, it has then only a microscopic meaning. A nonequilibrium preparation of the system will "amplify" this arrow of time and bring it to the macroscopic level.

Moreover, the collision process as described here provides us with a nice illustration of a process of transfer of information from the two-body system formed by two colliding particles, to the medium. It would be most interesting to repeat these simulations in a quantum mechanical context, as this loss of information would provide us with a dynamical mechanism for the collapse of the wave function which occurs in dissipative quantum mechanical systems /10/.

In summary, far from being an illusion, the arrow of time corresponds to a ubiquitous phenomenon present even at equilibrium, to emerge to the surface whenever the prevailing conditions permit.

ACKNOWLEDGEMENTS

One of us (E.K.) wants to thank the financial support of the "Instituts Internationaux de Physique et de Chimie fondés par E. Solvay". We also acknowledge fruitful discussions with Dr. M. Malek Mansour.

REFERENCES

1. Quoted by Weyl H., <u>Philosophy of Mathematics and Natural Science</u>, Princeton University Press, Princeton, N.J. (1949).
2. Chandrasekhar S., Rev. Mod. Phys. <u>15</u>, 3-91 (1943).
3. Prigogine I. and Stengers I., <u>Order out of Chaos</u>, Heinemann, London (1984).

4. Lighthill, Sir J., Proc. R. Soc. Lond. A407, 35-50 (1986).
5. Alder B.J. and Wainwright T., in <u>Proceedings of the International Symposium on Transport Processes in Statistical Mechanics</u>, ed. by I. Prigogine, Interscience, New York (1958).
6. Orban J. and Bellemans A., Phys. Lett. 24A, 620 (1967).
7. Prigogine I. and Résibois P., <u>Atti del Simposio Lagrangiano</u>, Academia delle Scienze di Torino (1964).
8. Balescu R., Physica 36, 433 (1967).
9. Résibois P. and Mareschal M., Physica 94A, 211-253 (1978).
10. Prigogine I. and Petrosky T., Physica A (in press, 1987); Petrosky T. and Prigogine I., Physica A (in press, 1987).

Nonlinear Dynamics, Self-Organization and the Symbolic Representation of Complexity

G. Nicolis

Faculté des Sciences, Université Libre de Bruxelles,
Campus Plaine, CP. 231, Bd. du Triomphe, B-1050 Brussels, Belgium

1. Introduction

Several contributors to this volume produce in their chapters convincing evidence about the usefulness of nonlinear dynamics and self-organization, in the modelling of a variety of concrete problems of relevance in chemistry and biology. In the present chapter we focus on a different aspect. Specifically, we show how nonequilibrium physics and nonlinear dynamics allow us to go one step further than the traditional description of physico-chemical and biological systems and set up a <u>symbolic representation</u> in which such concepts as attractors, predictability, probability and information play a prominent role and allow us to better grasp the nature of complexity.

We shall first present, in section 2, a succinct overview of dynamical systems theory. In section 3 we show how, out of the nonlinear evolution laws of physico-chemical systems, a symbolic description can emerge. We shall end (section 4) by a few thoughts on the biological interest of these considerations.

2. Dynamical Systems Theory: Thinking with Attractors

The evolution equations of the state variables $\{X_i\}$ of a physico-chemical or biological system can be written in the general form

$$dX_i/dt = F_i(\{X_j\},\{\lambda_k\}) \quad i, j = 1, \ldots n$$
$$k = 1, \ldots n . \tag{1}$$

Here F_i represents the overall effect of the rate laws responsible for the evolution of X_i and are, typically, <u>nonlinear</u> and <u>dissipative</u> operators. The quantities λ_k, referred to as <u>control parameters</u>, describe the ways in which the system is coupled with the outside world.

Because of the nonlinearities inherent in (1) a full quantitative analysis is out of the question. One therefore resorts to a qualitative approach [1] at the basis of which is the embedding of the dynamics in <u>phase space</u>, the space spanned by the full set of variables participating in the dynamics. By construction, there is a one-to-one correspondence between a point in this space and an instantaneous state of the underlying dynamical system. Consequently, the time evolution of such a system will be represented by a curve in phase space, the phase space trajectory. In a dissipative system, as time grows and transients die out, the phase space trajectory tends to a privileged part of phase space whose dimensionality is smaller than that of the full space and whose volume is therefore zero. We call this set the <u>attractor</u>, and we may regard it as the archetype for a particular type of behavior that may arise in our system, whatever its details might be.

Figure 1 depicts some possible attractors. The 0-d (zerodimensional) point attractor of Fig. 1a represents a steady state regime. The 1-d attractor of

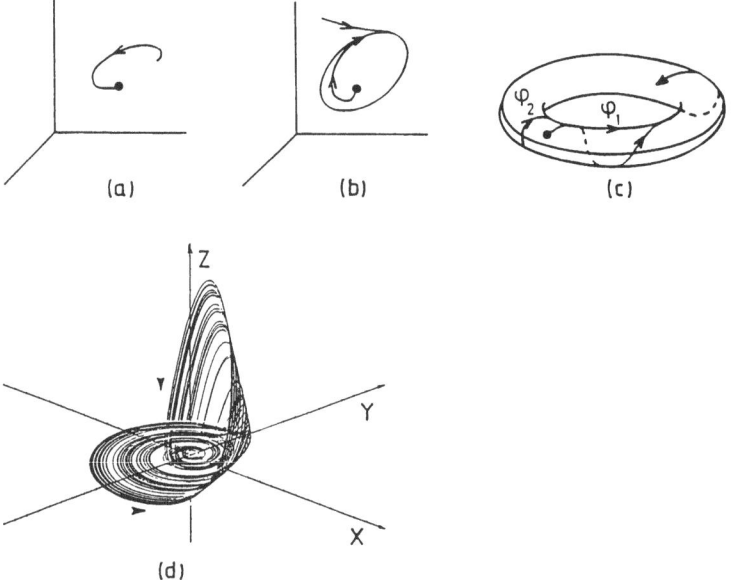

Fig. 1: Possible attractors of a dissipative dynamical system: (a) point attractor; (b) limit cycle; (c) torus; (d) chaotic attractor

Fig. 1b, the familiar limit cycle, represents a time periodic regime whose amplitude and period are intrinsically determined by the system. The 2-d torus of Fig. 1c represents a multiperiodic regime which, in the case of incommensurate frequencies, gives rise to aperiodic behavior. A different kind of aperiodic behavior is depicted in Fig. 1d [2]. Contrary to Fig. 1c, in which the motion on the attractor is stable, we observe now two opposing trends. On the one side (see horizontal arrow in the Figure) an instability of the motion, tending to remove the phase space trajectory away from the "reference" state X = Y = Z = 0; and on the other side (see vertical arrow) the bending on the outgoing trajectories followed by their reinjection back to the vicinity of this state.

The unstable motion on the attractor (as opposed to the stability in the directions transverse to the attractor) is reflected by the sensitivity of the trajectories on the attractor to small changes in the initial conditions. As a result two initially nearby states can diverge, for a certain period of time, in an exponential fashion whose rate is measured by the (largest positive) Lyapounov exponent σ of the system. This leads immediately to two important conclusions. First, it is impossible to make predictions on this type of system beyond a lapse of time of the order of σ^{-1}. And second, if an observer still wants to characterize the state of the system for t > 0 as sharply as for t = 0, he will need an increasing amount of data or, put differently: during its motion in phase space the system increases in variety and complexity, thereby creating information. These qualitative statements can be further implemented, as we shall see in more detail in section 3. It can indeed be shown that certain kinds of chaotic attractors can be mapped into well-defined stochastic processes like, for instance, Markov chains. In as much as a stochastic process can be characterized by its information theoretic entropy and algorithmic complexity [3] we have therefore at our disposal a dynamical model for two key concepts of science, which so far had been discussed only at the level of abstract definitions or static prescriptions.

One of the most exciting aspects of nonlinear phenomena is that the same system can show a great variety of behaviors, each corresponding to a different attractor. The mechanism which is at the origin of this diversification is the bifurcation of new branches of states, as the parameters built in the system are varied. The simplest bifurcations are those leading from a point attractor to a limit cycle (Hopf bifurcation), or from one point attractor to another point attractor (pitchfork and limit point bifurcations). One also observes complex bifurcation cascades culminating in certain cases to the bifurcation of chaotic attractors. This raises naturally the question of universal description of bifurcation diagrams. We discuss this matter in the next section.

3. Toward a Symbolic Description of Complexity

A most remarkable property of the transition phenomena summarized in the preceding section is that in the vicinity of a bifurcation point the dynamics of the system is entirely determined by a limited number of key variables, in terms of which all others can be expressed. For instance, near a pitchfork bifurcation, there exists a combination of the initial n variables X_i which we denote by z, obeying the universal equation

$$dz/d\tau = (\lambda - \lambda_c)z - uz^3 \tag{2}$$

in which τ is a scaled time and u a real parameter whose value depends on the detailed structure of the system. As for the individual X_i's, they are given in terms of z by equations of the form $X_i = g_i(z)$, in which the functions g_i are again system-dependent.

Equation (2) is known in the literature as normal form [1], and the one-dimensional subspace of the initial phase space on which z is defined is known as the center manifold. The variable z itself is referred to as the order parameter [4]. Turning now to the bifurcation leading to periodic attractors, known as the Hopf bifurcation, one can again establish the existence of a universally valid normal form. This time there exists a pair of real valued order parameters (the center manifold is two-dimensional) or, equivalently, one complex-valued order parameter. The normal form equation reads

$$dz/d\tau = [(\lambda - \lambda_c) + i\omega_c] z - u |z|^2 z , \tag{3}$$

where u is in general complex-valued, $|z| = zz^*$, z^* being the complex conjugate of z, and ω_c is a combination of the system's parameters giving the frequency of the oscillations at the bifurcation point $\lambda = \lambda_c$.

More intricate situations arise in the presence of interaction between bifurcations, generating secondary or even tertiary bifurcation phenomena. The above results carry through, in the sense that one can guarantee that the part of the dynamics that gives information on the bifurcating branches takes place in a phase space of reduced dimensionality. The explicit construction of the normal forms becomes, however, much more involved. In addition, their universality can no longer be guaranteed. For instance, in addition to fixed points or limit cycles, new attractors can be generated by global bifurcation mechanisms which have nothing to do with the critical value $\lambda = \lambda_c$. If the system contains at least three coupled order parameters, these global bifurcations may lead to chaotic dynamics.

The above results open the tantalizing possibility to model the complex phenomena observed in real-world multivariable systems in terms of a limited number of variables or, equivalently, in terms of low-dimensional attractors. What is more, starting from data pertaining to the evolution of one variable only, one can develop algorithms to actually compute the dimensionality of the attractor, the Lyapounov exponents and the minimum number of variables participating in the dynamics, independent of any modelling. We refer to the chapter in this volume by A. BABLOYANTZ and A. DESTEXHE for an illustration of these possibilities.

We now show that in certain classes of systems it becomes natural to introduce a still higher level of abstraction by mapping the dynamics into a succession of information-rich strings of symbols. More specifically, we indicate some general mechanisms allowing to cast chaotic dynamics in the form of stochastic processes of varying complexity.

3.1. Master Equation for Chaotic Attractors and Markov Partitions

To fix ideas we consider one-dimensional iterative maps, like the logistic map in the fully chaotic region:

$$x_{n+1} = f(x_n, \lambda)$$
$$= 4x_n(1 - x_n) \ , \ 0 \leq x \leq 1 \ . \tag{4}$$

The probability density $\rho(x)$ of this dynamical system obeys the Perron-Frobenius equation

$$\rho_n(x) = U \ \rho_{n-1}(x)$$
$$= \int_0^1 dy \ \delta(x - f(y))\rho_{n-1}(y) \ , \tag{5}$$

which fails to define a physically meaningful stochastic process owing to the singular character of the "transition probability" $\delta(x-f(y))$. In order to map the dynamics into a regular stochastic process we therefore partition the state space of the system into a set of N non-overlapping cells C_i (i = 1,...,N) chosen to define a Markov partition [5,6]. The main feature of such a partition is that the boundaries between cells are kept invariant by the dynamics. This allows us to define properly the states of the underlying process. Introducing the probability vector

$$\underline{P}_n = \left(\int_{C_1} \rho(x)dx \ , \ \ldots, \ \int_{C_N} \rho(x)dx \right) \tag{6}$$

and assuming "coarse grained" initial conditions $\rho_0(x)$ taking a constant value in each of the cells of the partition, one can then deduce from (5) a master equation of the form [7]

$$\underline{P}_n = \underline{W} \ \underline{P}_{n-1} \ . \tag{7}$$

The transition probability matrix $\underline{W}$ is a positive stochastic matrix. Moreover, under some additional conditions (which turn out to be fulfilled for the logistic map) the process described by (7) is a first order Markov process compatible with all the constraints imposed by the deterministic dynamics, (4)-(5).

As an example, consider the three-cell Markov partition provided by the end points 0 and 1 of the interval and by the points $x_1 = 0.345$, $x_2 = 0.905$ belonging to the period two orbit of the system. The probability vector

$$(P_1, \ P_2, \ P_3) = \left(\int_0^{0.345} \rho dx, \ \int_{0.345}^{0.905} \rho dx, \ \int_{0.905}^1 \rho dx \right)$$

obeys the Markovian master equation

$$(P_1, \ P_2, \ P_3)_{n+1} = (P_1, \ P_2, \ P_3)_n \begin{pmatrix} 1/2 & 1/2 & 0 \\ 0 & 1/2 & 1/2 \\ 1 & 0 & 0 \end{pmatrix} . \tag{8}$$

A symbolic description of the dynamics incorporating in a natural way the concept of information can now be achieved. For instance, the average amount of information needed to locate the system in state space is given by the information theoretic entropy

$$S_I = -\sum_i P_s(i) \ln P_s(i)$$

in which P_s stands for the invariant probability. Its value is $S_I \simeq 1.06$. Moreover, the average amount of information created by the system in one unit of time is given by the Kolmogorov-Sinai entropy

$$S_K = -\sum_{ij} P_s(i) W_{ij} \ln W_{ij}$$

and turns out to be equal to $S_K = 4/5 \ln 2 \simeq 0.55$.

3.2. Level Crossing Probabilities

We now turn to a second mechanism producing information-rich sequences of symbols [8]. We consider a dissipative dynamical system whose state variables X, Y, Z, ... perform sustained oscillations in time, not necessarily periodic. We assume that when a variable crosses a certain level L with a positive slope (Fig. 2), a new process is switched on, as a result of which the "symbol" X, Y or Z is collected on a "tape". For instance, if the variables represent the concentrations of chemicals in a reagent, one may imagine that beyond the threshold the substance is rapidly precipitating. One obtains in this way a one-dimensional pattern of digits generated by the dynamics.

Let us illustrate this procedure on the Rössler attractor (Fig. 1d) described by the following set of equations [9]:

$$
\begin{aligned}
dX/dt &= -Y - Z \\
dY/dt &= X + aY \\
dZ/dt &= bX - cZ + XZ \ .
\end{aligned}
\tag{9}
$$

For a = 0.38, b = 0.3 and c= 4.5 this system gives rise to the chaotic attractor depicted in Fig. 1d. Choosing the threshold values $L_x = L_y = L_z = 3$ and the initial conditions X = Y = Z = 1 one obtains then a sequence of the form [8]

$$
\begin{aligned}
&\text{ZYXZXYXZXYXZYXZXYXZYX} \\
&\text{ZYXZXZYXZYXZYXZXYXZYX} \quad ....
\end{aligned}
\tag{10a}
$$

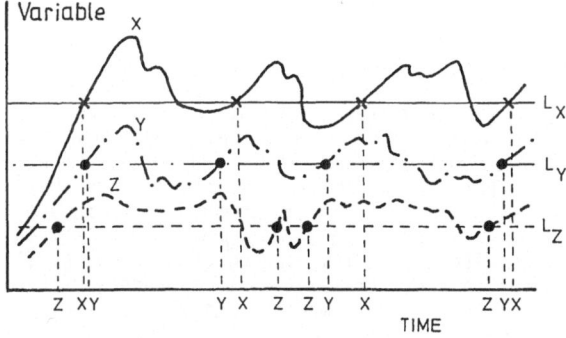

Fig. 2: Generation of a one-dimensional aperiodic string of symbols forming an information-rich spatially asymmetric structure

It can be checked that this sequence can be entirely reformulated by introducing the hypersymbols

$$\alpha = ZYX, \quad \beta = ZXYX, \quad \gamma = ZX .$$ (10b)

The result reads

$$\alpha\beta\beta\alpha\beta\alpha\alpha\gamma\alpha\alpha\beta\alpha \dots$$ (10c)

This interesting property has to be attributed to the deterministic origin of the mechanism giving rise to the sequences. It suggests the existence of strong correlations in the succession of symbols. Actually it turns out that (10a) defines a fifth order Markov chain:

$$P(ABCDEFG) = P(ABCDEF).W(F|ABCDE).W(G|BCDEF)$$

whose information theoretic and Kolmogorov-Sinai entropies can again be defined and evaluated explicitly. In addition, the sequences generated by this chain are asymmetric, in the sense that the probability of a particular "word" formed by our 3-letter "alphabet" is different from the probability of the reverse "word".

4. Implications in Biology

Much of what we call "information" is carried in nature by a one-dimensional aperiodic and asymmetric structure, the DNA. The origin of this primordial information raises two issues, which appear to be mutually contradictory. On the one side information must be associated with some kind of randomness. Thus, although perfectly reproducible from one generation to the next, the codon sequence along the DNA is basically unpredictable in the sense that its global structure cannot be inferred from the knowledge of a part of it, however large that part might be. But on the other side, the astronomically large number of random sequences makes it extremely improbable to select, on a priori grounds, the particular class of sequences that is likely to play the major role in a given phenomenon. For instance, as pointed out by EIGEN and SCHUSTER [10], the number of random polypeptides of a moderate length, say 100 units, which have to be scanned to select a biologically meaningful protein, is of the order of 10^{130}. On these grounds the spontaneous origin of information as a result of a purely probabilistic game, like the game that would take place around the state of thermodynamic equilibrium, must be ruled out.

We suggest that informationally meaningful structures can be generated from an underlying mechanism which is nonlinear, dissipative, and operating in the far-from-equilibrium chaotic region. In this way randomness and asymmetry - two prerequisites of information - are incorporated from the outset in the resulting structure. In addition, being the result of a mechanism - which has ultimately to do with the existence of molecules endowed with suitable catalytic properties - these structures automatically overcome the difficulty of the tremendous "thermodynamic improbability" characterizing random sequences.

Because of the dissipative character of the dynamics, the system can run on an asymptotically stable attractor, a situation which would be impossible had the dynamics been conservative. As a result, the sequences generated by the mechanisms discussed in subsections 3.1 and 3.2 are structurally stable towards small changes of parameters. Specifically, although the particular succession of symbols may be modified considerably, their statistical properties remain essentially unchanged. This provides us with a dynamical analog of a mutation. It also suggests that it is more meaningful to speak of families of sequences rather than of a preferred one, a situation which is somewhat reminiscent of the quasispecies concept of Eigen and Schuster.

Naturally, in order that this scheme be tested against real-world biopolymers carrying biological information, it will be necessary to identify the series of reactions which may result in such polymers of precursors thereof, for instance in the context of prebiotic chemistry. In this respect, it is worth noting that the distribution of configurational sequences in vinyl polymers obtained from polymerizing monomers like $CH_2 = CHR$ or even the symmetrically substituted monomers such as $CHF = CHF$, can be accounted for by a third order irreversible Markov chain [11]. More to the point, the analysis of DNA sequences of various organisms [12] suggests that there exist correlations between the nucleotides. Finally, evidence has recently accumulated that tRNA and rRNA share deep structural affinities [13], suggesting that these two species may have derived from a common group of ancestral molecules. A primordial chaotic dynamics of a suitable set of chemical reactions would provide a mechanism for such an ancestor, the first molecule ever to have captured irreversibility and nonlinearity and to have used them in a constructive manner.

References

1. J. Guckenheimer, P. Holmes: Nonlinear Oscillations, Dynamical Systems and Bifurcations of Vector Fields (Springer, Berlin, Heidelberg 1983)
2. P. Gaspard, G. Nicolis: J. Stat. Phys. 31, 499 (1983)
3. J.S. Nicolis: Dynamics of Hierarchical Systems (Springer, Berlin, Heidelberg 1986)
4. H. Haken: Synergetics (Springer, Berlin, Heidelberg 1977)
5. R. Bowen, D. Ruelle: Inven. Math. 29, 181 (1975)
6. Ya Sinai: Introduction to Ergodic Theory (Princeton Univ. Press, Princeton, N.J. 1977)
7. G. Nicolis, C. Nicolis: Phys. Rev. A (1987), submitted
8. G. Nicolis, G. Rao, S. Rao, C. Nicolis: In Coherence and Chaos in Dynamical Systems (Manchester Univ. Press, 1987)
9. O. Rössler: Ann. New York Acad. Sci. 316, 376 (1979)
10. M. Eigen, P. Schuster: The Hypercycle (Springer, Berlin, Heidelberg 1979)
11. H. Frisch: Adv. Chem. Phys. 55, 201 (1984)
12. D. Lipman, W. Wilbur: J. Mol. Biology 163, 363 (1983)
13. A. Nazarea, D. Bloch, A. Semrau: Proc. Natl. Acad. Sci. USA 82, 5537 (1985)

Spectral Kinetics and the Efficiency
of (Bio)Chemical Reactions

J. Ross[1], S. Pugh[1;*], and M. Schell[1;**]

[1]Department of Chemistry, Stanford University,
Stanford, CA 94305, USA

Abstract

In this article we discuss some aspects of physical chemical non-linear systems, in particular the effect of external periodic perturbations on non-linear chemical reactions, which we call spectral kinetics. As an application we analyze the efficiency of a biological energy transduction system, a proton pump. The organizers of the conference requested qualitative summaries and some speculation, and we indulge so in this regard here and there.

1. External Periodic Perturbations of Non-linear Systems

The study of the response of non-linear systems to external periodic perturbations leads to interesting and useful information [1-7]. We have investigated the cool flame oxidation of acetaldehyde under various conditions of reactant flow rates, temperature of the baths in which the reaction vessel is immersed, stirring rates, and average pressure in the reaction vessel. The autonomous system, that is the system in the absence of external perturbations, shows a variety of dynamic behavior including stationary states, hysteresis, and diverse types of oscillations [8,9]. We have observed complex oscillations in a limited range of external constraints [10] and have studied the measured time series of light emission by means of statistical distribution of periods, Poincaré phase portraits, correlation function techniques, power spectra, Lyapunov numbers coefficients, and dimension with a method due to GRASSBERGER and PROCCACIA [11-15]. The response of this system to external periodic perturbations has shown [16,17] a number of new phenomena, and deserves more study. The reaction is run in a continuous stirred tank reactor, a glass vessel, which is maintained at a constant external temperature (Fig. 1). We choose as an external perturbation a sinusoidal component to the input flux of one or both of the reactants and vary both the amplitude and the frequency of perturbations. The response of oscillatory combustion systems to such perturbations may be periodic, in which case the system is said to be entrained, or it may be quasiperiodic or chaotic. The range of values for the frequency and amplitude of perturbation in which the response is entrained constitutes an entrainment band. A summary of measured entrainment bands is given in Fig. 2. At zero amplitude of perturbation entrainment bands emerge from points that are equal to the ratio of integers n/m, where m is the number of cycles that occur in the response for every n cycles of the perturbation. In this system the entrainment bands are rather narrow and as the integers n and m increase the band width decreases [3].

A convenient way to study the response of the system is by means of Poincaré next phase plots [16,18]. We pick a characteristic feature in the perturbation, such as the maximum amplitude, and a characteristic feature of the response, let it be the maximum amplitude of the light emission of the combustion system. The

---

* Present address: Photographic Research Labs., Eastman Kodak Company, 1669 Lake Avenue, Rochester, N.Y. 14650

**Present address: Department of Chemistry, Southern Methodist University, Dallas, TX 75275

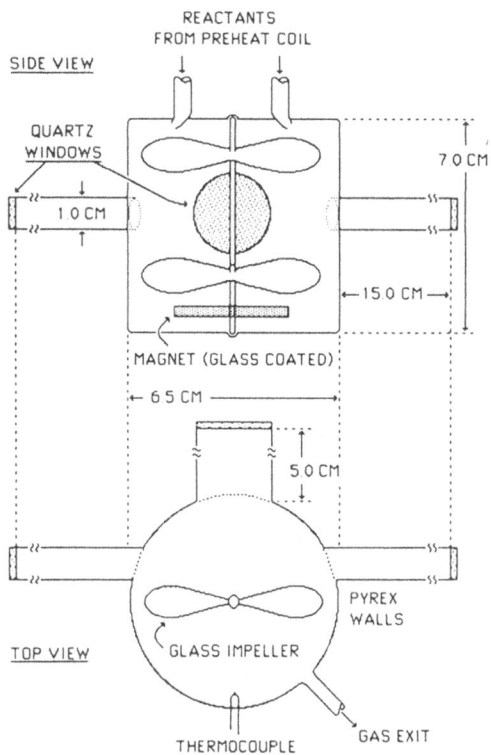

SIDE VIEW

REACTANTS
FROM PREHEAT COIL

QUARTZ
WINDOWS

1.0 CM

7.0 CM

15.0 CM

MAGNET (GLASS COATED)

6.5 CM

5.0 CM

PYREX
WALLS

TOP VIEW   GLASS IMPELLER

THERMOCOUPLE   GAS EXIT

Fig. 1: Schematic of continuous-flow stirred tank reactor (CSTR), side view and top view. The preheated reactants enter at the top of the reactor and exit from the side of the reactor near the bottom [9]

temporal difference between these chosen features in the perturbation and the response, divided by the period of the perturbation, is a phase difference. Within an entrainment band the phase difference remains constant in time but in a quasiperiodic response the phase difference varies in time. In Fig. 3 we show a

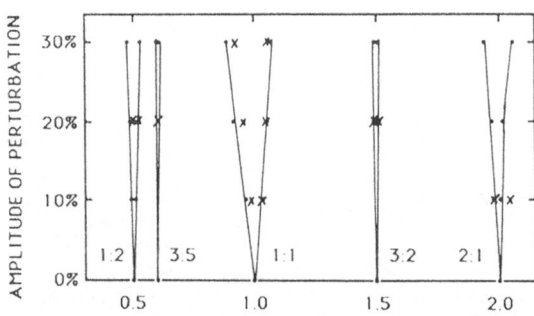

AMPLITUDE OF PERTURBATION

30%

20%

10%

0%

1:2   3:5   1:1   3:2   2:1

0.5   1.0   1.5   2.0

FREQUENCY OF PERTURBATION/AUTONOMOUS FREQUENCY

Fig. 2: Experimentally determined edges of five entrainment bands (crosses) arising from the periodic perturbation of the cool flame oscillation of acetaldehyde. The amplitude of the acetaldehyde flow rate perturbation is given as a percent of the average flow rate and the perturbation frequency is scaled by the frequency of the autonomous system. We also show calculated edges of five entrainment bands (lines) arising from the periodic perturbation of the cool flame oscillation of acetaldehyde [16]

35

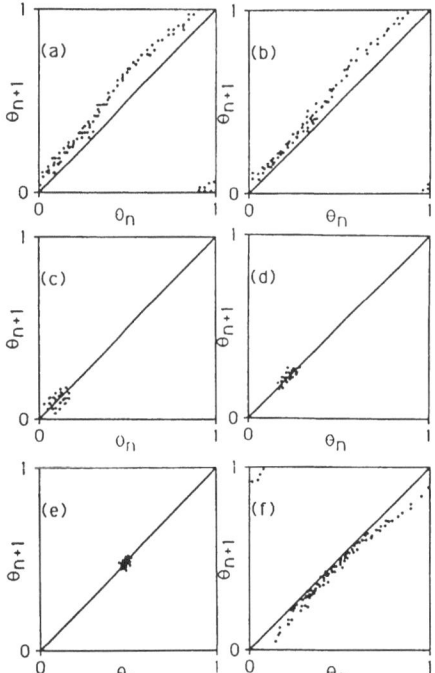

Fig. 3: Measurements of the response of the oscillatory cool flames of acetaldehyde combustion in the region of the fundamental entrainment band, displayed on next-phase maps. The amplitude of the periodic perturbation of the acetaldehyde flow rate is 20% of the average flow rate. The frequencies of the perturbations, scaled by the frequency of the autonomous oscillation, are (a) 0.848, (b) 0.886, (c) 0.925, (d) 1.00, (e) 1.025, and (f) 1.033 [16]

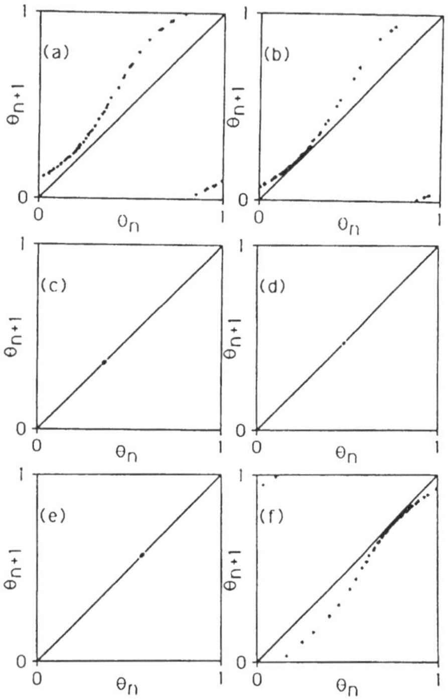

Fig. 4: Calculations of the response of the oscillatory cool flames of acetaldehyde combustion in the region of the fundamental entrainment band, displayed on next-phase maps. The amplitude of the periodic perturbation of the acetaldehyde flow rate is 20% of the average flow rate. The frequencies of the perturbations, scaled by the frequency of the autonomous oscillation, are (a) 0.85, (b) 0.90, (c) 0.95, (d) 1.00, (e) 1.05, and (f) 1.10 [16]

sequence of next-phase plots constructed from experiments: all are at a constant amplitude of external perturbation but each is for a different frequency. The choices of frequencies of perturbation correspond to the traversal of the fundamental entrainment band, from a quasiperiodic region to entrainment and continuing with increases in frequency to quasiperiodicity again. Within an entrainment band all the phase differences are the same at a given frequency of perturbation, and the experiment shows a clustering of points on the diagonal. However, it is important to note that as the entrainment band is traversed at constant amplitude of perturbation the phase shift changes from the low frequency side, on entering the entrainment band, to the high frequency on departing the entrainment band, by π [19]. This phase shifting is important and we shall return to it later. A calculation made with a five-variable model of the complex reaction mechanism of the combustion of acetaldehyde reproduces the measured phase plots quite well (Fig. 4).

The next phase plot is very sensitive to the change of transition from quasiperiodicity to entrainment so that entrainment bands, even narrow ones, are easily found. If we look at measurements of the time series, that is the response

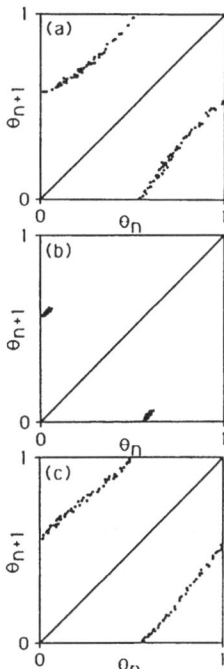

Fig. 5: Measurements of the response of the combustion of acetaldehyde to the periodic perturbation of the acetaldehyde input rate, in the region of the 1:2 entrainment band, displayed in next-phase maps. The amplitude of the perturbation is 20% of the average flow rate. The perturbation frequencies (scaled by the autonomous frequencies) are (a) 0.480, (b) 0.503, and (c) 0.525 [16]

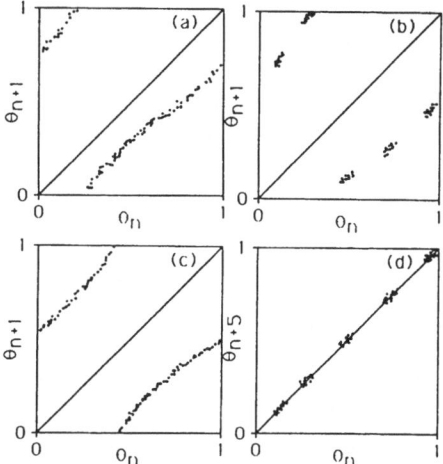

Fig. 6: Measurements of the response of the combustion of acetaldehyde to the periodic perturbation of the acetaldehyde input rate, in the region of the 3:5 entrainment band, displayed in next-phase maps. The amplitude of the perturbation is 20% of the average flow rate. The perturbation frequencies (scaled by the autonomous frequencies) are (a) 0.555, (b) 0.610, and (c) 0.775. (d) Response of the periodically perturbed combustion of acetaldehyde in the 3:5 entrainment band is displayed in a $\theta(n+5)$ vs. $\theta(n)$ plot. This higher-order map is constructed from the data used to plot the next-phase map shown in (b). All five sets of points fall on the bisectrix, indicating entrainment [16]

of the system as a function of time, it may be difficult to distinguish between periodicity and quasiperiodicity, and thus it is easy to miss an entrainment band.

Traversals of the 1:2 band, where the system has two responses for every perturbation, are shown in Fig. 5. The three to five entrainment band in which there are five responses for every three perturbations is shown in Fig. 6.

The measurements shown in Fig. 2 are an example of what we shall call spectral kinetics, the response of a complex kinetic system far from equilibrium to external perturbations. Consider the analogy with optical spectroscopy. Structures of polyatomic molecules are studied by means of the external periodic perturbation of light and in an absorption band the frequency of the molecular system has an integral relation to the frequency of the light, normally 1:1. From such, and other, optical spectroscopic measurements, one attempts to infer the structure of polyatomic molecules and in principle an inversion procedure may be possible, in which the structure is not guessed but deduced from sufficient optical measurements. In a complex kinetic system the chemical species are the analogs of the atoms, the reaction mechanism which gives the connections among the chemical species is the analog of the chemical bonds, and the rate coefficients are the analogs of the force constants. In spectral kinetics we have the advantage that we may determine phase relations among the oscillations of the various species and the perturbation, and may perturb one species or constraint at a time. The difficult problem of the interpretation of spectral kinetic measurements such as in Fig. 2, which may lead to some information about the reaction mechanism, remains to be developed. But we may speculate that there may exist the possibility of the deduction of the reaction mechanisms from sufficient spectral kinetic measurements in juxtaposition to the usual approach in which one guesses a reaction mechanism and sees if, as a sufficient condition, it satisfies the observations.

We may of course consider more than one perturbation on the system. The measurements in Figs. 2 to 6 show the response of the system to the single perturbation of a sinusoidal variation imposed on the flux of acetaldehyde. Suppose, however, we attempt two simultaneous perturbations, that of the flux of acetaldehyde and the flux of oxygen. At once a substantial number of new kinds of measurements becomes possible. For two perturbations we may choose their frequencies to be the same but then still have a choice of the phase between them. In Fig. 7 we show the response of the combustion system to a sinusoidal perturbation in the flux of acetaldehyde and a perturbation in flux of oxygen both at the same frequency but at different relative phase. If the two perturbations are in phase then the response of the system is, within the fundamental entrainment band, periodic and entrainment exists. The same occurs when the perturbation in the oxygen flux is advanced by 90° ahead of that of the perturbation in the acetaldehyde flux. However, if the two perturbations are applied with a phase difference of 180° then the system is no longer entrained. Furthermore, on return to a 90° phase difference we see that the system remains quasi-periodic (not entrained); at 0° phase difference the system is of course entrained again. We thus have what appears to be a new type of "phase" hysteresis on application of two perturbations.

If the frequency of the two perturbations is not the same we have observed a number of new phenomena including entrainment without phase locking [17].

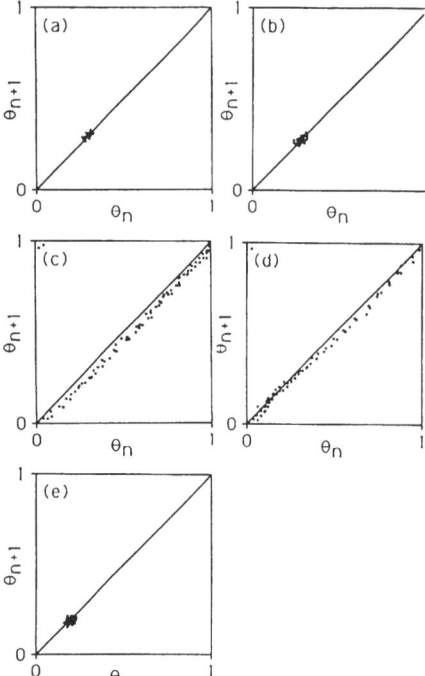

Fig. 7: Response of the oscillatory combustion of acetaldehyde, as measured by light emission, to the simultaneous perturbation of acetaldehyde and oxygen flow rates at the same frequency but different phases. The plots are next-phase plots, in which the (n+1)th phase relation of the acetaldehyde perturbation to the response in light emission is plotted against the nth phase relation. The succession of phase differences with which the oxygen perturbation lags the acetaldehyde perturbation, (a) 0°, (b) 90°, (c) 180°, (d) 90° and (e) 0°, shows hysteresis in entrainment as a function of that phase difference [17]

In seeking new measurements of responses of the system to external periodic perturbations, there is a dual search: for general types of behavior, that is universal relations, and for responses specific to a particular reaction mechanism. Universal relations please the physicist and mathematician in particular, and of course the chemist and biologist as well. If we think about the periodic table all the disciplines are delighted with the similarities among the alkali atoms and ions, and much diverse knowledge is thus categorized. However, to the chemist and biologist the deviations from universal behavior (or the commonalty) among the alkali metals and ions is equally important, as follows for example from the empirical finding that lithium carbonate effectively controls manic depressive illness whereas the other alkali carbonates do not (nor do other alkali salts other than lithium salts). It is similar with complex reaction mechanisms. The existence of bifurcations, limit cycles, period doubling etc. have common features in all reaction mechanisms, and universal laws, such as the occurrence of critical slowing down in the transition from entrainment to quasi-periodic behavior [16] provide powerful insight into commonalties of different reaction mechanisms. The chemist and biologist, however, will seek, in addition, properties which are specific to a reaction mechanism and thus help to distinguish one mechanism from another and help to establish complex reaction mechanisms.

The entrainment bands shown in Fig. 2 give no indication of the complexity of the structure within such bands. A calculation of a large number of different types of dynamical responses within the 1:2 entrainment band for the external periodic perturbation of the reaction of an epoxide with water in the presence of acids [20] is shown in Fig. 8. We have made a beginning on measuring some transitions in such bands from quasi-periodic to entrained responses. The number and variety of dynamic structures are complex and we speculate that there is a need for a mathematical theory of classification, analogous to the application of group theory to the interpretation of the spectra.

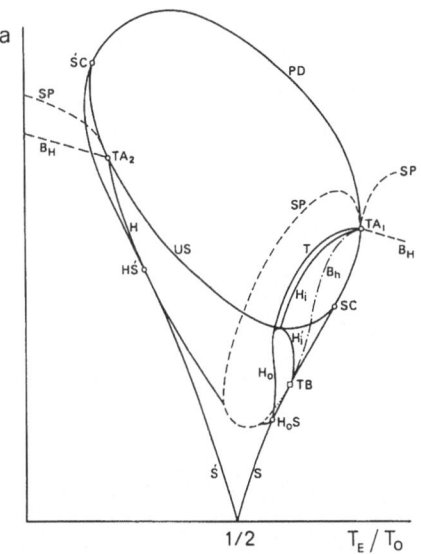

Fig. 8: Plot of calculation of amplitude of perturbation vs. period of external perturbation divided by period of autonomous system. The lines within the entrainment band separate different dynamical regimes. The symbols denote types of transitions (for details see [20])

2. Efficiency of Energy Transduction: The Proton Pump

In a proton pump the chemical energy obtained from the hydrolysis of adenosine triphosphate (ATP) is used to transport protons across a membrane against both a concentration gradient and an electrical potential difference. The energy transduction, from chemical energy to energy storage in a proton gradient, is the essence of the chemi-osmotic hypothesis [21]. This topic is in fact closely related to the discussion external periodic perturbations of chemical reactions, as we shall see by considering the dissipation of a chemical system.

For an isothermal chemical reaction the dissipation is defined as

$$D = \Delta G \ (rate), \tag{1}$$

where D is the dissipation, ΔG is the Gibbs free energy change of the reaction under consideration, and the (rate) is the rate of disappearance of reactants. For a spontaneous reaction ΔG is negative, the rate of disappearance of reactants by definition is negative, and hence the dissipation is positive, as is to be expected from the second law of thermodynamics. For an isothermal system the dissipation equals also the product of the temperature times the rate of entropy production. If the chemical reaction is in a stationary state, a stable node or a stable focus [22], then both ΔG and the rate are fixed quantities and the dissipation therefore is determined and constant. Let us look at the dissipation in a model oscillatory system (the Sel'kov reaction) [23] in which the reactants and products are perturbed periodically, Fig. 9. The dissipation is approximately constant except in regions of frequency marked on the abscissa by dark bands. These bands demarcate the width of the entrainment bands for the given amplitude of perturbation. Within entrainment bands the dissipation undergoes considerable variation, particularly within the fundamental entrainment band; a maximum and minimum are observed and thus by varying the frequency of the perturbation we may vary the dissipation both above and below the approximate average. (The magnitude of the variation of the dissipation depends on the amplitude of perturbation and the particular chemical system.)

If we impose a periodic variation of reactants and (or) products then we vary ΔG periodically, see (1), and we vary the rate of disappearance of reactants periodically. The variation of dissipation now depends on the phase shift of the flux, which is the rate of the reaction, and the thermodynamic force, that is ΔG, of the reaction. This phase shift is dictated by the nonlinearities in the system, and the frequency and amplitude of the external periodic perturbation. The reduction of dissipation is optimized if the rate is a maximum at minimum ΔG and a minimum at maximum ΔG, see (1). Thus we see that external periodic perturbations on nonlinear systems allow a control of the dissipation which is not available for linear systems (where an external periodic perturbation always increases the dissipation beyond that in a stationary state) [2].

Variability of dissipation is also possible in autonomous oscillatory systems, as in glycolysis [24], compared to such reactions in stationary states. The in-

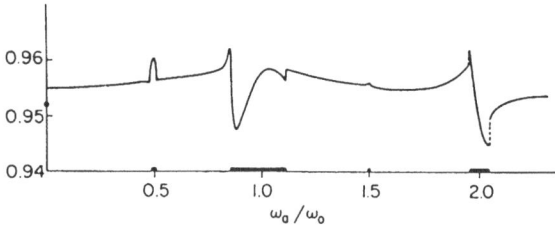

Fig. 9: Dissipation in the Sel'kov model [23] plotted against the ratio of an applied external perturbation of frequency ω_a, divided by the autonomous frequency ω_0 of the Sel'kov model, for a small amplitude of perturbation. The entrainment bands are marked by solid lines in the abscissa [23]

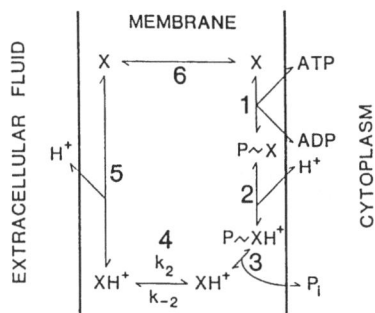

Fig. 10: Sketch of model-reaction sequence for the proton pump [28]

troduction of oscillations in chemical reactions, either imposed or inherent, has some interesting analogies to the transition from D.C. to A.C. electrical circuitry.

A decrease in dissipation in a chemical reaction which acts as an energy transduction mechanism may lead to increased efficiency of that mechanism. However, by means of external periodic perturbation we may also alter the final state of the reacting system and changes in efficiency may thus come about as well.

Consider now a simple model of a proton pump [25] as shown in Fig. 10 as for instance in the plasma membrane of neurospora or yeast cells [26,27]. The enzyme ATPase, denoted simply as X in Fig. 10, accepts an ATP molecule and chemical reaction, the hydrolysis of ATP, occurs. Both ADP and phosphate (P_i) are rejected from the membrane back to the inside of the cell, the cytoplasm, and a proton is accepted from there. The enzyme complex now has both a proton and energy from the hydrolysis of ATP. That energy is used to transport the proton against a concentration gradient and a potential difference, so that a proton may be delivered to the outside of the membrane bordering the extra-cellular fluid. Thereupon the enzyme completes the shuttle back to the side of the membrane facing the cytoplasm and the process may be repeated.

We wish to consider the issue of the thermodynamic, nonequilibrium efficiency of such a pump in two modes of operation: In the first mode the ATP concentration is held fixed and in the second mode the ATP concentration varies with a sinusoidal component of a given amplitude and frequency [28]. We may of course inquire about the work necessary to produce the oscillatory concentration of ATP. In glycolysis under anaerobic conditions ATP is produced such that its concentration oscillates with substantial variations in concentration, up to factor of 4 [29]. In earlier work we have discussed the efficiency of oscillatory production of ATP in glycolysis versus such production with all concentrations in fixed stationary states and found that for appropriate frequencies of autonomous oscillations (without any external periodic perturbations and for fixed input flux of glucose) the oscillatory production of ATP is more efficient [24,30].

We define the efficiency of the proton pump as the ratio of power output of the pump to that of the power input. The power output of the pump is the chemical potential difference of protons on the inside and the outside of the membrane multiplied by the proton flux produced on the outside of the membrane to which has to be added the product of the potential difference across the membrane multiplied by proton flux within the membrane. The power input is the product of the Gibbs free energy change of the hydrolysis of ATP times the rate of hydrolysis.

We digress for a remark on the definition of the thermodynamic efficiency. In classical thermodynamics the efficiency of a Carnot engine is defined as the work produced by the engine in the surroundings per cycle divided by the heat withdrawn from the reservoir at a high temperature per cycle. if the engine operates reversibly, then the efficiency is maximum. Proton pumps in living systems, and many other energy transduction systems, cannot run reversibly because such processes in general take an infinite time [31]. Living systems require energy transduction to be accomplished in finite time and under those circumstances irreversible processes are an essential component of energy transduction which cannot be removed by idealization, as for instance in dealing with friction. In living systems as well as in engines we are interested frequently in the power output and hence we choose a definition of efficiency in terms of power and not work [32,33].

The chemical kinetic equations describing the proton pump shown in Fig. 10 are given elsewhere in detail [28,34]; here we present only a qualitative description in terms of the feedback loops in the reaction mechanism Fig. 11. A periodic variation in ATP concentration affects the proton current. The total ion current in the system is the sum of the currents due to protons, calcium ions and potassium ions. Periodic variation in the proton current affects the time rate of change of the potential across the membrane and hence the potential $\Delta\psi$ itself. In turn, the potential across the membrane determines the fraction of open K^+ channels, M, and their time variation. The same happens for the fraction of open calcium channels. With all these interactions and feedbacks the autonomous system of the proton pump, in the absence of external periodic perturbations, is in a stable focus for lower values of the conductivities of calcium and potassium, and approaches a bifurcation to oscillations at higher values of those conductivities. The first case is shown in Fig. 12 in which we plot the response of the autonomous system to a pulse at zero time and note the oscillatory decay to a stationary state in the proton current, the potential and the total current. This oscillatory decay has been found experimentally [26].

The kinetic equations describing the proton pump are highly nonlinear and can be solved exactly only numerically. In Fig. 13 we show the results of such calculations for the ratio of the thermodynamic efficiency of the pump in the mode of oscillatory ATP divided by the efficiency of the pump in the mode of constant concentration of ATP. For a set of lower values of the ionic conductivities of calcium and potassium there is only a single peak in the ratio of efficiencies as a function of frequency of the oscillatory perturbation. For low amplitudes of that oscillatory perturbation the increase in efficiency above unity is small and

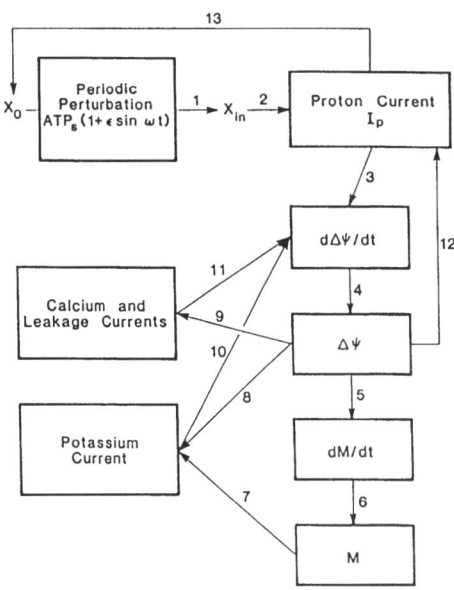

Fig. 11: Sketch of the feedback loops in the system consisting of a proton pump, calcium and potassium ion currents, and a leakage current [28]

42

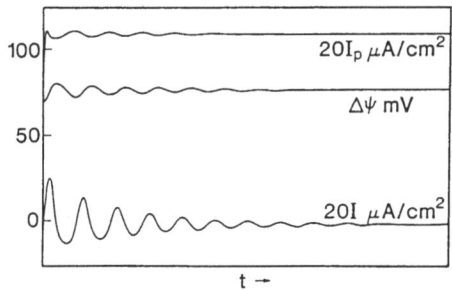

Fig. 12: The proton current I_p, total current I, and membrane potential $\Delta\psi$ plotted as functions of time. At $t = 0$ an impulse of excess ATP is imposed on the system in the steady state mode [28]

for most of the range of frequencies the ratio is below unity. As the amplitude of perturbation is increased, say to 60%, the ratio of efficiency can be as large as 1.08. For a set of ionic conductivities for calcium and potassium ions somewhat larger than the first set we obtain two maxima in the ratio of efficiencies as a function of frequency of variation of ATP. For this second set the autonomous system is close to a bifurcation to oscillations. We observe that the peaks occur at values of the frequency of perturbation divided by the frequency of the autonomous system close to 2 and 1/2, that is close to entrainment bands characterized by those ratios of integers.

The possibility of variable efficiency in the case of a proton pump due to the nonlinearities in kinetic system coupled with oscillatory variation of constraints and in autonomous glycolysis (not discussed here) leads to the possibility of control and evolutionary advantages. Increased efficiency in ATP production (in glycolysis) and ATP utilization (in the proton pump) has advantages in situations of short food supply, which always occurs at some point in the increase of the population of a species. On the other hand a decrease in efficiency, coupled to a higher dissipation, may be desirable under conditions requiring increased heat production.

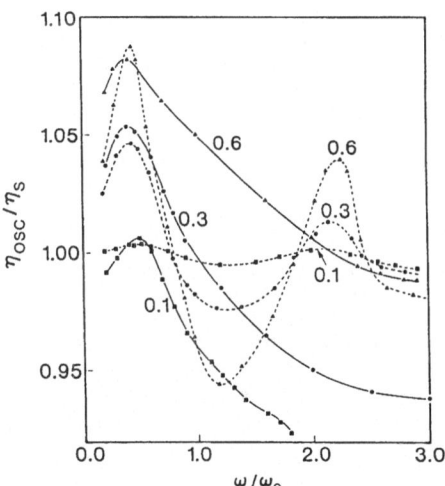

Fig. 13: The ratio of the efficiency in the oscillatory mode to that in the steady mode, η_{osc}/η_s, is plotted against the ratio ω/ω_0, the frequency of ATP oscillations to the relaxation frequency of the autonomous system. The amplitudes of perturbation are $0.1\cdot ATP_s$ (squares), $0.3\cdot ATP_s$ (circles), and $0.6\cdot ATP_s$ (triangles). The solid lines correspond to a set of lower ionic conductivities for Ca^{2+} and K^+, and the dashed lines to a set of higher values (see [28] for full details)

43

3. Delay Kinetics

The systems discussed in the prior sections are describable by ordinary differential equations as is the usual case for chemical kinetics in a spatially homogeneous system. Consider, however, a set of chemical reactions indicated symbolically as n steps in Fig. 14, in which the product of the end step influences the first step after some delay of time, τ. This delay may come about because of a spatial separation between the sites of the nth and the first step in the reaction mechanism, and biological reactions with delayed feedback have received some attention [35]. A similar but somewhat different delay effect may occur in a continuous stirred tank reactor in which a part of the output of the reactor is recycled into the input stream to the reactor, with a certain delay. Here the delay is imposed and the system with multiple stationary states in the reactor, for example, can be studied in much additional detail, in particular in regard to a precise determination of the branch of unstable states of the system [36]. Finally in Fig. 14C an optical delay system is sketched [37]. Physical and chemical systems may have natural or imposed delays and in either case such systems show the variety of dynamic behavior observed with complex reaction mechanisms and no delay, that is multiple stationary states, oscillations, quasi periodicity and chaos. Systems with time delay are equivalent to systems with an infinite number of degrees of freedom describable by ordinary differential equations [38]. Hence, even more complex behavior is possible in systems with time delay [39].

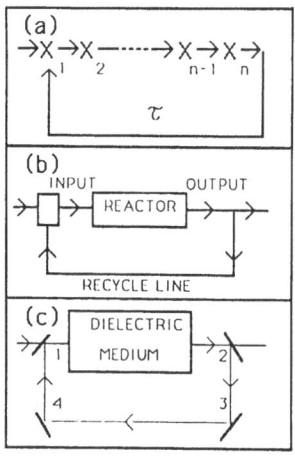

Fig. 14: Three examples of processes involving a delayed feedback. (a) A chain of chemical reactions in which the end product inhibits an earlier reaction; X_i denotes only the step. Each step involves an arbitrary number of species. The time required for the end product to move from its site of the production to the inhibition site is the origin of a time delay. (b) Diagram of a reactor with recycling which introduces a time delay. (c) A ring cavity containing a nonlinear medium. Mirrors labeled 1,2 (partial reflectivity), 3 and 4 (complete reflectivity) direct part of the transmitted light back into the medium [38]

Imposition of a delay can be utilized to stabilize unstable stationary states and to study properties of other nonequilibrium states. Biological organisms may use natural time delays, e.g. by turning on and off the mechanism for the delay, to charge and control the dynamic behavior of complex biochemical processes [34,38]. Therefore, it may be important in the study of complex systems to determine if delay plays an essential role.

4. Summary

External periodic variations of constraints of a non-linear system provide interesting and useful information on the dynamics of the system and the reaction mechanism; such variations further provide elements of control over the system and affect reaction efficiencies.

Dedication and Acknowledgement

It is a pleasure to devote this article to Professor Benno Hess in honor of the celebration of his 65th birthday. His science and his friendship have been an inspiration and joy.

This work was supported by the National Science Foundation, the Air Force Office of Scientific Research and the Department of Energy/BES Engineering Research Program.

References

1. (a) K. Tomita, T. Kai, F. Hikami: Prog. Theor. Phys. **57**, 1159 (1977)
   (b) T. Kai, K. Tomita: Prog. Theor. Phys. **61**, 54 (1979)
2. (a) P.H. Richter, J. Ross: J. Chem. Phys. **69**, 5521 (1978)
   (b) P.H. Richter: Physica **10D**, 353 (1984)
3. (a) P. Rehmus, J. Ross: J. Chem. Phys. **78**, 3747 (1983)
   (b) P. Rehmus, J. Ross: In _Oscillations and Travelling Waves_, ed. by R.J. Field and M. Burger (Wiley, New York 1985) p.287
   (c) P. Rehmus, W. Vance, J. Ross: J. Chem. Phys. **80**, 3373 (1984)
4. M.R. Guevara, L. Glass: J. Math. Biol. **14**, 1 (1982)
5. (a) A. Boiteux, A. Goldbeter, B. Hess: Proc. Natl. Acad. Sci. USA **72**, 3829 (1975)
   (b) M. Markus, S.C. Müller, B. Hess: Ber. Bunsenges. Phys.Chem. **89**, 651 (1985)
   (c) M. Markus, B. Hess: Proc. Natl. Acad. Sci. USA **81**, 4394 (1985)
   (d) M. Markus, D. Kuschmitz, B. Hess: FEBS Lett. **172**, 235 (1984)
6. (a) M. Dolnik, I. Schreiber, M. Marek: Phys. Lett. A **100**, 316 (1984)
   (b) T.W. Taylor, W. Geiseler: Ber. Bunsenges. Phys. Chem. **89**, 441 (1985)
   (c) F.W. Schneider: Ann. Rev. Phys. Chem. **36**, 347 (1985)
7. F. Bucholtz, F.W. Schneider: J. Am. Chem. Soc. **105**, 7450 (1983)
8. (a) P.G. Felton, B.F. Gray, N. Shank: In _Second European Symposium on Combustion_ (The Combustion Institute, Orleans, France 1975)
   (b) P. Gray, J.F. Griffiths, S.M. Hasko, P.-G. Lignola: Combust. Flame **43**, 174 (1981)
   (c) P. Gray, J.F. Griffiths, S.M. Hasko, P.-G. Lignola: Proc. R. Soc. London Ser. A **374**, 313 (1981)
   (d) B.F. Gray, J.C. Jones: Combust. Flame **57**, 3 (1984)
9. (a) S.A. Pugh, H.-R. Kim, J. Ross: J. Chem. Phys. **86**, 776 (1987)
   (b) S.A. Pugh, J. Ross: J. Phys. Chem. **91**, 2178 (1987)
10. R. Harding, H. Sevcikova, J. Ross: to be submitted for publication
11. (a) J.C. Roux, R.H. Simoyi, H.L. Swinney: Physica **8D**, 257 (1983)
    (b) F. Argoul, A. Arneodo, P. Richetti, J.C. Roux: J. Chem. Phys. **86**(6) 3325 (1987)
12. (a) S. Grossman, B. Sonneborn-Schmick: Phys. Rev. A. **25**(4), 2371 (1982)
    (b) A. Babloyantz, A. Destexhe: Proc. Natl. Acad. Sci. USA **83**, 3513 (1986)
13. (a) A. Wolf, J.B. Swift, H.L. Swinney, J.A. Vastano: Physica **16D**, 285 (1985)
    (b) J.M. Greene, J.S. Kim: Physica **24D**, 213 (1987)
    (c) S. Sato, M. Sano, Y. Sawada: Prog. Theor. Phys. **77**(1), 1 (1987)
14. (a) P. Grassberger, I. Procaccia: Phys. Rev. Lett. **50**(5), 346 (1983)
    (b) R. Badii, A. Politi: J. Stat. Phys. **40** (5,6), 725 (1985)
    (c) J. Guckenheimer, G. Buzyna: Phys. Rev. Lett. **51**, 1438 (1983)
15. J.D. Farmer, E. Ott, J.H. Yorke: Physica **7D**, 153 (1983)
16. (a) S.A. Pugh, M. Schell, J. Ross: J. Chem. Phys. **85**, 868 (1986)
    (b) S.A. Pugh: Ph.D. thesis, Stanford University, Stanford, CA (1985)
17. S.A. Pugh, B. Dekock, J. Ross: J. Chem. Phys. **85**, 879 (1986)
18. (a) M.R. Guevara, L. Glass, A. Shrier: Science **214**, 1350 (1981)
    (b) L. Glass, R. Perez: Phys. Rev. Lett. **48**, 1772 (1982)
19. (a) G. Hayashi: In _Nonlinear Oscillations in Physical Systems_ (Princeton University Press, Princeton, N.J. 1985)
    (b) J. Guckenheimer, P. Holmes: _Nonlinear Oscillatory Dynamical Systems and Bifurcations of Vector Fields_ (Springer Verlag, New York 1983)

(c) V.I. Arnold: <u>Geometrical Methods in the Theory of Ordinary Diff.</u> (Springer Verlag, Berlin 1983)
20. W. Vance, J. Ross: to be published
21. P. Mitchell: <u>Chemosmotic Coupling and Energy Transduction</u> (Glynn Research, Bodmin 1968)
22. I.D. Huntley, R.M. Johnson: <u>Linear and Nonlinear Differential Equations</u>, Chap. 3 (Wiley, New York 1983)
23. P.H. Richter, P. Rehmus, J. Ross: Prog. Theor. Phys. <u>6</u>, 385 (1981)
24.(a) Y. Termonia, J. Ross: Proc. Natl. Acad. Sci. USA <u>78</u>, 2952 (1981)
    (b) Y. Termonia, J. Ross: ibid. <u>78</u>, 3563 (1981)
    (c) Y. Termonia, J. Ross: ibid. <u>79</u>, 2878 (1982)
25. C.L. Slayman: Membr. Transp. <u>1</u>, 485 (1982)
26. C.L. Slayman, W.S. Long, D. Gradmann: Biochim. Biophys. Acta <u>426</u>, 732 (1975)
27.(a) J.-P. Dufour, A. Goffeau: Eur. J. Biochem. <u>105</u>, 145 (1980)
    (b) J. Slavik, A. Kotyk: Biochim. Biophys. Acta <u>766</u>, 679 (1984)
    (c) B. Hess, M. Markus, D. Kuschmitz: <u>Progress in Bioorganic Chemistry and Molecular Biology</u>, ed. by Yu. A. Ovchinnikov (Elsevier Science Publishers, B.V. 1985) p.165
    (d) A. Pena, S. Uribe, J.P Pardo, M. Borbolla: Arch. Biochem. Biophys. <u>231</u>, 217 (1984)
28. M. Schell, K. Kundu, J. Ross: Proc. Natl. Acad. Sci. USA <u>84</u>, 424 (1987)
29. K. Tornheim, J.M. Lowenstein: J. Biol. Chem. <u>250</u>, 6304 (1975)
30. P.H. Richter, J. Ross: Biophys. Chem. <u>12</u>, 285 (1980)
31.(a) V. Fairen, J. Ross: J. Chem. Phys. <u>75</u>, 5490 (1981)
    (b) V. Fairen, M.D. Hatlee, J. Ross: J. Phys. Chem. <u>86</u>, 70 (1982)
32.(a) A. Katchalsky, P.F. Curan: <u>Nonequilibrium Thermodynamics in Biophysics</u> (Harvard University Press, Cambridge 1965)
    (b) S. Minakami, H. Yoshikawa: Biochem. Biophys. Res. Comm. <u>18</u>, 345 (1965)
    (c) I.Z. Steinberg, A. Oplutka, A. Katchalsky: Nature <u>210</u>, 568 (1966)
    (d) A.L. Lehninger: <u>Biochemistry</u> (Worth, New York 1976)
    (e) L.O. Bjorn: Photosynthetica <u>10</u>, 121 (1976)
    (f) J.W. Stucki: In <u>Energy Conservation in Biological Membranes</u>, ed. by G. Schäfer and M. Klingenberg (Springer, New York 1978) p.264
    (g) L.A. Blumenfeld: <u>Physics of Bioenergetic Process</u> (Springer, New York 1983) Sect. 5.4.1
    (h) J.W. Stucki, M. Compianim, S.R. Caplan: Biophys. Chem. <u>18</u>, 101 (1983)
    (i) P.D. Weer: In <u>Electrogenic Transport Fundamental Principles and Physio-logical Implications</u>, ed. by M.P. Blaustein and M. Lieberman, Vol. 38 (Society of General Physiologists Series, Raven Press, New York 1984) p.1
33.(a) F.L. Curzon, B. Ahlborn: Am. J. Phys. <u>43</u>, 22 (1975)
    (b) J.W. Warner, R.S. Berry: J. Phys. Chem. <u>91</u>, 2216 (1987)
34. J.Ross, M. Schell: Ann. Rev. Biophys. Chem. <u>XVI</u> (1987)
35.(a) N. MacDonald: <u>Time Lag in Biological Models</u> (Lect. Notes in Biomath., Vol. 27 (Springer, New York 1978)
    (b) U. an der Heiden: J. Math. Anal. Appl. <u>70</u>, 599 (1979)
    (c) P.E. Rapp: In <u>Mathematical Models in Molecular and Cellular Biology</u>, ed. by L.A. Segel, Chap. 3 (Cambridge University, New York 1980) p.146
    (d) P.E. Rapp, A.I. Mees, C.T. Sparrow: J. Theor. Biol. <u>90</u>, 531 (1981)
    (e) L. Glass, M.C. Mackey: Ann. N.Y. Acad. Sci. <u>316</u>, 214 (1979)
    (f) J.J. Tyson: J. Theor. Biol. <u>103</u>, 313 (1983)
    (g) U. an der Heiden, M.C. Mackey: J. Math. Biol. <u>16</u>, 75 (1982)
    (h) M.C. Mackey, U. an der Heiden: J. Math. Biol. <u>19</u>, 211 (1984)
36. E.C. Zimmermann, M. Schell, J. Ross: J. Chem. Phys. <u>81</u>, 1327 (1984)
37.(a) K. Ikeda: Opt. Commun. <u>30</u>, 257 (1979)
    (b) H.M. Gibbs, F.A. Hopf, D.L. Kaplan, R.L. Schoemaker: Phys. Rev. Lett. <u>46</u>, 474 (1981)
    (c) P. Nardone, P. Mandel, R. Kapral: Phys. Rev. A <u>33</u>, 2465 (1986)
38. M. Schell, J. Ross: J. Chem. Phys. <u>85</u>, 6489 (1986)
39. J.D. Farmer: Physica <u>4D</u>, 366 (1982)

Part II

Chemical Organization

Distinction Between Amplified Noise and Deterministic Chaos by the Correlation Dimension

A. Freund, Th.-M. Kruel, and F.W. Schneider

Institute of Physical Chemistry, University of Würzburg,
Marcusstr. 9/11, D-8700 Würzburg, Fed. Rep. of Germany

1. Introduction

All experimental data are subject to noise due to a variety of sources depending on the nature of the experiment. In a continuous flow stirred tank reactor (CSTR), for example, relatively large concentration gradients may exist in the immediate neighborhood of the in-flow tube where the feed stream comes into first contact with the contents of the reactor. Due to stirring there will be concentration gradients variable in time which will decrease with increasing distance from the point of entry of the feed stream. Since mixing is not instantaneous, these concentration gradients may represent macroscopic fluctuations in concentration (and temperature) for time periods which also depend on the geometry of the reactor. Thus the CSTR represents a fluctuation or noise generator. Reactant streams may be important [1].

Nonlinear chemical reactions may be quite sensitive to local concentration fluctuations. Amplification of the macroscopic fluctuations is expected to occur through the non-linear mechanism. The problem arises of how the resulting macroscopic motions (due to these amplified fluctuations) may be distinguished from any deterministic chaos present in a particular system. As a solution we have previously suggested observation of the dependence of the largest Lyapounov exponent (λ_{max}) on the rate of stirring at a constant flow rate [2,3]. At high stirring rates the average size of the macroscopic fluctuations in the reactor will be smaller than at low stirring rates. Therefore any kinetic effects due to macroscopic fluctuation will be enhanced at low stirring rates. In all of our following considerations the quantitative size of the fluctuations will not be important. In previous work [3] we have done numerical integrations of the periodically perturbed Brusselator mechanism by superimposing statistical fluctuations of various amplitudes (multiplicative noise) on the concentration variables. As a result the extrapolation of λ_{max} to zero fluctuations led to positive Lyapounov exponents for chaos and zero values for periodic motions, respectively, in the driven Brusselator. Interestingly, all λ_{max} values were positive for sufficiently large noise even when the motions were periodic. Thus we conclude that a single positive value of λ_{max} cannot be regarded as a reliable indication for deterministic chaos in a noisy experiment. We have tested this assertion in the experimental Belousov-Zhabotinsky (BZ) reaction in a CSTR of 1.61 ml volume where the dependence of λ_{max} on the rate of stirring between 175 and 2000 rpm was measured. An extrapolation to highly effective stirring at high stirring rates showed positive values of λ_{max} indicating the presence of deterministic chaos at this particular flow rate. On the other hand, experiments in the periodic range of the BZ reaction lead to an extrapolation for λ_{max} of zero at high stirring rates in agreement with the general theoretical predictions based on the Brusselator which, however, should not be regarded as a specific model for the Belousov-Zhabotinsky reaction.

2. Correlation Dimension

In this work we investigate the use of the correlation dimension D_{corr} according to GRASSBERGER and PROCACCIA [4] with the intention of providing another tool to distinguish between amplified noise and deterministic chaos in a general theoretical model (Brusselator) and in the BZ reaction in a CSTR. The "correlation integral" $C(r)$ is defined as

$$C(r) = \frac{1}{N^2} \sum_{i=1}^{N} \sum_{\substack{j=1 \\ i \neq j}}^{N} H\left(r - |\vec{x}_i - \vec{x}_j|\right),$$

where H is the Heaviside function. $C(r)$ is the number of pairs of points $\vec{x}_i$ and $\vec{x}_j$ in a sphere of radius r about a point $\vec{x}_i$ on the attractor. The correlation dimension D_{corr} is represented by a slope according to [4]

$$D_{corr} = \lim_{N \to \infty} \lim_{r \to 0} \frac{d \log_2 C(r)}{d \log_2 r}.$$

The construction of attractors may be carried out by two methods which we compare in this work. Most authors use the method of delays by TAKENS [5]. One successively applies a delay time τ to the digitized data of one single variable to generate all higher dimensions of the attractor according to $x(t_k)$, $x(t_k + \tau)$, $x(t_k + 2\tau)$, ... , $x(t + (n-1)\tau)$, where $t_k = k\Delta t$, $k = 1, 2, \ldots \infty$ and n is the embedding dimension. The resulting multidimensional curve is called the attractor of the system. It is usually observed in a two-dimensional projection. For the determination of τ we used the first zero of the calculated autocorrelation of the time series, although other procedures have been proposed such as the mutual information [6] or the test for stationarity of D_{corr} [7]. Another method for the construction of an attractor from a time series of a single variable makes use of the derivatives of the system, x', x'', ... $x^{(n)}$. This method was originally proposed by PACKARD et al. [8], but to our knowledge it has not been used in the calculation of the correlation dimension of experimental systems so far. The method according to PACKARD et al. is characterized by the following points: the choice of a good value of the delay time becomes unnecessary; the coordinates obtained by this procedure have no linear interdependence, and finally, there is no need to follow the reconstruction to 2m+1 coordinates as postulated by Takens' embedding theorem, where m is the effective number of independent variables.

The procedure is carried out numerically as follows: the experimental data are interpolated by a natural cubic spline fit; then the derivative is taken from the spline coefficients. The derivative obtained is treated in the same manner to give the second and higher derivatives. This procedure is called "spline on spline". Numerical errors accumulate at higher derivatives of the time series. Therefore one has to increase the frequency of data sampling in order to obtain reliable results at higher dimensions. It is important that all derivatives are normalized to unity, since their absolute values depend on the time scale of the dynamic process. Thus only normalized derivatives guarantee an equal contribution of every coordinate to the correlation integral in the derivative method.

50

3. Brusselator

In order to learn from the behaviour of a model, we present some Brusselator calculations that were performed in two regions of the phase diagram [9] of the driven oscillator, namely at the point of known chaos where A = 0.4, B = 1.2, α = 0.08 and ω_p = 0.852 rad/time and at a point of periodic motion for A = 0.4, B = 1.2, α = 0.04 and ω_p = 0.290 rad/time where A, B, α and ω_p represent reactant concentrations A, B, the amplitude of perturbation and the frequency of periodic perturbation, respectively. The periodic motion was chosen to be close to a quasi-periodic boundary in view of the possibly strong effects of large amplitude fluctuations to be superimposed later on the concentration variables. We used both Takens' delay method and the method of derivatives for the reconstruction of the attractors.

3.1. Chaos

For chaos in the driven Brusselator the correlation integral showed approximately a uniform scaling region for both methods when its logarithm was plotted versus the logarithm of the sphere radius r (Fig.1a,c). With increasing embedding dimension the slope approached a constant but fractal value of $D_{corr} \approx 2.26$ for the delay method. This represents deterministic chaos (Fig.1b). The derivative method gave D_{corr} = 2.05 in good agreement with calculations done by HAO

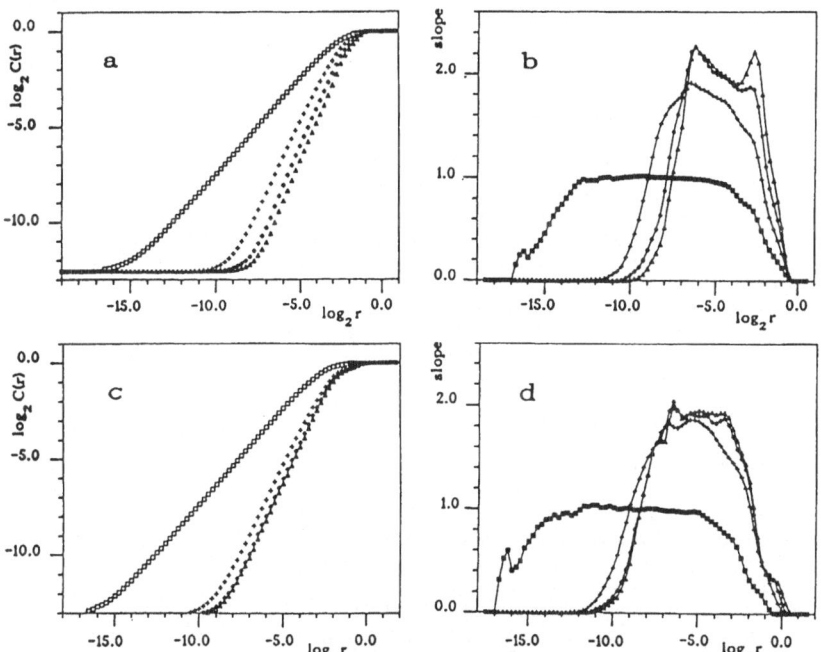

Fig.1.   Chaos in the driven Brusselator (ω_p = 0.852; α = 0.080): delay method (τ = 2.5) (a,b) and derivative method (c,d); correlation integral versus sphere radius r (a,c) and slope versus radius (b,d).
$D_{corr} \approx 2.26$ (at maximum) for the delay method (b)
$D_{corr} \approx 2.05$ (at maximum) for the derivative method (d)
$\square$ dimension 1, + dimension 3, $\diamond$ dimension 5, $\triangle$ dimension 7

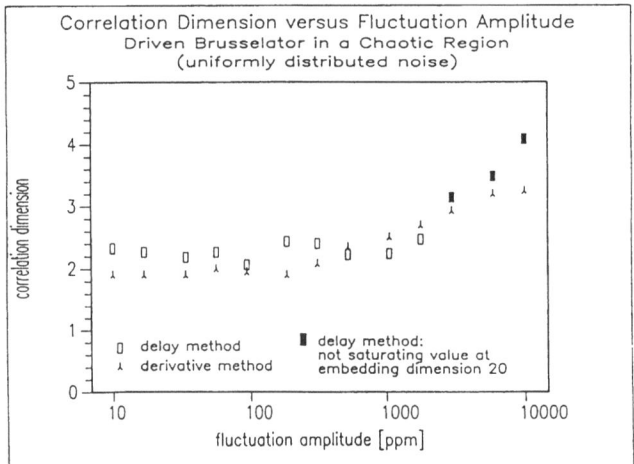

Fig.2.   Chaos in the driven Brusselator (ω_p = 0.852; α = 0.080)
          as a function of uniformly distributed noise:
          correlation dimension versus noise amplitude for the delay
          method (τ = 2.5); D_{corr} increases with noise amplitude;
          values at embedding dimension 20 did not saturate in the
          delay method for high fluctuation amplitudes

and ZHANG [10] (D_{corr} = 2.18) who used the boxcounting method [11]
which is considered to be reliable only for low dimensions as
encountered here [12]. The maximum dimension of the driven Brusselator is equal to three which is equal to its number of independent
variables. In addition we superimposed fluctuations of Gaussian
distribution on the reactant concentrations A and B with relative
standard deviations σ (in ppm). This simulates the effect of multiplicative white noise on the driven Brusselator. One average
oscillation period was approximated by about 3000 integration steps.
As the amplitude σ of the superimposed fluctuations increased, D_{corr}
for chaos also increased for both methods (Fig.2). Both methods
display the best agreement at low fluctuation amplitudes whereas
they diverge at higher amplitudes. The reason for this behaviour may
be the loss of precision of the higher derivatives as calculated by
the cubic spline procedure for "noisy" quasi-experimental data.   We
intend to investigate this effect further.

3.2. Periodic Oscillations

For superimposed Gaussian fluctuations in periodic motion, the correlation curves show characteristic breaks at r_{crit} (Fig.3a) whose
location shifts to higher r values with increasing fluctuation size
(Fig.4). A plot of the slope versus log r shows a maximum at low r
and a lower plateau (or a minimum) at higher r (Fig.3b). Breaks in
the correlation curves have been observed earlier [13]. They are
considered to separate the correlation dimension of the superimposed
noise at small r from the correlation dimension of the attractor
itself at high r. This is borne out nicely in our derivative calculations as seen in Fig.5 which shows a slight increase of D_{corr}
for the attractor from unity (corresponding to a limit cycle) to
1.07 at 10^4 ppm in parallel with an increase of D_{corr} for the lower
scaling region from 1.75 at 10 ppm to approximately 2.0 at 10^4 ppm.
Here the derivative method gave consistent results with only a small

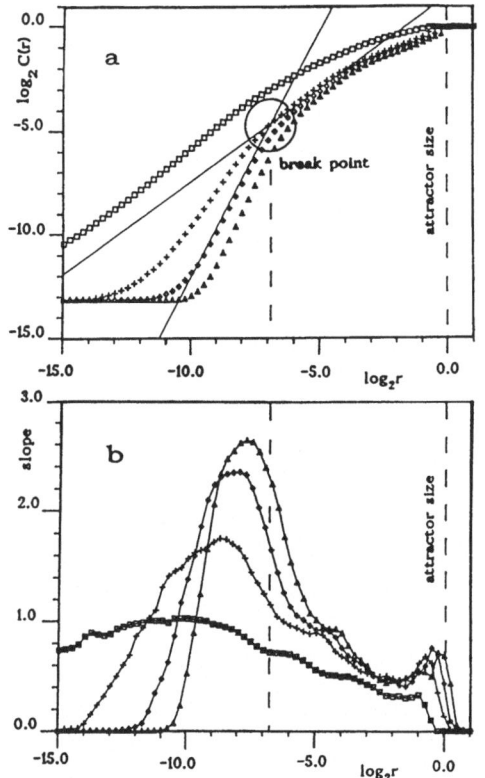

Fig.3. Periodic motion in the driven Brusselator for Gaussian fluctuations (σ = 1000 ppm):
a) correlation integral C(r) versus sphere radius r in the derivative method. Notice the break point at r_{crit};
b) slope of above curves versus r,
□ dimension 1,
+ dimension 3,
◇ dimension 5,
△ dimension 7.

spill-over of the noise into the D_{corr} of the attractor. At very high fluctuation sizes (>10^3 ppm) a break point could no longer be observed in either method. Here it becomes increasingly difficult to determine any definite values whatsoever for D_{corr} of the attractor. This indicates that the structure of a periodic attractor may be blurred by large external noise in such a fashion that it may not be

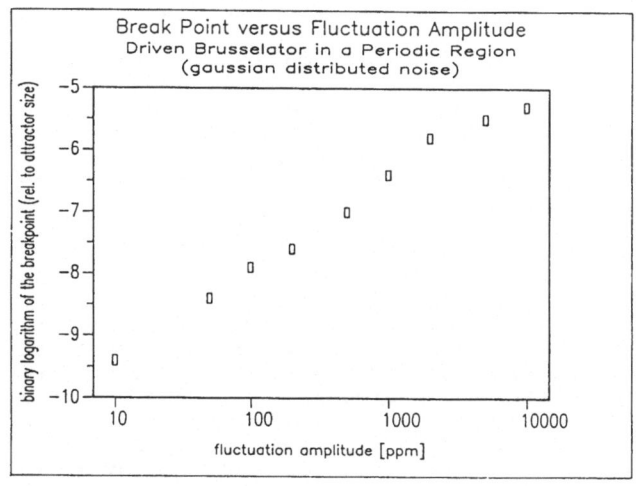

Fig.4. Break point (r_{crit}) versus fluctuation amplitude for Gaussian fluctuations in the driven Brusselator in a periodic region (ω_p = 0.290; α = 0.040; derivative method)

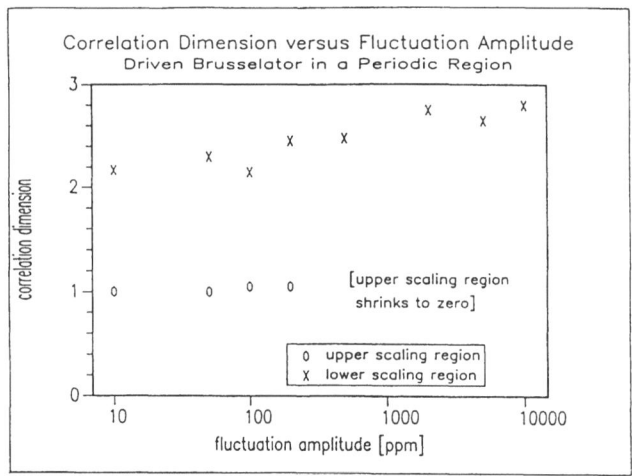

Fig.5. Periodic motion in the driven Brusselator for Gaussian
        fluctuations   in   the   driven   Brusselator $(\omega_p$ =   0.290;
        α = 0.040; derivative method); D_{corr} versus fluctuation
        amplitude; D_{corr} for the attractor cannot be determined
        above 1000 ppm, since the upper scaling region disappears

distinguishable from a chaotic one. These results indicate that the
application of the above methods to experimental data should be
carried out with great care as demonstrated in the following experiments.

4.Experiments

4.1.Nonperiodic Oscillations in the Belousov-Zhabotinsky Reaction

Our experimental set-up for the BZ experiments uses a syringe pump
(Infors Precidor) that drives three 50 ml syringes to deliver three
solutions   into a 1.61 ml CSTR at constant flow rate via a stepping
motor which is regulated by a desk computer. For the experimental
investigation of nonperiodic oscillations we use concentrations and
residence times similar to those of HUDSON and coworkers [14] in the
BZ reaction. Our piston pump and reactor construction allowed only 3
solutions to be delivered simultaneously. Syringes 1, 2 and 3 contain 0.9 M malonic acid and 0.003 M $Ce_2(SO_4)_3$; 0.6 M H_2SO_4; and
0.42 M $NaBrO_3$, respectively. The whole set-up is thermostatted at
25.00 ± 0.02°C. Digital data (absorbance of Ce^{+4} at 350 nm) are
taken every 2 sec for oscillation periods of 45 to 150 sec; 3,600
data points are usually taken per run. It is known that periodic
windows may be found in a region of chaotic motion, when a control
parameter, such as the flow rate, is altered. In the present BZ
experiments a high flow rate of 0.225 min^{-1} was used similar to
HUDSON and coworkers [14]. This produced a nonperiodic appearance in
the time series and in the broad Fourier spectra. In order to test
whether the present conditions are characteristic of a true chaotic
attractor we carried out a series of experiments at various stirring
rates. The rate of rotation of a relatively large magnetic stirrer
could be adjusted with high precision in the range of 175-2000 rpm.
Time series of ·100 oscillations were taken for about 2 hours at

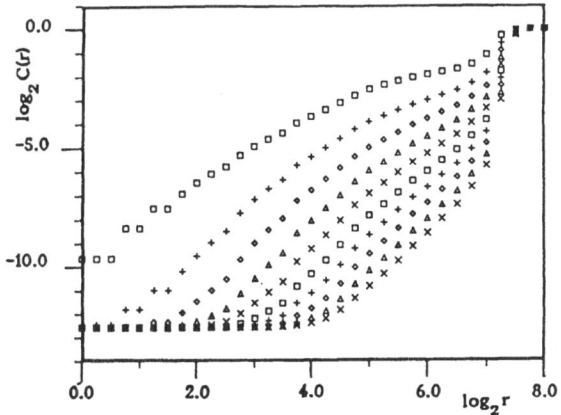

Fig.6. BZ reaction at k_f = 0.225 min⁻¹: correlation integral versus r (delay method, τ = 30 sec), breaks disappear at high embedding dimensions, □ dimension 2, successive curves differ by 2 dimensions, × (right side) is dimension 20

each stirring rate. The resulting time series were subjected to the same statistical analyses as the above Brusselator calculations. Takens' delay method for the calculation of D_{corr} showed definite breaks in the log C versus log r plots which, however, tended to disappear at high embedding dimensions (Fig.6). The upper scaling region showed D_{corr} values around two with larger scatter at low stirring rates (Fig.7).

This is usually interpreted as chaos. The lower scaling region consistently led to D_{corr} values (below three) characteristic of higher dimensional noise. Saturation occurred in both scaling regions at an embedding dimension of about 16, indicating that about 7-8 independent variables are necessary to describe this reaction. On the other hand, the application of the derivative method to the same data clearly showed the presence of periodic motion with lower values of D_{corr} ($\approx$ 1.10) than those of the Takens method (Fig.7). The dimensionality was consistently below two. Extrapolation to high stirring rates retained the low dimensionality. This behaviour is characteristic of a "noisy" limit cycle. The lower scaling range

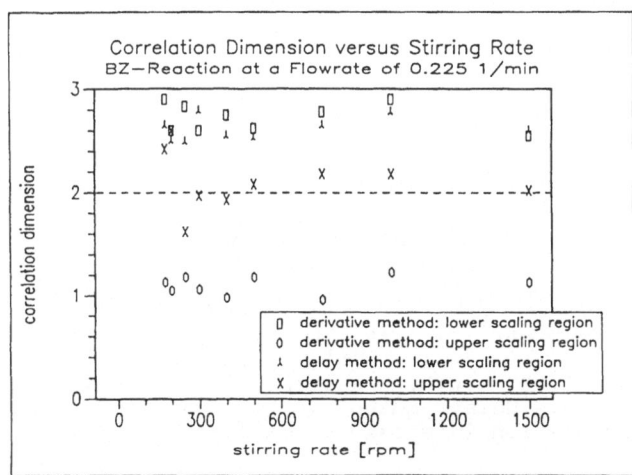

Fig.7.   BZ reaction at k_f = 0.225 min⁻¹:
           D_{corr} versus stirring rate, τ = 30 sec (delay method)

showed a higher dimensionality with values comparable with those of the Takens method. Saturation occurred at a reconstruction dimension between 5 and 7 in good agreement with the Takens method. Obviously, in the presumed chaotic region, a contradiction exists between the results of the two methods, about which we can only speculate. It seems that the low density of the data points together with the high experimental "noise" are partially responsible for this difference.

4.2.Periodic Oscillations in the BZ Reaction

In order to observe periodic motion in the BZ reaction several flow rates were used. We report here on experiments at two flow rates. The experiments at the low flow rate (= 0.087 min^{-1}) produced a time series of almost perfect periodic appearance whose Fourier spectrum showed a relatively sharp fundamental frequency. For this time series we show the curves of the correlation integral versus log r and the calulated slopes for the delay and the derivative methods (Fig.8). It is seen that a break occurs and that saturation in the slopes is achieved for 5-7 independent variables where $D_{corr} \approx 1.22$ for the Takens method and 1.47 for the derivative method in the plateau region. This result is expected for a noisy limit cycle. Interestingly, the D_{corr} values for the lower scaling regions are significantly higher, indicating the presence of substantial experimental noise. From the time series (not shown) this noise level is upon casual observation not immediately obvious.

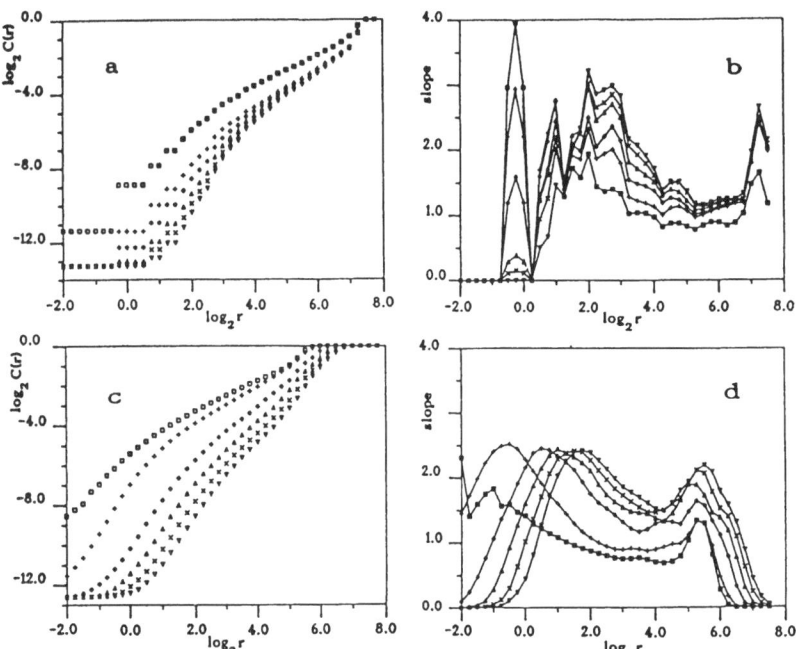

Fig.8. BZ reaction at k_f = 0.087 min^{-1}, stirring rate = 1000 rpm, periodic oscillations: delay method (a,b) and derivative method (c,d); correlation integral versus r (a,c) and slope versus r (b,d).
$D_{corr} \approx 1.22$ (delay method τ = 30 sec); $D_{corr} \approx 1.47$ (derivative method); □ dimension 2, + dimension 3, ◇ dimension 4, ▲ dimension 5, × dimension 6, ▼ dimension 7

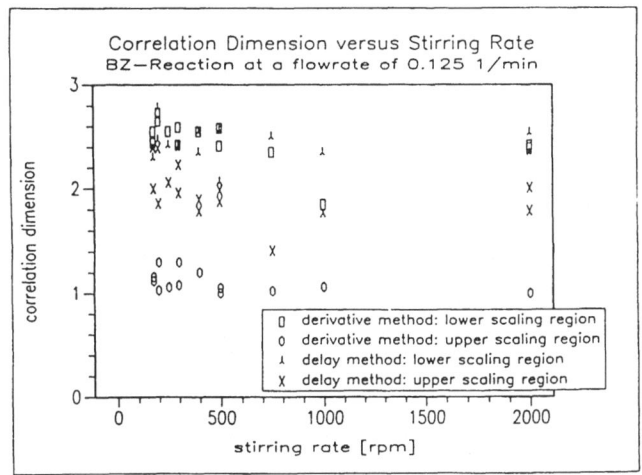

Fig.9. BZ reaction at k_f = 0.125 min^{-1}:
     D_{corr} versus stirring rate, (delay method, τ = 25 sec)

A dependence on the stirring rate was carried out for periodic motion at a medium flow rate of 0.125 min^{-1} (Fig.9). The experimental conditions for Figs. 8 and 9 were 0.90 M malonic acid and 0.003 M Ce(NO$_3$)$_3$; 1.12 M H$_2$SO$_4$; and 0.26 M KBrO$_3$ in syringes 1, 2, and 3, respectively. Breaks in the experimental correlation curves (not shown) are evident in both methods, in agreement with the above mentioned Brusselator calculations for periodic motion with superimposed fluctuations. The upper scaling region produces D_{corr} values above two for the delay method at low stirring rates, indicating the presence of amplified fluctuations, since at higher stirring rates D_{corr} declines below two and rises again at 2000 rpm. For the derivative method D_{corr} is significantly below two in the upper scaling region, indicating the presence of "noisy" periodic oscillations. In the lower scaling region the D_{corr} values obtained from the two methods substantially agree at all stirring rates. They range between 2 and 3 with larger scatter at low stirring rates. Formally speaking, chaos is indicated in the Takens method at low stirring rates whereas an indication for fractal D_{corr} values above two is not obtained in the derivative method.

5. Discussion

To obtain the true dimension of an attractor, it is desirable to extrapolate all D_{corr} values to vanishing fluctuations or, in a CSTR experiment, to very high stirring rates [2]. A comparison of D_{corr} as calculated from Takens' delay method and the derivative method showed good agreement in the case of chaotic and periodic attractors in the driven Brusselator model. Moreover, both methods produced the literature values for Lorenz and Rössler chaos. It should be kept in mind that these methods are restricted to low dimensionalities [12]. For high amplitudes of superimposed fluctuations the D_{corr} values consistently increase above their values at zero fluctuations, although the latter increase is difficult to determine for periodic motions for which the upper scaling region eventually shrinks to zero (Fig.5). This increase of D_{corr} at large fluctuations provides the basis for the experimental stirring rate dependence which we propose here [2,3].

In our BZ experiments we found a large difference between the Takens and the derivative method, particularly in the case of presumed chaos (Fig.7). While the Takens method gave indications of chaos by fractal D_{corr} values above two for most but not all points, the derivative method showed D_{corr} values consistently close to unity as characteristic for a "noisy" limit cycle. Part of the reason for this difference is probably the low sampling frequency of data points (~50) per average oscillation. Brusselator calculations of D_{corr} in the chaotic region as a function of sampling frequency clearly show that the sampling frequency has to be sufficiently high (> 100) to obtain the literature value of D_{corr} in the derivative method (Tab.1). Therefore it is necessary to always ensure that the experimental sampling frequency is appropriately high as well. This may be achieved by calculating D_{corr} at various experimental sampling frequencies.

Table 1

Derivative Method : Effect of Sampling Rate

Brusselator in a Chaotic Region:
$\omega_p = 0.852$ rad/sec; $\alpha = 0.080$

sampling rate in data/mean period	correlation dimension	literature values [10]	
		from Lyapunov exponents[1]	2.15
		from boxcounting algorithm	2.18
17	1.84		
35	1.90	[1] via Kaplan-Yorke conjecture [18]	
70	1.96		
140	2.12		

It has been shown that Takens' delay method depends on the correct choice of the delay time. We chose the first zero of the calculated autocorrelation function of a given data set. In the test for stationarity of D_{corr} one calculates D_{corr} for a series of τ values and chooses the τ range at which D_{corr} remains effectively constant. This method turned out to be extremely time consuming and it did not lead to a satisfactory stationarity in the D_{corr} values for our BZ system. Problems may arise if the data are strongly autocorrelated as THEILER [15] has pointed out. Our method of calculation avoids the possible pitfalls due to these autocorrelation effects. Moreover, for most τ values a definite D_{corr} value (saturation) could not be obtained in our experiments except for the τ values that correspond to the first zero of the autocorrelation function. It is possible that the noise level in our CSTR experiments is correspondingly large. In fact the CSTR is an excellent fluctuation generator at all stirring rates, particularly at low stirring rates. High experimental noise may distort the actual structure of a periodic attractor whereas chaotic attractors are more robust towards noise [16,17].

If D_{corr} is found to be higher than 2.0 in an experimental system, a stirring rate dependence must be performed. The presence of amplified fluctuations is indicated if the D_{corr} values decline below two at increasing stirring rates. Amplified fluctuations may produce aperiodic trajectories and their invariant measures may formally resemble chaos [2,3]. However, amplified fluctuations should not be confused with chaos, which is deterministic in nature. The presence of deterministic chaos is indicated if the D_{corr} values at low stirring rates retain fractal values higher than 2.0 as the stirring rate is increased. The range of stirring efficiency should be as large as possible in order to obtain sufficiently large variations in the size of the macroscopic fluctuations.

Noise - induced chaos may be obtained if the fluctuations push a system beyond its bifurcation point into a preexisting chaotic region. For this case a stirring rate dependence should show a decline in D_{corr} from a value which is substantially higher than 2.0 to values below 2.0 as the stirring rate is enhanced.

According to Fig.7 some indication of chaos is found for Takens' delay method where most but not all of the points show fractal D_{corr} values above two. In the periodic range of the BZ reaction we find D_{corr} values around two only at low stirring rates (150-300 rpm). At high stirring rates the D_{corr} values decline as expected for amplified fluctuations. We conclude from the positions of the break points (not shown) that mixing is less efficient at extremely high stirring rates of about 2000 rpm in our reactor. Therefore we regard the D_{corr} values as not very meaningful above 1500 rpm. The derivative method led to D_{corr} values that were consistently around unity in the BZ reaction (Figs.7 and 9) in contrast to the higher values found for the Takens method. We ascribe this difference to the low sampling frequency and the high dimensionality of the BZ reaction which may lead to an underestimation of D_{corr} by the derivative method.

On the basis of Takens' delay method we find that amplified fluctuations (Fig.9) may be distinguishable from deterministic chaos (Fig.7) in the BZ reaction by measuring D_{corr} as a function of stirring rate. This result is in accord with similar predictions made on the basis of the largest Lyapounov exponent [3].

Acknowledgements

We thank Mr.D.Lisch for experimental assistance, the VW Stiftung and the Fonds der Chemischen Industrie for financial support and the Computer Center of the University of Würzburg for a generous supply of computer time.

References

1. Horsthemke,W. and Hannon,L.: In Non-Equilibrium Dynamics in Chemical Systems, ed. by C.Vidal and A.Pacault, (Springer Berlin, Heidelberg, New York, Tokyo 1984) p.178
2. Freund,A., Kruel,Th. and Schneider,F.W.: Ber.Bunsenges.Phys. Chem. 90, 1079 (1986)
3. Freund,A., Kruel,Th.-M. and Schneider,F.W.: In Proceedings from MIDIT 1986 Workshop, ed. by R.D.Parmentier and P.L.Christiansen, (Manchester University Press, 1987) in press
4. Grassberger,P. and Procaccia,I.: Phys.Rev.Lett. 50, 346 (1983) Grassberger,P. and Procaccia,I.: Physica 9D, 189 (1983)
5. Takens,F.: In Lecture Notes in Mathematics, ed. by D.A.Rand and L.S.Young, Vol.898 (Springer 1981) p.366
6. Fraser,A.M. and Swinney,H.L.: Phys.Rev. 33A, 1134 (1986)
7. Wolf,A., Swift,J.B., Swinney,H.L. and Vastano,J.A.: Physica 16D, 285 (1985)

8. Packard,N.H., Crutchfield,J.P., Farmer,J.D. and Shaw,R.S.: Phys. Rev.Lett. $\underline{45}$, 712 (1980)
9. Kai,T. and Tomita,K.: Prog.Theor.Phys. $\underline{61}$, 54 (1979)
10. Hao,B.-L. and Zhang,S.-Y.: J.Stat.Phys. $\underline{28}$, 314 (1982); Hao,B.-L.: "Bifurcation and Chaos in a Periodically Forced Limit Cycle Oscillator", preprint
11. Russel,D.A., Hanson,J.D. and Ott,E.: Phys.Rev.Lett. $\underline{45}$, 1175 (1980); Froehling,H., Crutchfield,J.P., Farmer,D. and Packard,N.H.: Physica $\underline{3D}$, 605 (1981)
12. Greenside,H.S., Wolf,A., Swift,J. and Pignataro,T.: Phys.Rev. $\underline{25A}$, 3453 (1982)
13. Ben-Mizrachi,A., Procaccia,I. and Grassberger,P.: Phys.Rev. $\underline{29A}$, 975 (1984); Atten,P., Caputo,J.G., Malraison,B. and Gagné,Y.: In Special Issue of the Journal de Mécanique: Bifurcations and Chaotic Behavior, (1984), p.133
14. Hudson,J.L. and Mankin,J.C.: J.Chem.Phys. $\underline{74}$, 6171 (1981)
15. Theiler,J.: Phys.Rev. $\underline{34A}$, 2427 (1986)
16. Schaffer,W.M., Ellner,S. and Kot,M.: J.Math.Biol.$\underline{24}$, 479 (1986)
17. Herzel,H., Ebeling,W. and Schulmeister,Th.: Z.f.Naturforsch. $\underline{42a}$, 136 (1987)
18. Kaplan,J. and Yorke,J.: In Lecture Notes in Mathematics, ed.by H.O.Peitgen and H.O.Walter, Vol. 730 (Springer 1978) p.228

Exotic Chemical Reactions with Cu(II) Catalyst

M. Orbán

Institute of Inorganic and Analytical Chemistry, L. Eötvös University,
Múzeum krt. 4/B, P.O. Box 123, H-1443 Budapest, Hungary

Abstract. Trace amount of copper ion catalyst induces many exotic
phenomena in the chemistry of the elements in group VI A. Simple
and complex periodic oscillations as well as multistability have
been observed in the oxidation of SCN^- by H_2O_2, in the oxidation
of $S_2O_3^{2-}$ by H_2O_2, in the oxidation of $S_2O_3^{2-}$ by $S_2O_8^{2-}$ and in the
reaction between ClO_2^- and S^{2-}. Some preliminary mechanistic suggestions are offered to explain these behaviours.

1. INTRODUCTION

Exotic chemical phenomena are referred to as various kinds of long-time, nonmonotonic , dynamical behaviour which can be expected to appear in far-from-equilibrium systems if the kinetics governing the chemical processes contains at least
one nonlinear step. These behaviours include temporal and spatial periodicities,
multiple steady states, excitability, phase synchronisation, chemical chaos, etc.
The exotic phenomena occur frequently in biological systems and in nature but more
and more chemical reactions are also found to exhibit simple or complex or chaotic oscillations, bi- and tristability, birhythmicity in stirred reactors or may
form spatial patterns in one, two or three dimensional space in unstirred solutions.

The chemical composition of the known exotic reactions mainly involve oxyhalo-
gen compounds. Several dozens of bromate, iodate and chlorite driven oscillatory
and multistable systems have been described [1]. The groups of VI A, V A and IV A
elements also support a wide range of oxidation states which makes them the most
likely candidates to constitute new inorganic homogeneous oscillatory and multistable systems. During the last years efforts have been made to test the compounds of oxygen, sulfur, nitrogen and carbon, and some successful experiments
have been reported such as the cobalt and bromide ion catalysed air oxidation
of benzaldehyde, the methyleneblue catalysed air oxidation of sulfide-sulfite
system and the hydrogen peroxide oxidation of sulfide ion without any added
catalyst. A recently published review article [2] attempts to classify the oscillatory reactions discovered so far and presents a taxonomy showing the family
links between the groups.

The mechanistical studies of these reactions and model calculations revealed
that autocatalysis is one of the requirements for exotic behaviour. In many
cases metal ion catalysts like Ce^{4+}, Mn^{2+}, Co^{2+} and ferroin participate in this
step. In the examples described here, Cu^{2+} ions were applied to catalyse the
oxidation reaction between inorganic sulfur compounds and oxidants, such as hydrogen peroxide, persulfate and chlorite ions in neutral, alkaline and acidic
media. Depending upon the experimental conditions large amplitude oscillations,
many steady states and bistabilities between them have been observed.

2. EXPERIMENTAL

The far-from-equilibrium requirement for long-term exotic behaviours in chemical systems is most conveniently met in continuously stirred flow reactor (CSTR). In

stirred batch reactors oscillations may also appear if the critical distance from
equilibrium is ensured by using excess of reactants. In our work both techniques
were applied. In batch experiments the components in appropriate concentrations
were mixed in a thermostated reaction vessel and visual and instrumental observations were made. In flow configuration the reactants were separately introduced into the CSTR and the excess volume of reaction mixture was continuously
drained off. The experimental variables or constraints were the input concentrations, flow rates (residence time) and temperature. The following responses were
monitored: potential of Pt electrode $\underline{vs}$ $Hg|Hg_2SO_4|K_2SO_4$ reference electrode, pH,
optical density and the evolution rate of gaseous product. Hysteresis loop in
the constraint-response phase diagrams was searched by observing the responses
of the system when the control parameter was first decreased and then – after
transition of the state occurred – increased. From a set of constraint-response
diagrams the constraint-constraint phase plane cuts were constructed.

3. RESULTS AND DISCUSSIONS

3.1 H_2O_2 – SCN^- – Cu^{2+} – OH^- system [3]

The copper-catalysed reaction between H_2O_2 and SCN^- gives rise to a remarkable
variety of dynamical phenomena.

Oscillations appear both in batch and flow reactors. The oscillatory responses
are the colour change between yellow and colourless, the periodic change in the
potential of Pt electrode and the oscillatory rate in the oxygen gas evolution.
No pH oscillations occur. For the oscillations the pH must be above 9. A typical
batch experiment is presented in Fig. 1.

The damped oscillations become sustained in CSTR.

Under certain sets of external variables the system can show three different
types of bistability in CSTR in which two stationary states and one oscillatory

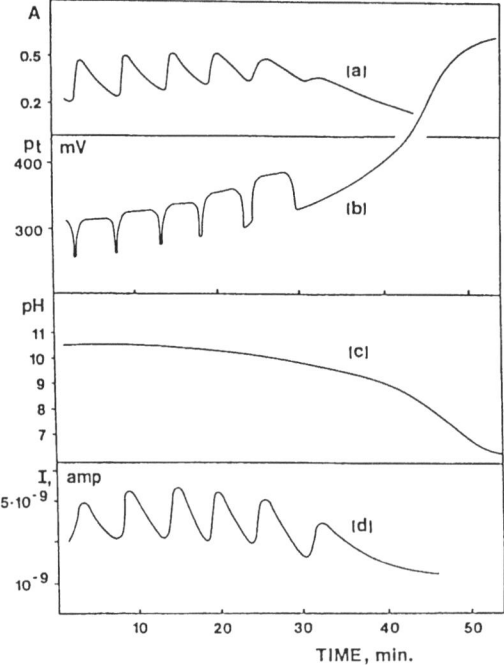

Fig. 1.
The oscillatory responses vs.
time in the Cu^{2+}-catalysed
reaction between H_2O_2 and KSCN.
The concentrations (M) are:
$[H_2O_2]$ 0.25, $[KSCN]$ 0.0375,
$[CuSO_4]$ 7.5×10^{-5}, $[NaOH]$ 0.025.

Curves: (a) absorbance at 375
nm; (b) potential of Pt; (c)
pH; (d) ionic current in mass
spectrometer which is proportional to the product oxygen

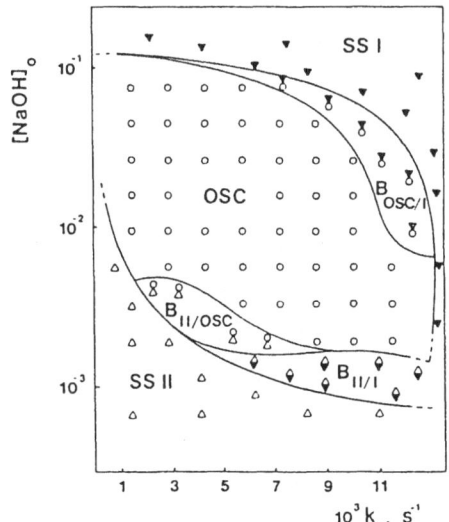

Fig. 2. Phase diagram for the H_2O_2 - KSCN - $CuSO_4$ - NaOH system. Fixed constraints $[H_2O_2]$ 0.25 M; $[KSCN]$ 0.025 M; $[CuSO_4]$ 5×10^{-5} M; temperature: 25 °C. Symbols: ▼ SS I, △ SS II; O oscillations. Letter B indicates the bistabilities among the corresponding modes

state are involved. Figure 2 gives the boundary lines and ranges of oscillations, steady states and bistabilities in the input NaOH concentration vs flow rate Phase plane.

Systematic research has been carried out with the purpose to replace Cu^{2+} and SCN^- ions with species of similar chemical properties. The majority of the transitional metal ions as well as Br^-, I^- and S^{2-} were unsuccessfully tested. In contrary to Cu^{2+} ions the other metal ions showed no significant effect on the H_2O_2 - SCN^- reaction but many of them accelerated the decomposition of the H_2O_2.

It is interesting to note that in the absence of Cu^{2+} catalyst in neutral solution the sulfide ion undergoes oscillatory oxidation by H_2O_2 in CSTR. The enhanced pH oscillations indicate an entirely different mechanism in the H_2O_2 - S^{2-} reaction from that responsible for the periodicity in the alkaline H_2O_2 - SCN^- reaction.

Mechanistical considerations to explain oscillations and bistability in the Cu^{2+}-catalysed H_2O_2 - SCN^- reaction led us to the following conclusions. The overall reaction between the two reactants can be expressed by (1)

$$4 H_2O_2 + SCN^- = HSO_4^- + NH_4^+ + HCO_3^- + H_2O \qquad . \tag{1}$$

First order kinetics in each reactant and a relatively simple mechanism is characteristic for the reaction in alkaline media. A moderate catalytic effect on the brutto process was found in the presence of small amount of copper ions. The two stationary states observed in CSTR are due to the different extent of the overall process. State SS I in Fig. 2 represents a small extent of the reaction, whereas in state SS II the reaction is more complete and the stoichiometry may be approximated by (1). The experimental data suggest that the brutto process hardly plays a role in bringing about the oscillations and the periodicity here cannot be the result of a feedback mechanism on the SS I - SS II bistable system. The oscillations in the alkaline H_2O_2 - SCN^- - Cu^{2+} reaction are the result of a chemistry in which the Cu^{2+}-catalysed autocatalytic decomposition of H_2O_2 takes place according to the cycle of (2a), (2b), (3a) and (3b):

$$Cu^{2+} + HOO^- \rightarrow Cu^+ + HO_2^{\cdot} \tag{2a}$$
$$Cu^{2+} + HO_2^{\cdot} \rightarrow Cu^+ + H^+ + O_2 \tag{2b}$$

$$2 \ Cu^{2+} + H_2O_2 + 2 \ OH^- \rightarrow 2 \ Cu^+ + O_2 + 2 \ H_2O \qquad (2)$$

$$\begin{array}{ll} Cu^+ + H_2O_2 \rightarrow Cu^{2+} + OH^- + HO^\cdot & (3a) \\ \underline{Cu^+ + HO^\cdot \rightarrow Cu^{2+} + OH^-} & (3b) \\ 2 \ Cu^+ + H_2O_2 \rightarrow 2 \ Cu^{2+} + 2 \ OH^- & (3) \end{array}$$

Equations (2) and (3) result in the decomposition of H_2O_2:

$$2 \ H_2O_2 \rightarrow 2 \ H_2O + O_2 \ . \qquad (4)$$

In process (2) the competition for $HO^\cdot$ by Cu^{2+} and H_2O_2, in process (3) the competition for $HO^\cdot$ by Cu^+ and SCN^- ($SCN^- + OH^\cdot \rightarrow SCN^{\cdot 2-} + OH^-$) may be important in switching between processes (2) and (3). The role of the SCN^- in the oscillatory cycle is not clear in every detail. The periodic formation of the colloidal CuSCN may also contribute to the separation in time of reactions (2) and (3).

3.2 $H_2O_2 - S_2O_3^{2-} - Cu^{2+}$ system [4]

In the presence of catalytic amount of Cu^{2+} ions the reaction between H_2O_2 and $S_2O_3^{2-}$ results in rich dynamical behaviour when it is carried out in CSTR. Oscillations occur in the pH, in the potential of Pt electrode and of Cu^{2+}-selective electrode. No batch oscillations appear. The oscillatory responses of the system are presented in Fig. 3. No oscillations and only one steady state are observed without Cu^{2+} catalyst. Three stationary states and bistabilities between two stationary states as well as between the oscillatory and one stationary state are established. An example of the many possible phase diagrams is depicted in Fig. 4.

The chemistry responsible for the exotic phenomena is totally different from that which governs the oscillations in the $H_2O_2 - SCN^- - Cu^{2+}$ reaction. In the $S_2O_3^{2-}$ containing system at a slightly acidic pH, the catalyst readily reacts with $S_2O_3^{2-}$ rather than with H_2O_2 which occurred in the previous system at high pH. The catalytic cycle can be described by (5)-(7):

$$2 \ Cu^{2+} + 2 \ S_2O_3^{2-} \rightarrow Cu(S_2O_3)_2^{2-} \qquad (5)$$

$$2 \ Cu(S_2O_3)_2^{2-} \rightarrow 2 \ Cu^+ + S_4O_6^{2-} + 2 \ S_2O_3^{2-} \qquad (6)$$

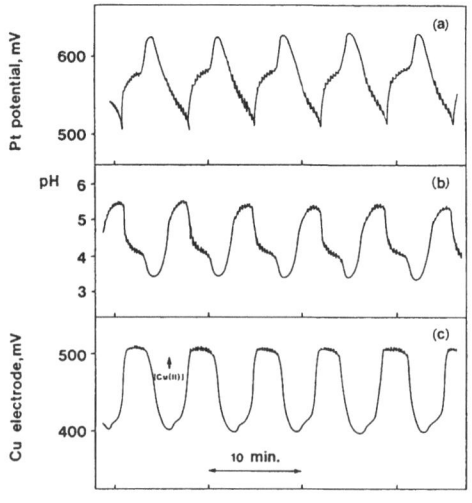

Fig. 3. Oscillations in (a) potential of Pt; (b) pH; (c)Cu(II) selective electrode in the H_2O_2 (0.25 M) – $S_2O_3^{2-}$ (0.025 M) – $CuSO_4$ (2.5×10^{-5} M) – H_2SO_4 (0.0075 M) system. Flow rate $k_o = 3.4 \times 10^{-3} \ s^{-1}$

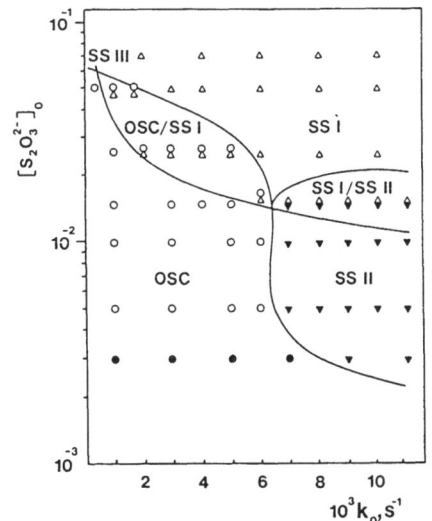

Fig. 4. Phase diagram of the H_2O_2 (0.25 M) $- S_2O_3^{2-} - Cu^{2+}$ (2.5×10^{-5}) $- H_2SO_4$ (0.001 M) system in the $[S_2O_3^{2-}]$ vs flow rate plane.
Symbols: o high amplitude oscillations; • low amplitude oscillations, Δ SS I; ▼ SS II. OSC/SS I and SS I/SS II mean bistability in the regions marked in the figure

$$2 \, Cu^+ + H_2O_2 \to 2 \, Cu^{2+} + 2 \, OH^- \quad . \tag{7}$$

Equations (5)-(7) yield the component reaction (8) which leads to the production of $S_4O_6^{2-}$ and consumption of H^+:

$$H_2O_2 + 2 \, S_2O_3^{2-} \to S_4O_6^{2-} + 2 \, OH^- \quad . \tag{8}$$

As a consequence of (8) the pH rises and new reaction path opens near and above the neutral pH. Acid production and total oxidation of $S_2O_3^{2-}$ to SO_4^{2-} follow reaction (9):

$$4 \, H_2O_2 + S_2O_3^{2-} \to 2 \, SO_4^{2-} + 2 \, H^+ + 3 \, H_2O \quad . \tag{9}$$

Reaction (9) is also catalysed by Cu^{2+}.

The oscillations in the $H_2O_2 - S_2O_3^{2-} - Cu^{2+}$ flow system can be the result of cross-coupling in which the component reactions (8) and (9) separately produce and consume $[H^+]$, switch off themselves by product inhibition and give free way for each other to start again. The stationary states are apparently approximated by reaction (8) (SS I) and reaction (9) (SS II). From the measured responses it is obvious that the oscillations here occur between steady states SS I and SS II.

3.3 $S_2O_8^{2-} - S_2O_3^{2-} - Cu^{2+}$ system

The reaction between $S_2O_8^{2-}$ and $S_2O_3^{2-}$ is very slow and shows nonmonotonic behaviour neither in batch nor in flow reactor. The rate of the reaction is markedly increased by catalytic amount of Cu^{2+} ion. In the presence of traces of Cu^{2+} ions high amplitude oscillations in the potential of Pt electrode and low amplitude oscillations in pH appear when the reactants are introduced into CSTR. The oscillations always happen in the acidic pH range. Two oscillatory responses vs time are shown in Fig. 5. The interesting feature of the system is that no bistability has been found so far on the wide variation of the constraints.

The overall reaction is represented by (10)

$$S_2O_8^{2-} + S_2O_3^{2-} \to S_4O_6^{2-} + 2 \, SO_4^{2-} \quad . \tag{10}$$

The Cu^{2+} catalyst does not change the stoichiometry. The Cu^{2+} ions act probably

65

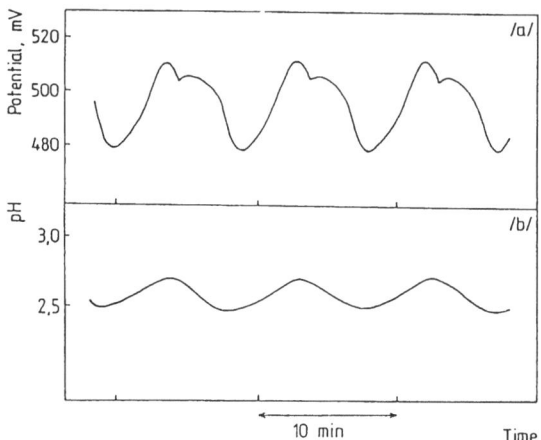

Fig. 5. Oscillation in (a) the potential of Pt; (b) pH in the $S_2O_8^{2-}$ (0.025 M) – $S_2O_3^{2-}$ (0.005 M) – Cu^{2+} (2.5×10^{-5} M) system. Flow rate $k_o = 4.0 \times 10^{-3}$ s^{-1}; temp.: 25 $^\circ$C

in a similar way as in (5)–(7) with the exception that in (7) the peroxo sulfate is involved. There was no trace of colloidal sulfur in any of our experiments. However, formation of some H_2S was noticed. The pH changes considerably during the course of reaction which is unexpected from the stoichiometric equation. Side reactions are supposed to be responsible for the pH changes. When buffer is introduced into an oscillatory mixture, the oscillations stop. It is premature to suggest a mechanism which explains the dynamical behaviour but beside the main reaction the role of the side reactions may also need to be considered.

3.4 ClO_2^- – S^{2-} – Cu^{2+} – H^+ system

A large number of chlorite oscillators have been discovered during the last 6 years. The substrates are mainly iodine compounds and occasionally sulfur species. Simple, complex periodic and aperiodic oscillations and travelling waves have been observed. No catalyst was required to be applied in any known chlorite system.

The reaction between chlorite and sulfide ions does not give rise to exotic behaviour. However, in the presence of trace amount of copper ions simple and complex oscillations appear in CSTR. The appropriate buffer is a pH 1.9–3.0 sodium acetate-acid mixture. The periodic responses are the potential of Pt electrode and – in some experiments – the precipitation and dissolution of the colloidal sulfur. No pH oscillations occur. At low concentrations and lower flow rates the prevailing wave form is a complex, spike-containing signal (see striped area in Fig. 7) which turns into simple oscillations with increasing flow rates and input concentrations.

A phase diagram in the input sulfide vs input chlorite concentrations and some oscillatory traces as a function of flow rates are presented in Figs 6 and 7. The intensive search for finding bistability has not been successful so far.

The mechanism and the role of the catalyst in the ClO_2^- – S^{2-} – Cu^{2+} – H^+ oscillator is not known. The periodic appearance and disappearance of sulfur suggest that the two kinetic states may be given by reactions (11) and (12)

$$HS^- + ClO_2^- + H^+ \rightarrow S + ClO^- + H_2O \tag{11}$$
$$HS^- + 2\ ClO_2^- \rightarrow SO_4^{2-} + 2\ Cl^- + H^+ . \tag{12}$$

The copper ions probably slow down process (11) through formation of CuS precipitate and/or speed up process (12) through complex formation with a transient sulfur species allowing the composite reactions to happen separately.

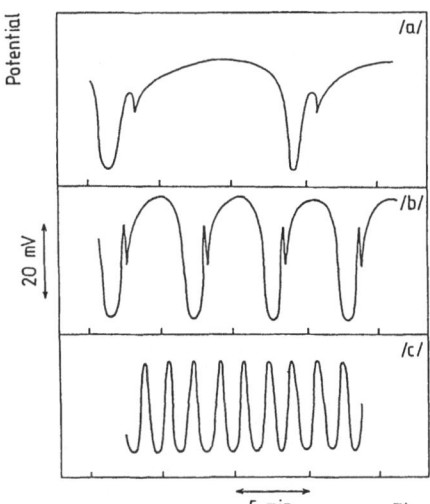

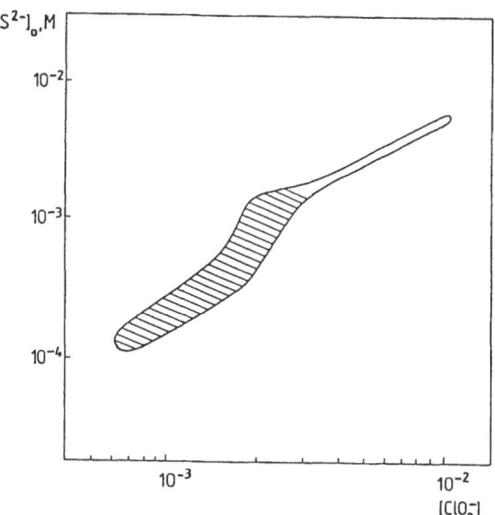

Fig. 6. Oscillations in the ClO_2^- (0.002 M) – S^{2-} (0.0013 M) – Cu^{2+} (5×10^{-5} M) system at flow rates of (a) 0.0022 s^{-1}; (b) 0.0044 s^{-1}; (c) 0.0066 s^{-1}; pH 2.55; temp.: 25 $^{\circ}C$

Fig. 7. Phase diagram for the ClO_2^- – S^{2-} – Cu^{2+} (5×10^{-5} M) system. pH: 2.55; k_o = 0.0022 s^{-1}. temp.: 25 $^{\circ}C$

4. CONCLUSIONS

In the present paper we have focused on the role of copper ions in some exotic reactions. The Cu^{2+}/Cu^+ redox couple seems to be an important constituent in many nonmonotonic chemical systems. Because of the wide range of its redox potential due to precipitation and complex formation, the copper ion catalysis may manifest itself in many ways: it can induce either the periodic decomposition of the oxidant or the oscillatory oxidation of the substrate and it is able to establish many stationary states. The copper ions mainly catalyse the chemical reactions of the elements in group VI A. Since the oxygen, sulfur as well as the copper belong to the elements of vital importance in many living systems, the discovery of copper catalysed exotic phenomena in biochemistry and biology would not be a surprise.

Acknowledgement. The work has been supported by a U.S.-Hungarian cooperative grant from the National Science Foundation (INT 8217658) and the Hungarian Academy of Sciences. The fruitful discussions with I.R. Epstein are gratefully acknowledged.

References

1. R.J. Field, M. Burger, eds: Oscillations and Travelling Waves in Chemical Systems (Wiley, New York 1985)
2. I.R. Epstein: Chem. Eng. News 65, 24 (1987)
3. M. Orbán: J. Am. Chem. Soc. 108, 6893 (1986)
4. M. Orbán, I.R. Epstein: J. Am. Chem. Soc. 109, 101 (1987)

Bifurcation, Periodicity and Chaos
by Thermal Effects in Heterogeneous Catalysis

E. Wicke and H.U. Onken

Institut für Physikalische Chemie, Universität Münster,
D-4400 Münster, Fed. Rep. of Germany

1. Introduction

In the case of reactions far from equilibrium under isothermal conditions in homogeneous liquid media the interaction of a non-linear reaction mechanism with diffusion of reaction components determines the dynamical behaviour of the reacting system and the formation of dissipative structures. Contrary to this, in the case of the heterogeneous gas-solid media of catalysis that will be dealt with in the following, the reaction is restricted to the surface of (usually porous) pellets, and their superheating by an exothermic reaction gives rise to non-isothermal conditions. The dynamical behaviour of such heterogeneous reacting systems far from equilibrium is determined by the interaction of the non-linearities of the reaction mechanism at the catalyst surface with the processes of heat transfer, predominantly conduction and convection. The strongest non-linearity is the exponential-like increase of the reaction rate with increasing catalyst temperature due to the Arrhenius law. Besides this, other types of non-linearities, too, may occur in the reaction mechanism at the surface, controlling the dynamical behaviour under isothermal conditions, and leading in some cases to isothermal oscillations of the reaction rate as they have been observed at single catalyst pellets [1,2], at samples of catalyst powder [2-4], and at specimens of single crystal planes of the catalytically active metal [5-8].

When a reacting system is "moved" along the thermodynamic path away from equilibrium - by suitably changing the parameter values - the first effect of the growing influence of non-linearities is a bifurcation, i.e. a splitting up of the unique and stable state of reaction to several states, in part stable or unstable, or oscillatory. In the systems discussed in the following the bifurcation occurs as a splitting up into three steady states, one of which is unstable, the others are stable: an ignited state of reaction with rather high catalyst temperature and high conversion, and a quenched state where the superheating of the catalyst and the conversion are low. In the region of parameter values near this "thermal" bifurcation the system displays a high parametric sensitivity of the reaction rate, and hence of the catalyst temperature. This will be demonstrated in the next two sections by a model calculation in the case of an adiabatic CSTR, and by experiments with the oxidation of ethane in an adiabatic packed bed tubular reactor. Subsequently the oxidation of carbon monoxide will be investigated within this region of high sensitivity in the tubular reactor, and the effects of self-sustained oscillations of the reaction rate at single catalyst pellets on the dynamics of the temperature pattern and of the conversion in the packed bed will be presented and discussed.

2. A Model Case in an Adiabatic CSTR

2.1. The Thermal Bifurcation

Continuous stirred tank reactors are standard for homogeneous reactions in liquids, their dynamical behaviour, however, can be applied likewise as a model for heterogeneous catalytic reactions, visualizing for instance an open catalytic loop reactor with gradientless operation [9]. The dynamics of a reaction in a CSTR is governed by the material balance

$$V \frac{dc}{dt} = q(c_0 - c) - Vkc \qquad (1)$$

and the heat balance

$$V\rho c_p \frac{dT}{dt} = -q\rho c_p(T-T_0) - \Delta H \cdot Vkc \qquad . \qquad (2)$$

Here V and q are the reaction volume within and the volume flow rate through the CSTR, respectively, (both taken as time invariant), c_0 and T_0 are the concentration of the relevant reaction component and of the temperature in the feed, respectively, c and T the same within the reaction vessel and in the effluent. ρc_p is the heat capacity per unit volume of the reaction mixture. k is the rate constant of the reaction (taken for simplicity as first order) and temperature dependent according to

$$k = k_0 \exp(-E/RT) \qquad (3)$$

with k_0 = frequency factor and E = activation energy. ΔH, finally, is the reaction enthalpy. A term for external cooling of the reaction vessel is omitted in the heat balance with regard to the adiabatic conditions presupposed. By introduction of the mean residence time, τ, and the adiabatic temperature rise, ΔT_{ad}, for complete conversion, according to

$$V/q \equiv \tau - \Delta H c_0/(\rho c_p) \equiv \Delta T_{ad} \qquad (4a,b)$$

eqs. (2,3) change to

$$\frac{dc}{dt} = \frac{c_0 - c}{\tau} - kc \qquad (5)$$

$$\frac{dT}{dt} = -\frac{T-T_0}{\tau} + \frac{\Delta T_{ad}}{c_0} kc \qquad . \qquad (6)$$

In the steady state, where $dc/dt = d\tau/dt = 0$,

$$\frac{c_s}{c_0} = \frac{1}{1+k_s\tau} \quad ; \qquad \frac{T_s-T_0}{\Delta T_{ad}} = k_s\tau \frac{c_s}{c_0} \qquad (7a,b)$$

with $k_s \equiv k(T_s)$.

The possible steady states can be obtained by plotting the chemical heat production per unit time in the CSTR:

$$\dot{Q}_r = (-\Delta H) \cdot Vkc = q\rho c_p \cdot \Delta T_{ad} \frac{k\tau}{1+k\tau} \qquad , \qquad (8)$$

(valid for steady state conditions) as well as the heat removal by convection

$$\dot{Q}_a = q\rho c_p(T-T_0) \qquad (9)$$

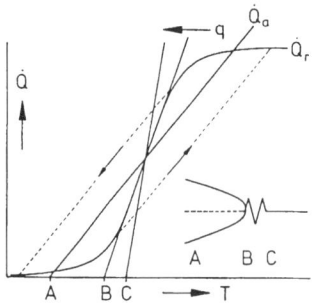

Fig. 1: Heat balance diagram of an exothermic reaction with chemical heat production, $\dot{Q}_r$, and convectional heat removal, $\dot{Q}_a$, vs. reaction temperature T in the bifurcation region A-B-C, schematically. Dotted lines: ignition and quenching with hysteresis. Details in the text

versus the reaction temperature T. As Fig. 1 shows, $\dot{Q}_r$ increases with T like an s-shaped curve: at low temperatures, where in eq.(8) $k\tau \ll 1$, $\dot{Q}_r$ increases exponentially-like according to the Arrhenius law, eq. (3), at higher temperatures it levels out due to complete conversion (c→o) in the vessel. The heat removal, $\dot{Q}_a$, is represented by a straight line with a slope proportional to the flow rate q and starting from the abscissa at the feed temperature $T = T_0$. The intersection points $\dot{Q}_r = \dot{Q}_a$ represent the steady states.

There are three characteristic cases shown in Fig. 1. At high flow rates, case C, one intersection only occurs, indicating a unique (and stable) steady state; at low flow rates, case A, three intersections occur, indicating a stable state at low temperature and reaction rate (quenched state), another stable state at high temperature and almost complete conversion (ignited state), and an intermediate (unstable) steady state. The transitions of ignition and quenching occur with hysteresis as shown by the dotted lines in Fig. 1. Between A and C the singular case B occurs where the straight $\dot{Q}_a$ line touches the $\dot{Q}_r$ curve as the tangent at the inflection point. This is the critical point of bifurcation, i.e. of splitting up into three steady states (when q is slightly decreased) as shown by the sequence C → B → A on the right-hand side of Fig. 1. In this situation $\dot{Q}_a$ and $\dot{Q}_r$ coincide to a common straight line within a certain range of reaction temperatures; there is no definite intersection, and hence no definite steady state. The reaction rate and temperature, accordingly, are subject to fluctuations that may originate from external or internal sources. The common straight line of $\dot{Q}_r$ and $\dot{Q}_a$ acts similarly to the characteristic of a transistor amplifier, magnifying strongly such external or internal perturbations (this is symbolized by the zig-zag-line at B in Fig. 1)

The transition from case A via B to C, demonstrated in Fig. 1 by increase of the flow rate q (with suitable adjustments of the feed temperature T_0), but invariant $\dot{Q}_r$ curve, can likewise be obtained when the slope of the $\dot{Q}_a$ line is kept constant (for instance according to case A) and the feed concentration c_0 is diminished. Then the ordinates of the s-shaped $\dot{Q}_r$ curve are correspondingly compressed, and the sequence A→B→C is passed through.

2.2. Model Calculations in the Bifurcation Region

For integration of the balance eqs. (5,6) in form of difference equations by means of time steps [10] the following basis data were chosen: c_0 = 1 vol. %, ΔT_{ad} = 98 K, k_0 = 1.64 × $10^9 s^{-1}$, E = 80 kJ/mol. The steady state reaction temperature was calculated as function of the feed temperature T_0 for three fixed values of the reciprocal residence time, τ^{-1}, as indicated in the legend of Fig. 2. The values of τ^{-1}, i.e. of the flow rate q, increase from case a) via b) to case c); they are chosen in such a way that case c) corresponds to case C in Fig. 1 - one unique steady state only at every feed temperature T_0 - and case a) to A in Fig. 1: two stable steady states with hysteresis between "ignition" and "quenching". Case b) represents the bifurcation point, situated here at τ^{-1} = 4.78s^{-1}. The co-

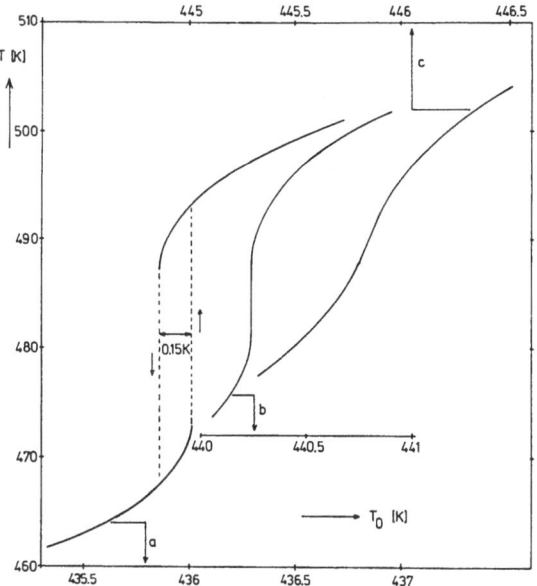

Fig. 2: Steady-state temperatures in a CSTR model calculation vs. feed temperature T_0 at different reciprocal residence times: a) $\tau^{-1} = 4$ s^{-1}, b) $\tau^{-1} = 4.78$ s^{-1}, c) $\tau^{-1} = 6$ s^{-1}

incidence of $\dot{Q}_r$ and $\dot{Q}_a$ about the inflection point in case B, Fig. 1, is testified here by an almost vertical increase of the curve $T = f(T_0)$, indicating a very high amplification factor $\Delta T/\Delta T_0$ of fluctuations that might occur in the feed temperature T_0.

In order to introduce self-sustained oscillations into this model, a sequence of alternating blocking and reactivation processes of the catalyst surface was simulated. To this purpose the frequency factor - proportional to the active surface area - was assumed to decrease when the reaction temperature exceeded the steady-state value:

$$k_0(T) = k_s(T_a) + \left(\frac{dk_0}{dT}\right)_s (T-T_s) \; ; \qquad \left(\frac{dk_0}{dT}\right)_s < 0 \quad . \tag{10}$$

On the other hand, a "reactivation" of the frequency factor with a relaxation time τ_k was supposed:

$$\frac{dk_0}{dt} = \frac{k_0(T_s) - k_0(t)}{\tau_k} \quad . \tag{11}$$

Fig. 3 shows on the left-hand side the temperature oscillations obtained with the parameter set:
$T_0 = 435.5$ K, $\tau^{-1} = 4$s^{-1}[case a in Fig.2], $(dk_0/dT)_s = -0.005$ s^{-1}K^{-1}, $\tau_k = 100$s. The oscillations disappear when τ^{-1} is reduced below 3s^{-1}; then the hysteresis becomes too broad for the reaction to "jump over" between states of low and high reaction rate. The oscillations also disappear if τ^{-1} exceeds the critical value 4.78 s^{-1}, because then one unique stable state only establishes. In order to demonstrate the high sensitivity of the system to external perturbations in this region, the feed temperature T_0 was varied by small statistical fluctuations. The effect is to be seen in Fig. 3 to the right, where the fluctuations ΔT_0 are shown 10-fold extended: the oscillations are disturbed by irregularities that exceed the small fluctuations by an amplification factor of $\Delta T/\Delta T_0 \approx 30$ (nearer to the critical value $\tau^{-1} = 4.78$s^{-1} the amplification factor would even be higher).

71

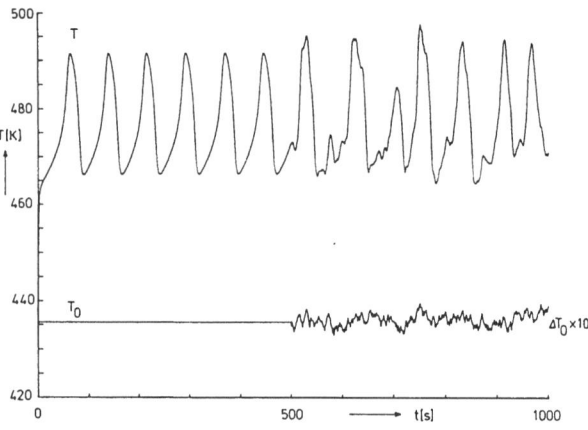

Fig. 3: Simulated self-sustained temperature oscillations in a CSTR with feed temperature T_0 constant (left hand) and fluctuating at random (right hand). $T_0 = 435.5$ K (on the average) $\tau^{-1} = 4$ s^{-1}

3. Measurements on the Ethane Oxidation in an Adiabatic Packed Bed Reactor

3.1. The Thermal Bifurcation of Reaction States at Catalyst Pellets and in Packed Catalyst Beds

Whereas for homogeneous exothermic reactions bistability in a CSTR was discussed first in the fifties by C. VAN HEERDEN [11] (see also [12]), the concept of "ignition" and "quenching" of exothermic reactions at catalyst pellets was developed even earlier by C. WAGNER [13]. His basic consideration was - in modern terms - the following. The material and the heat balance for the steady state of a first order reaction at a catalyst pellet in the gas flow are, in fair analogy to the steady state forms of eqs. (1,2),

$$0 = \beta F(c_g - c) - V \, k\eta c \tag{12}$$

$$0 = -\alpha F(T - T_g) - \Delta H \, V k\eta c \quad . \tag{13}$$

Here β and α are the mass and the heat transfer coefficient between the gas flow, by passing the pellet, and the external surface of area F of the pellet, c, c_g and T, T_g are the concentrations and temperatures at this surface and in the gas flow, respectively, V is the volume of the pellet, ΔH the reaction enthalpy and k the reaction rate constant. The factor $\eta < 1$ takes account of the depletion of the relevant reaction component within the porous interior of the pellet; the temperature T, on the other hand, can be taken as homogeneous throughout the pellet. By elimination of c from eqs. (12,13) one obtains

$$(-\Delta H) \, \frac{k\eta \, \beta}{k\eta / F + \beta / V} \, c_g \; = \; \alpha F(T - T_g) \quad . \tag{14}$$

Visualizing that the quantity on the left is the chemical heat production in the pellet per unit time, and that on the right the convectional heat removal from the pellet, the analogy to eqs. (8) and (9) is obvious. Actually, the heat balance diagram Fig. 1 can be applied schematically also for a catalyst pellet. The heat production here also follows an s-shaped curve; at high temperatures, when $Vk\eta \gg \beta F$, it is limited in this case by the external mass transfer (complete conversion at the external pellet surface, $c \to 0$). The straight line of heat removal starts on the abscissa at $T = T_g$, and its slope also increases with increasing flow rate, although not proportional to it ($\alpha \sim v^n$ with $n < 1$ where v = linear gas velocity in the empty tube under standard conditions: the use of v instead of the volume flow rate q is customary in the case of tubular reactors).

The considerations concerning the transition C → B → A from monostable to bistable behaviour via the bifurcation case B in Fig. 1 can therefore be applied also to catalyst pellets.

Now we will pass over from a single catalyst pellet to the behaviour of a packed bed of catalyst pellets in a tubular reactor. At sufficiently high gas velocity case C occurs throughout the packed bed, i.e. a unique stable state of reaction. With decreasing gas flow rate the catalyst temperature will increase most strongly at the outlet of the catalyst bed (adiabatic conditions presupposed), hence, the bifurcation point B will be attained first in the exit cross section. The same holds for the first occurrence of bistable behaviour when the flow rate is reduced further. The correlate to Fig. 2 for the CSTR can therefore be obtained for a tubular reactor best by measurements at the end of the packed bed.

3.2. The Tubular Reactor

The adiabatic packed bed reactor, constructed for the experimental investigations [14] is demonstrated in Fig. 4. The core of the apparatus is a vertical glass tube with a double mantle that is silver-plated inside and evacuated to 10^{-5} mbar to ensure adiabatic conditions. The packing of catalyst pellets of 3 to 4 mm diameter is positioned between layers of inactive support pellets that provide for homogeneous distribution of the gas flow and for adiabatic conditions also in vertical direction. The feed flow, air with about 1 vol.% of C_2H_6 or CO, was adjusted and kept constant carefully by electronically controlled flow meters. Quick and homogeneous mixing was achieved by introducing the small flow of C_2H_6 or CO through a porous filter tube into the large air flow. The feed temperature T_0 was established by preheating the air flow; electronic control, directed by a NiCr/Ni thermocouple near the top of the upper inert layer provided for strict temporal constancy. In view of the high sensitivity of the system in the bifurcation region to external irregularities, high precision and constancy of the feed conditions and avoidance of control fluctuations was a predominant effort of the equipment.

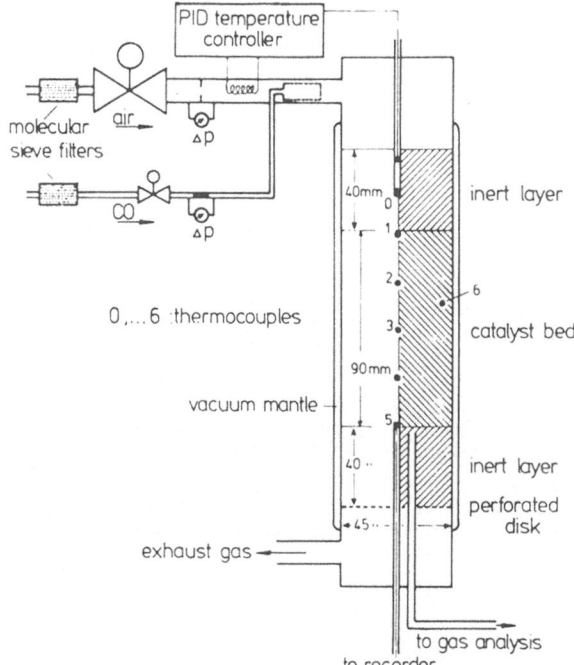

Fig. 4: The adiabatic packed bed tubular reactor

A number of thermocouples was positioned along the axis of the catalyst bed (the number changed with changing the bed height); one thermocouple was placed near the wall in order to indicate radial inhomogeneities. From the effluent gas flow a small part could be separated for continuous recording of the CO_2 content by non-dispersive IR analysis.

3.3. Measurements and Results

The catalyst used for the ethane oxidation to CO_2 and H_2O was a carrier catalyst with 1 wt. % of Pd on porous, spherical Al_2O_3 pellets, 3 mm ϕ, where an external shell only was impregnated. A packing of 110 mm height was placed into the reactor between 50 mm layers of inert pellets, and three thermocouples, NiCr/Ni in a thin-walled ceramic capillary, were positioned along the axis in the inlet and outlet cross section and in the midst of the catalyst bed.

The transition from monostable to bistable behaviour of the reaction was observed by measuring the CO_2 content in the effluent as function of the feed temperature T_0 at different fixed values of flow rate and ethane content in the feed. The results are shown in Fig. 5: to the right at high gas flow rates monostability, to the left at lower flow rates bistability with hysteresis (in this case the feed temperature was first increased for "ignition", and then decreased again for "quenching" at the exit of the catalyst bed). In the midst of Fig. 5 two curves are shown, the left-hand one of which with the steepest slope nearly represents the bifurcation point. The confirmation of the schematic concept in Fig. 1 by these results is obvious, the narrow similarity between Figs. 2 and 5 even surprising. The small difference of the parameter values between the two curves near the bifurcation - ethane content in the feed 0.74 and 0.75 vol.% - indicates the reliability of the control mechanisms applied. No irregularity, originating from external perturbations, could be detected in this region of high sensitivity of the system.

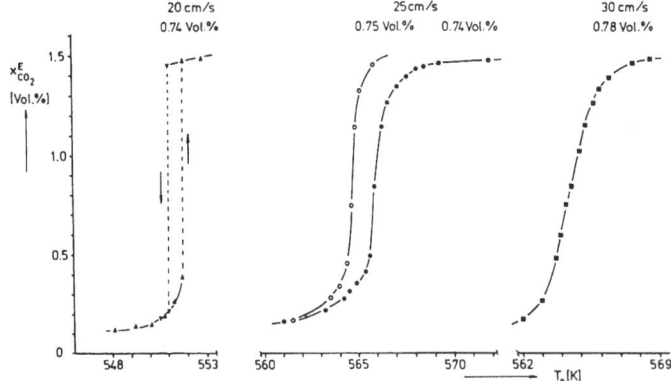

Fig. 5: Ethane oxidation in the tubular reactor, showing transition through the bifurcation region with varying gas flow rate and ethane content in the air/ethane feed

Fig. 6 demonstrates this high sensitivity by showing the response of the system to small periodic modulations of the feed temperature at fixed values of the gas flow rate and the ethane content in the feed corresponding to the bifurcation case in Fig. 5. The temperature oscillations increase strongly from the inlet to the outlet of the catalyst bed (thermocouples T_1-T_3) by altogether a factor of about 50. The CO_2 content in the effluent changes in phase with the catalyst temperature at the exit. - A similar pattern is obtained by modulation of the ethane content in the feed.

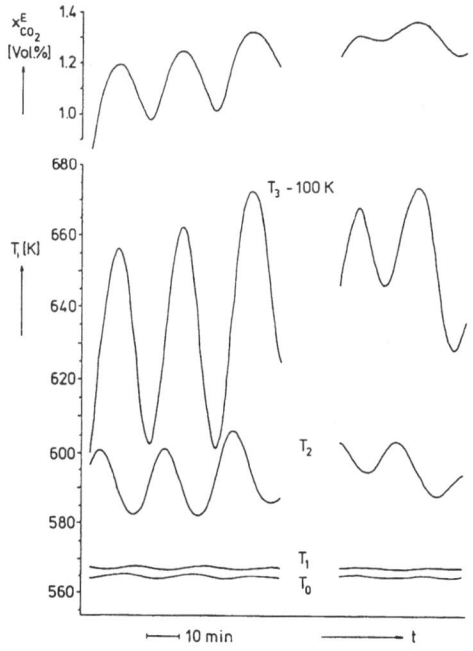

Fig. 6: Response of temperature evolution along the catalyst bed and of CO_2 production to periodic modulation of the feed temperature T_0 near the bifurcation point. $v = 25$ cm/s, 0.75 vol% ethane. Period 1200 s; amplitude $\pm$ 0.6 K (left), $\pm$ 0.3 K (right)

4. The Catalytic Oxidation of Carbon Monoxide

4.1. The Dynamic Behaviour in the Tubular Reactor

A carrier catalyst with 0.3 wt.% of Pt on Al_2O_3 pellets of cylindrical shape (3x3mm) was used for the reaction. The same catalyst had been applied by P. FIEGUTH [15] in 1970 already for the CO oxidation in a tubular reactor with a more simple equipment. He observed in the transition region between bistable and monostable behaviour a certain range of feed parameter values where the temperature at the exit of the catalyst bed fluctuated irregularly and without hysteresis between quenching and ignition, and the CO_2 content in the effluent fluctuated strongly in a similar irregular way. The intention to clarify the nature of these fluctuations and to examine if they represent deterministic chaos was the impetus for the new investigation of the problem with modern means.

The height of the catalyst bed in the reactor tube and the positions of the thermocouples were such as shown in Fig. 4. In this case the welded joints of the thermocouples 1-6 from 0.1 mm NiCr and Ni wire were placed in 0.2 mm channels drilled through single catalyst pellets. The fluctuations of exit temperature and CO_2 content were observed in a similar range of parameter values as found by FIEGUTH. Fig. 7 presents an example of still regular fluctuations and their evolution along the catalyst bed. The thermocouple in the entrance cross section, T_1, indicates practically no temperature perturbations. The thermocouple T_2, 22 mm downstream, however, shows humps of up to 5 K that occur in quasiperiodic sequence every 10 to 15 min. These humps travel along the catalyst bed with an average speed of about 0.2 mm/s (see dotted lines in Fig. 7). Some of them increase to steep peaks of up to 100 K in the exit (T_5) and produce maxima in the effluent CO_2 (Fig. 7 above). Some of the humps, however, remain small or even fade away; it seems that these do not reach the threshold value necessary to trigger the ignition of the reaction.

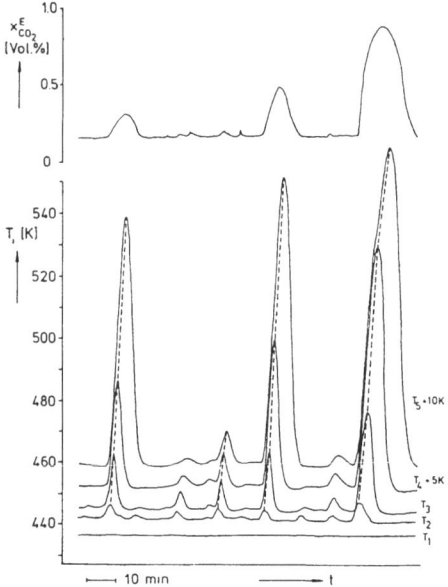

Fig. 7: CO oxidation in the tubular reactor. Evolution and movement of temperature peaks along the catalyst bed and CO_2 production. v = 21 cm/s, 1.0 vol.% CO in air, T_0 = 431 K

Travelling reaction waves are a well-known type of dissipative structure in homogeneous reaction-diffusion systems; in heterogeneous heat transfer-reaction systems, as dealt with here, they have been observed and investigated for a long time [16,17].

The reaction pattern shown in Fig. 7 changes rapidly with feed temperature and gas flow rate. A bit higher values of these parameters give rise to a much more irregular pattern, Fig. 8, with higher frequencies and amplitudes of the reaction waves. There is only little correspondence left between the temperature peaks and the CO_2 maxima at the exit. It seems that the CO_2 maxima do no more extend uniformly over the whole cross section, but that in radial direction regions of lower and higher temperature and reaction rate alternate locally and temporally. Accordingly, the course with time of the temperature at the eccentric thermocouple T_6 (drawn in Fig. 8 above) shows only little correlation with the signals of the thermocouple near the axis.

Since these fluctuations are not initiated by external disturbances - the inlet temperature, T_1, shows no perturbations also in this case - the fluctuations obviously originate in the system itself. This augments the suspicion that not random noise but deterministic chaos is involved in the reacting system.

4.2. Check for Deterministic Chaos

Deterministic chaos in the behaviour of a system means that its temporal evolution, represented by a trajectory in phase space, loses gradually all correlation with earlier states of the system, and finally becomes unpredictable. The reason is that in the case of chaos minor uncertainties in the initial conditions of a trajectory are amplified with time by the non-linear dynamics of the system. If there are two trajectories with almost identical initial conditions, i.e. starting from narrowly adjacent points in phase space, they will diverge with time, accordingly. The increase of their distance can be described by an exponential, $\exp(\lambda_1 t)$, where λ_1 is called the positive Lyapunov exponent. If we have not only two but a whole bundle of trajectories, the diameter of the bundle may in some directions remain constant ($\lambda_i = 0$), in other directions decrease ($\lambda_i < 0$). If

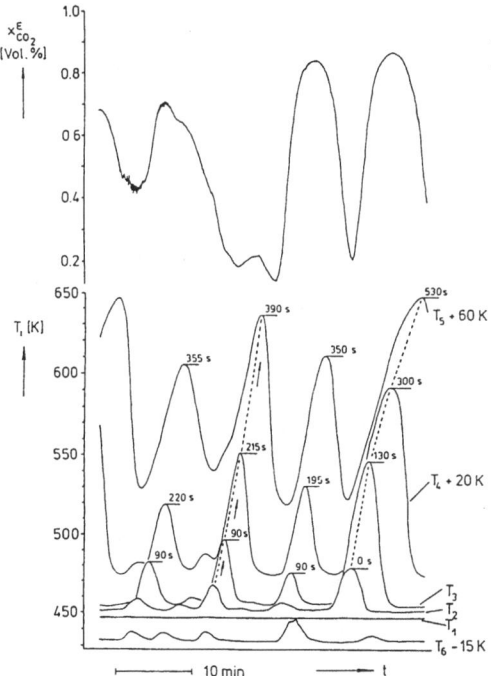

Fig. 8: The same as in Fig. 7 at v = 36.7 cm/s, T_0 = 440.5 K. T_6 : thermocouple in excentric position

there are only zero and negative, or only negative Lyapunov exponents, the bundle proceeds towards an attractor, i.e. to a periodic orbit or a fixed stable point, respectively. If there occurs, however, one positive value among the Lyapunov exponents of the different directions, the trajectories belong to a "strange" or "chaotic" attractor.

In order to check if the measurements described above represent deterministic chaos the method of A. WOLF et al. [18] to determine the largest (positive) Lyapunov exponent was applied. The method is based on long-time series of measurements of a single state variable of the reacting system. The CO_2 content in the effluent was chosen as this test variable. Several long-time measurements with 10,000 data points (3 hours) each were carried out, with an accuracy of the values of the CO_2 fluctuation amplitudes of 0.5%. Figure 9 shows part of such a measurement series with a 10-fold extended detail pattern above. Here the fluctuation peaks exhibit a fine structure of numerous short-time variations. Fourier power spectra of the long-time series display an accumulation of peaks in the range of 0.01 s^{-1} < ω < 0.03 s^{-1} (periods 600 to 200 s), underlaid by a continuous background without structure. This excludes the possibility that the fluctuations might represent a quasi-periodic superposition of a number of harmonic oscillators.

The long-time series of CO_2 data, x(t), have been evaluated first to "reconstruct" the phase space according to the method of F. TAKENS [19]. A time delay τ = NΔt was chosen of a number N of intervals Δt between successive data points - such that a correlation still exists between x(t) and x(t+τ) - and for each x(t) a point in n-dimensional space

$$x(t), x(t+\tau), \ldots , x[t+(n-1)\tau]$$

was determined. The sequence of these points forms a trajectory in n-dimensional phase space. If the "embedding dimension" n is of the order of 2d, where d is the number of state variables of the real system, this trajectory - according to

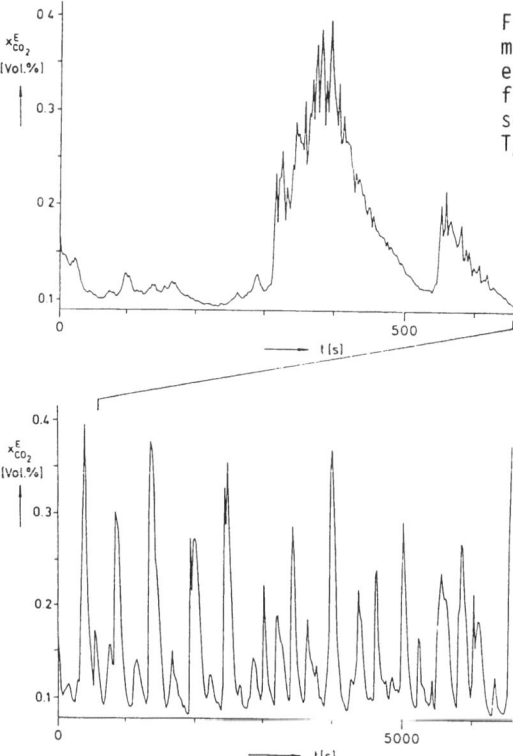

Fig. 9: Part of a long-time measurement series of CO_2 fluctuations in the effluent. Above: The region of the first peak with ten-fold expanded abscissa. $v = 36.7$ cm/s, 0.5 vol.% CO, $T_0 = 440$ K

TAKENS [19] - can be considered as representing the trajectory of the system in the real, d-dimensional phase space. The trajectory follows in the n-dimensional space a frequently wound and tortuous path, some segments of which are shown schematically in Fig. 10. The basic idea in the method of WOLF et al. is to evaluate the divergence of those segments as if they were separate trajectories. The procedure is to look for the increase of the distance between the trajectory A and the segment B from $r(t_0)$ to $r'(t_1)$ during a certain (arbitrary) evolution time τ_e, then to replace B by another trajectory segment C that is nearer to A and follow the increase of the distance from $r(t_1)$ to $r'(t_2)$, and so on along the whole trajectory A that was constructed from the series of measured data points. The Lyapunov exponent is then determined by

$$\lambda_1 = \frac{1}{t_m - t_0} \sum_{i=1}^{m} \ln \frac{r'(t_i)}{r(t_{i-1})} , \qquad (15)$$

where m is the total number of replacements along the trajectory A.

The results evaluated from 5 long-time series are displayed in Fig. 11. In order to check for the influence of external noise, the CO concentration in the feed was randomly disturbed with different noise amplitudes, plotted on the abscissa in Fig. 11 (the lowest value, $\Delta c = \pm 0.006$ % of CO, is the noise amplitude of the flow control system). The diagram yields $\lambda_1 = 0.013 \pm 0.002$ s^{-1}, independent of external noise. A change of the embedding dimension from n = 4 to n = 9 is without effect as shown by Table 1. A diminution of the evolution time τ_e, however, leads to larger values of λ_1. Presumably the fine structure of the fluctuations, shown in Fig. 9 above, simulates a larger divergence at smaller evolution times.

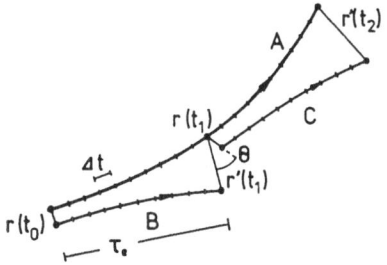

Fig. 10: Part of a "reconstructed" trajectory A and nearby situated segments B and C, displaying the divergence of trajectories, schematic. Δt = time interval between data points, τ_e = evolution time between replacements, θ = angle of readjustment at the replacement. Further details in the text

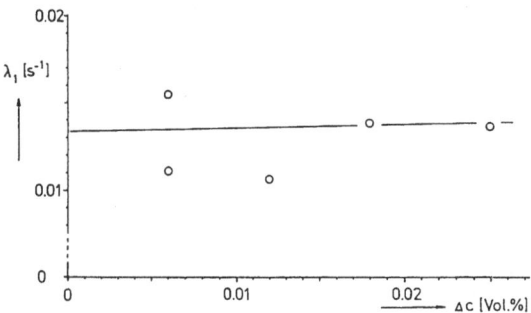

Fig. 11: Lyapunov exponent λ_1 from 5 long-time series (10,000 data points, separated by $\Delta t = 1.1$ s) at different noise amplitudes Δc of CO concentration in the feed (mean value c_0 = 0.5 vol.%). Time delay τ = 55 s, embedding dimension n = 4, evolution time τ_e = 220 s

These results , confirming the occurrence of deterministic chaos in the behaviour of the system, were obtained by making use of the computer program given by A. WOLF et al. in Appendix B of their publication [18]. The calculations, considerably time-consuming, were done on the central computer, IBM-4381, of the university.

As a second method to distinguish between deterministic chaos and random noise the procedure to determine the correlation exponent ν according to P. GRASSBERGER and I. PROCACCIA [21] was applied to a series of 2000 data points out of the long-time measurment presented in Fig. 9. The evaluation was done by means of a particular FORTRAN program [14] on the central university computer. The result was: ν= 2.1, independent of external noise superimposed on the measurements, and of the embedding dimension, n, when n $\geq$ 9. This confirms once more deterministic chaos and indicates by the finding ν > 2 that the real phase space is at least 3-dimensional. Actually it should be 4-dimensional according to the 4 state va-

---

Table 1: Lyapunov exponent λ_1 [s^{-1}] from measurement series 1 at different evolution times τ_e and embedding dimensions n.

τ_e[s]	n = 4	n = 9
330	0.010	
220	0.013	0.010
55	0.040	
22	0.063	0.060

riables decisive for the system: catalyst temperature and CO_2 effluent concentration, surface coverage by chemisorbed CO and by oxygen.

4.3. Response to Periodic Feed Modulation

With periodic modulation of the CO content in the feed all types of responses can be observed that are characteristic of reacting systems with nonlinear dynamics. With a modulation period of 480 s, that is nearly the quasiperiodicity of the unperturbed system shown in Fig. 7, harmonic entrainment has been observed of the sequence of temperature peaks at the entrance, and of reaction waves through the catalyst bed. At higher modulation frequencies, period 360 s, period doubling occurs, and still higher perturbation frequencies, period 180 s, create lower subharmonics and chaotic behaviour in the space-time pattern of temperatures in the system. Fourier power spectra of these patterns reveal subharmonic as well as superharmonic peaks with distinct sextet structure that is not yet understood. Typical examples of these phenomena have been published recently elsewhere [20].

4.4. The Origin of the Fluctuations

There remains the question of the origin of the temperature humps that occur near the bed entrance - thermocouple T_2 in Figs. 7 and 8 - and give rise to the evolution of the reaction waves. The answer is presented in Fig. 12 where the course with time of the temperature in the center of four single catalyst pellets a) - d) is shown that have been placed with 8 mm distance from one another in a cross section of an inert pellets packing in the tubular reactor. The temperatures reveal quasiperiodic sequences of ignition and quenching transitions between states of low and high reaction rate with jumps of 2 to 4 K. The ignited state displays a fine structure of short-time breaks-in of reaction rate that increase steadily until finally the rate falls down to the quenched state. In this state the jump back to ignition is gradually prepared, as shown by the small but distinct temperature rise in these regions in Fig. 12. With increasing temperature the residence time of the system in the ignited state extends and in the quenched state shortens; at T_0 = 465 K, already, the ratio of the residence time is reversed, compared to Fig. 12, and thereby resembles the pattern of the relaxation oscillations observed earlier by P. FIEGUTH [1].

These phenomena indicate a fluctuating deactivation of the catalyst surface in the ignited state, and a slow, continuous reactivation in the quenched state. Oscillation patterns that are in a surprising extent similar to those shown in Fig. 12 have been observed recently in the group of ERTL [8] with the CO oxidation of Pt(110) surfaces under high vacuum conditions. LEED investigations revealed that in connection with the change of CO and oxygen coverage of the sur-

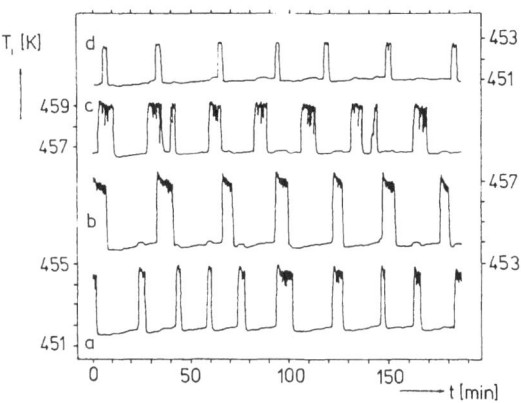

Fig. 12: Self-sustained oscillations of temperature (reaction rate) of 4 single catalyst pellets embedded in a packing of inert pellets. v = 25 cm/s, 1 vol.% CO in air, T_0 = 452 K

face severe alterations of the surface structure occur [22], much stronger than the mere reconstruction in the case of the Pt (100) surface [5-7]. It is most probable that the same happens at that part of the surface of the Pt crystallites that consists of (110) planes. The reaction rate oscillations at those planes initiate oscillations of temperature of the whole Pt crystallite and, due to heat conduction, of the whole catalyst pellet. (The rather good heat conductivity of the support material explains, furthermore, that the 10^9 to 10^{10} Pt crystallites in the porous interior of a single catalyst pellet "oscillate" so narrowly in phase that the pellet collectively displays an uniform oscillation pattern like in Fig. 12). It must be considered, further, that the dimensions of the crystal planes are of the order 100 to 1000 Å only, and therefore the regions at the borders - where the structure transformations used to start [22] - are of appreciable importance. In these regions the chemisorption of CO and of oxygen - due to structure defects and irregularities - are different from the chemisorption at smooth crystal planes; this presumably exerts substantial influences on the ignition and quenching processes of reaction rate at a catalyst pellet.

4.5. The Evolution of the Reaction Waves and of Chaos

Self-sustained oscillations of the reaction rate at single catalyst pellets in the entrance region of the packed bed turned out to be the sources of the reaction waves. Under the conditions of Fig. 7 - rather low feed temperature - there is obviously only one catalyst pellet in the region between bed entrance and cross section of thermocouple T_2 - 22 mm downstream - that "oscillates", similarly to the pellets presented in Fig. 12. By the local temperature peaks, although small, of the oscillation an ignition of the reaction rate, extending and growing on downstream, can be initiated. However, because the system is definitely in the quenched state, only a few of the oscillation peaks are successful for ignition - 3 during 100 min in Fig. 7 - most of them fade away. At higher temperature, Fig. 8, more pellets in the entrance region become active to oscillate, and more oscillation peaks are successful in the ignition of reaction waves. There appear now 5 reaction waves during 40 min with the result that the following reaction wave travels into the still non-uniform temperature field, left behind by the foregoing one. This leads to irregularities of temperature and reaction rate over the cross section of the packing, and thereby to loss of correlation between successive temperature peaks in the exit cross section and between CO_2 peaks in the effluent, i.e. to chaos. A real crowding of reaction waves with the result of chaos occurs when the 4 pellets, presented with their oscillations in Fig. 12, are placed in the entrance cross section of the catalyst packing and there initiate a dense sequence of reaction waves. Similar chaotic behaviour is observed when periodic modulations of the CO feed with short periodic (180s) trigger the initiation of densely successive reaction waves. Hence, deterministic chaos in this case does not evolve by a sequence of period doublings like in the Feigenbaum scenario [23], but by mutual interference of expanding instabilities that are initiated in short time intervals at different points of the reacting system.

The short-time variations that form a fine structure of the large CO_2 fluctuations in the effluent, as shown in Fig. 9, are , however, of different origin. In this fine structure, the numerous oscillations of the reaction rate at single catalyst pellets and at synchronized groups of pellets become apparent that are not successful in initiating reaction waves. With increasing distance of the parameter values from the bifurcation point, all these self-sustained fluctuations fade away, in the bistable region due to the hysteresis that stabilizes the quenched and the ignited state, and in the monostable region due to the autoregulation of the unique stable state.

References

1. H. Beusch, P. Fieguth, E. Wicke: Chem.-Ing.-Techn. $\underline{44}$, 445 (1972)
2. E. Wicke, P. Kummann, W. Keil, J. Schiefler: Ber. Bunsenges. Phys. Chem. $\underline{84}$, 315 (1980)
3. W. Keil, E. Wicke: Ber. Bunsenges. Phys. Chem. $\underline{84}$, 377 (1980)
4. D. Böcker, E. Wicke: Ber. Bunsenges. Phys. Chem $\underline{89}$, 629 (1985)
5. G. Ertl, P.R. Norton, J. Rüstig: Phys. Rev. Lett. $\underline{49}$, 177 (1982)
6. R. Imbihl, M.P. Cox, G. Ertl, N. Müller, W. Brenig: J. Chem. Phys. $\underline{83}$, 1578 (1985)
7. R. Imbihl, M.P. Cox, G. Ertl: J. Chem. Phys. $\underline{84}$, 3519 (1986)
8. M. Eiswirth, G. Ertl: Surf. Sci. $\underline{177}$, 90 (1986)
9. P. Hugo: Ber. Bunsenges. Phys. Chem. $\underline{74}$, 121 (1970)
10. E. Wicke, H.U. Onken: Chem. Engng. Sci. $\underline{41}$, 681 (1986)
11. C. van Heerden: Ind. Eng. Chem. $\underline{45}$, 1242 (1953); Chem. Engng. Sci. $\underline{8}$, 133 (1958)
12. W. Oppelt, E. Wicke: <u>Grundlagen der chemischen Prozeßregelung</u> (R. Oldenbourg Verlag 1964)
13. C. Wagner: Chemische Technik $\underline{18}$, 1, 28 (1945)
14. H.U. Onken: Dissertation, Münster (1986)
15. P. Fieguth, E. Wicke: Chem.-Ing.-Techn. $\underline{43}$, 604 (1971) (P.Fieguth, Dissertation, Münster 1971)
16. E. Wicke, D. Vortmeyer: Z. Elektrochem. Ber. Bunsenges. Phys. Chem. $\underline{63}$, 145 (1959)
17. G. Padberg, E. Wicke: Chem. Engng. Sci. $\underline{22}$, 1035 (1967)
18. A. Wolf, J.B. Swift, H.L. Swinney, J.A. Vastano: Physica $\underline{16D}$, 285 (1985)
19. F. Takens: In <u>Lecture Notes in Mathematics,</u> Vol. 898 (Springer, Berlin 1981) p. 366
20. P. Grassberger, I. Procaccia: Physica $\underline{9D}$, 189 (1983)
21. H.U. Onken, E. Wicke: Ber. Bunsenges. Phys. Chem. $\underline{90}$, 976 (1986)
22. G. Ertl: private communication
23. H.G. Schuster: <u>Deterministic Chaos</u> (Physik-Verlag, Weinheim 1984)

Spatial Patterns in (Bio)Chemical Reactions

S.C. Müller

Max-Planck-Institut für Ernährungsphysiologie, Rheinlanddamm 201,
D-4600 Dortmund 1, Fed. Rep. of Germany

1. Introduction

The study of spatiotemporal organization in reactive media due to the coupling of chemical reactions with transport processes has become, in recent years, a subject of intensive research in the field of non-equilibrium dynamics in chemical systems. The attention given to the evolution of spatial inhomogeneities on a macroscopic scale has been rapidly growing, ever since the remarkable phenomenon of chemical wave propagation in the Belousov-Zhabotinskii reaction was discovered almost 20 years ago [1,2]. However, as pointed out by TYSON and KAGAN in their historical review in this Volume, one should remember that numerous observations of chemical patterns in heterogeneous systems were reported before systematic research on temporal and spatial periodicities in homogeneous liquid-phase reactions began.

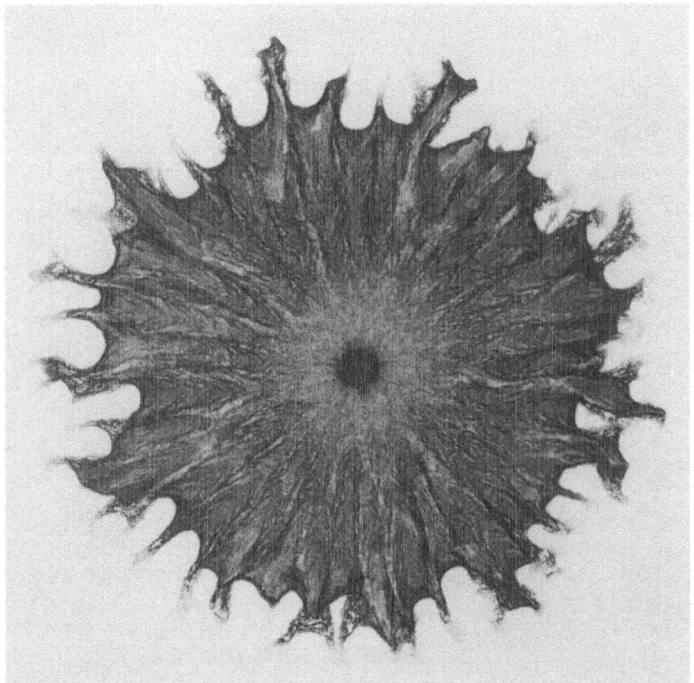

Fig. 1: A pattern produced according to a recipe from RUNGE [3]. Drops of a solution of $CuSO_4$ and $MnSO_4$ are repetitively placed on a suitable filter paper impregnated with KOH, K_2CrO_4, and $K_4Fe(CN)_6$. Diameter ≈ 10 cm

An early example of the investigation of spatially inhomogeneous reactions is the work of RUNGE around 1850 [3]. After impregnating a piece of filter paper with one chemical solution, he placed a sequence of drops of an appropriately chosen second reactant on this paper at regular time intervals. He thus discovered a large number of pairs of solutions for which the procedure results, upon spreading of the drops into the surrounding area, in quite spectacular and colorful spatial structures, which Runge called "paintings". The example in Fig. 1 was prepared according to one of the recipes given in his book "Der Bildungstrieb der Stoffe". Whereas the importance of Runge's methodological contribution to the development of chromatography is generally accepted, his more artistic approach to produce the "paintings" was not appreciated by the scientists. Nevertheless, one merit of this work should be emphasized: he was the first to observe ring-shaped regions of precipitated reaction product, many years before LIESEGANG rediscovered this phenomenon in 1896 to whom the first systematic study of periodic precipitation is credited [4].

Another early observation was reported by LIPPMANN in 1873 [5]. He described an electrochemical - mechanical oscillator, usually referred to as the "beating mercury heart". A mercury drop, placed in a watch glass and submerged in a pool of acidic solution of an oxidizing reagent, starts to pulsate with a frequency of about 1 Hz as soon as an iron needle touches the drop laterally. Depending on the details of the experimental setup the periodic geometrical distortions are isotropic or anisotropic, e.g. triangular as shown in Fig. 2.

Following these and other sporadic reports, patterns in heterogeneous systems received considerable attention during the early decades of our century, especially the Liesegang phenomenon and electrochemical oscillations. This is well documented in the book of HEDGES and MYERS on "The Problem of Physico-Chemical Periodicity" [6].

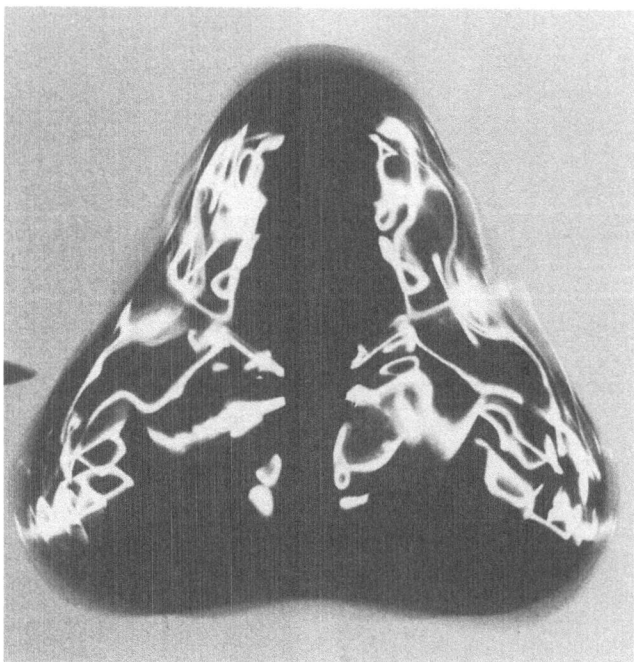

Fig. 2: The "beating mercury heart". A drop of mercury assumes a triangular shape during one cycle of periodic deformation. The tip of the iron needle can be seen at the left margin. Drop diameter approximately 1 cm

Nowadays, with the maturation of non-equilibrium thermodynamics and progress in understanding nonlinear system dynamics [7,8], the phenomenon of spatiotemporal organization in homogeneous liquid-phase reactions is also generally accepted, although a long time period of hesitation had to be overcome. Chemical waves belong now to the most frequently studied patterning processes. But "old" phenomena such as Liesegang rings have moved again into the focus of interest and new discoveries of pattern-forming systems are readily discussed in the light of the underlying mechanisms and the nonlinear coupling of complex reaction kinetics with diffusion or convection. A biochemical example is the periodic formation of stationary patterns during oscillating glycolysis, observed in cytoplasm extracted from yeast cells [9]. Photochemical patterns involving reactions at an interface were discovered a few years ago [10] and investigated by several authors [11,12]. Mosaic patterns were found in excitable media [13] and the role of convection for their occurrence discussed to some extent [14,15]. The interaction of chemical waves with hydrodynamic flows is frequently encountered and an increasing number of experiments in this area have been recently performed [16,17].

Progress in the theoretical understanding of non-equilibrium structures is closely connected with the development of modern methodologies that allow detailed and space-resolved measurements of the concentrations of relevant chemical variables. Of course, at the beginning of investigations phenomenological descriptions prevail. Thus, visual observations and photography are an essential first step for pattern characterization. For quantification many studies with detailed chemical analysis have been performed during the years and non-invasive optical methods, such as photometry, light scattering, Schlieren optics, have been used. In 1985, as a major improvement of the application of non-invasive physical techniques, spatially resolved spectrophotometric measurements combined with computerized data processing were introduced [18-20]. With help of one- or two-dimensional digital equipment operating in a wide spectral range of light, chemical and biochemical pattern analysis can now be performed with unprecedented precision and efficiency.

In this contribution a brief account is given of a number of recent experimental studies that resulted in a significant improvement in terms of the quantification of various system parameters. The systems are: periodic precipitation patterns, chemical waves in the Belousov-Zhabotinskii reaction, interaction of waves with hydrodynamic flow, and stationary patterns in oscillating glycolysis.

2. Periodic Precipitation Patterns

The periodic precipitation process known as Liesegang ring formation has been investigated - with varying activity - during the past 90 years. In his first experiment LIESEGANG placed a drop of silver nitrate on a gelatin layer containing potassium dichromate [4]. The subsequent process of diffusion led to the deposition of silver dichromate in concentric rings. More generally: if a soluble electrolyte is placed in contact with a second electrolyte and, on interdiffusion, both react to form a poorly soluble salt, a metastable solution can be produced. For many combinations of electrolytes at suitable concentrations the subsequent precipitation process does not occur homogeneously in the volume. Instead, well−separated bands of precipitate appear parallel to the diffusion front. They develop sequentially in a period of hours to several days. Usually, a gel-forming material is added to prevent sedimentation and convection, but structures can also be observed without a gel in very thin layers (films) or capillary tubes. For a review and bibliography concerning this subject (with a list of 800 references up to the year 1967) see [21].

Research on the Liesegang phenomenon clearly shows that, after its discovery, phenomenological descriptions were reported in abundance, while the obviously important quantification of spatial concentration distributions during the process of pattern formation was - and still is - quite rare. An early study yielded concentration data along a tube after performing a chemical analysis of thin cuts of

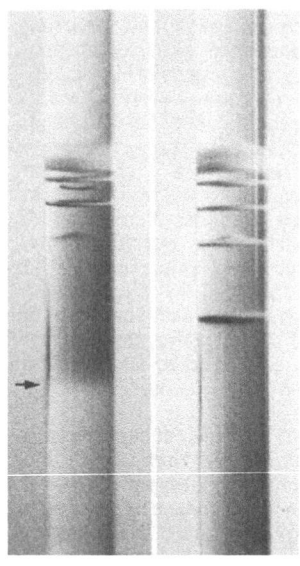

A   B

Fig. 3: Periodic precipitation patterns (Liesegang bands).
(A) Formation of parallel bands of $Mg(OH)_2$ with initially 5.5 M NH_4OH
in the upper and 0.4 M $MgSO_4$ in 9% gelatin in the lower portion of a
tube, after 6.5 h (left) and 16 h (right). The color change of added
metacresol purple indicates the location of pH 9 (arrow).
(B) Helicoidal precipitation band of PbI_2 in 1% agar with initially
0.240 M KI in the upper and 0.009 M $Pb(NO_3)_2$ in the lower portion of a tube.
Inner tube diameter: 5.5 mm (A), 13 mm (B)

the gel column in which the pattern of interest evolved [22]. In the subsequent
literature such local experimental procedures have only been reported in very few
cases, for instance in [23]. Other techniques for the investigation of local properties of precipitation patterns focussing on the early stages of their temporal
evolution, such as light scattering or light transmission of a focussed laser
beam, were applied quite recently [23]. For this purpose a one-dimensional system
was prepared in a test tube in which well-separated bands of $Mg(OH)_2$ appear sequentially in the course of time (Fig. 3A). The following properties were determined locally along the tube: the moving front of a color change of a suitable
indicator, corresponding to a specific value of OH^- concentration (pH front); the
front of a slight increase of turbidity (turbidity front) following a certain
distance apart and caused by growth of colloidal particles; the onset of localized, pronounced changes in refractive index, indicating a focussing of material
into a narrow spatial region; and finally, the formation of visible bands of
$Mg(OH)_2$ at these locations. The observations, which are summarized in Fig. 4,
were substantiated by optical methods, such as light scattering or the local
transmission and deflection of a thin laser beam, the deflection being a quantitative measure of refractive index gradients. (For details see [24].)

These and other experiments led to a theoretical discussion concerning the validity of either the long-existing supersaturation theory, originally proposed by
OSTWALD [26] and later described on the basis of reaction-diffusion equations
[27], or the hypothesis of a chemical instability due to competitive growth of
colloidal particles coupled with diffusion, first put forward by ROSS and
coworkers [28]. The basic difference between these two approaches is that the
former theory postulates pattern formation to occur at the nucleation stage,

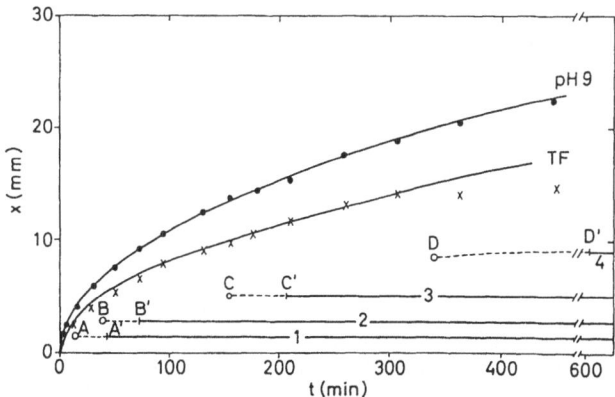

Fig. 4: Time dependence of the location of pH 9 front and turbidity front (TF) in the Mg(OH)$_2$ Liesegang system of Fig. 3A. The location of four precipitation bands is indicated by horizontal bars, starting with the onset of localized refractive index gradients (symbols A to D). The occurrence of visible bands is marked by symbols A' to D' (from [24])

while the latter assumes uniform nucleation and subsequent patterning due to the autocatalytic nature of colloidal particle growth. Actually, there are new re-sults, both in theory and in experiment, which indicate that a common basis may be found which could solve the controversy whether the patterns evolve in a pre- or a postnucleation stage [25,29].

A challenge remains the elucidation of the mechanisms responsible for precipi-tation patterns of higher complexity. These are found quite often but cannot yet be reproduced in a systematic fashion. The helix shown in Fig. 3B, which formed instead of a set of parallel bands, is one example; others, such as two-armed spirals instead of concentric rings and radial dislocations between neighboring rings, have been reported (see [30]).

3. Chemical Waves

Chemical gradients travelling through a thin solution layer have been described in several reaction mixtures showing temporal oscillations in batch systems [1,15,31]. Quantitative studies, however, have been mainly restricted to the classical Belousov-Zhabotinskii (BZ) reaction in which malonic acid is oxidized and decarboxylized by bromate in the presence of a catalyst, usually the redox couple ferroin (red)/ferriin (blue). The propagation of such waves of chemical activity is one of the most striking examples of pattern formation due to the coupling of reaction and diffusion. Wave patterns may evolve in an oscillating or in a quiescent, excitable medium. Here we are only concerned with the so-called trigger waves in an excitable reaction mixture.

The most common geometry of their fronts is a circle or a set of concentric circles ("target" patterns) triggered by, e.g., a hot platinum wire or evolving around impurities or small CO$_2$ bubbles. Because of the excitability of the sur-rounding regions this perturbation then travels outward, while excited solution volumes gradually return - through a refractory period of time - back to a quie-scent and newly excitable state.

Concentration profiles of a circular wave were first reported by WOOD and ROSS in 1985 [18]. They recorded light absorption at 490 nm on a one-dimensional pho-todiode array. Since then, much more work on concentration distributions has been done on the basis of computerized video techniques [20,32,33]. Recently, for instance, by using a UV sensitive video target, wave patterns could be investigated in the cerium-catalyzed BZ reaction that are invisible to the eye [34]. With this extended spectral range one can expect to detect waves in other chemical systems, as well.

In this section we will focus on two specific topics: the structure of spiral-shaped waves and the phenomenon of wave collision.

3.1. Spiral Waves

Spiral-shaped waves in an excitable medium [35,32] can be produced, for instance, by disrupting the front of an expanding circular wave with a gentle blast of air ejected from a pipette. At the irregularly shaped open wave ends spiral structures evolve with highly regular geometry. A digital image of one spiral wave is shown in Fig. 5. It was obtained with a two-dimensional (2D) spectrophotometer, an apparatus which consists of precision optics with interference filter (490 nm, corresponding to maximum ferroin absorption) for homogeneous illumination and imaging purposes, a video camera for the UV and visible range, a video frame buffer, and a fast, large memory computer for the storage of digital images and their further processing. The raster resolution is 512 x 512 picture elements each of which measures one out of 256 possible grey levels. Acquisition and storage of the 2D data are feasible at a frequency of 30 frames per minute. Elaborate software routines have been developed for further evaluation of the image data [17,36].

The quantitative information contained in an image can be used for the extraction of intensity profiles, as done in Fig. 6 along a vertical line passing through the tip of the spiral. From this a concentration profile can be easily computed. There is a remarkable difference between the shape of the individual

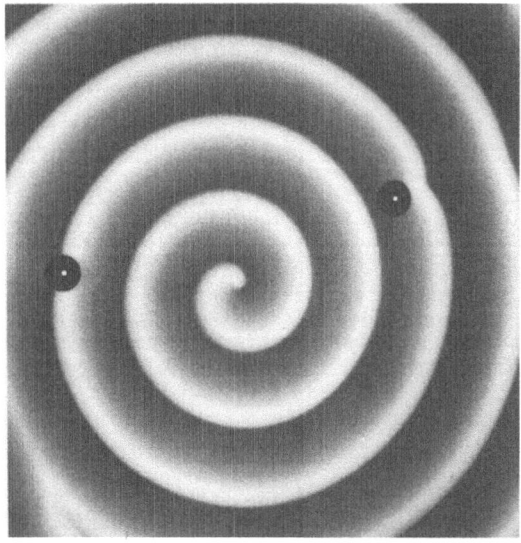

Fig. 5: Digital image of a spiral-shaped wave of chemical activity in an excitable medium of the ferroin-catalyzed BZ reaction. Transmitted light intensity (490 nm) at 450x450 pixels reflects the distribution of the ferroin/ferriin redox couple in a 9x9 mm^2 area of the dish. Initial concentrations: 0.34 M NaBrO$_3$, 0.048 M NaBr, 0.095 M malonic acid, 0.38 M H$_2$SO$_4$, 0.0035 M ferroin

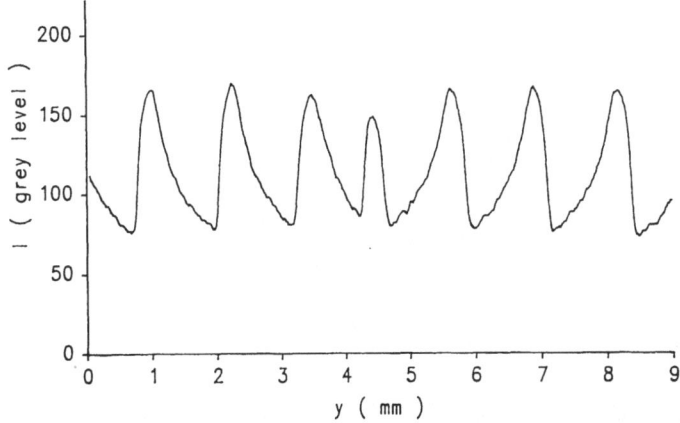

Fig. 6: Profile of transmitted light extracted from Fig. 5 along a vertical line (y-direction) and passing through the tip of the spiral

waves inside and outside the central region around y = 4.3 mm. An analysis of the profiles by fitting cubic splines yields concentration gradients of at least 5 mM/mm at the steep wave fronts and of the order of 1 mM/mm at the smooth backs [37].

Spirals are frequently arranged in pairs, but more complex systems with many spirals may be easily observed, as depicted in Fig. 7A. Each tip turns inward around a rotation center with a period of about 18 s, while the fronts propagate away from this center. (Note that the collision of two fronts leads to their mutual annihilation, as discussed below.)

In order to determine the location of the rotation center of the spirals in Fig. 7A and the structure of their immediate neighbourhood (the spiral cores) a logical overlay technique was applied to a sequence of six digital images recorded at 3s-intervals, thus covering just one revolution period. This procedure retains only the maximum of the six intensity values at any given spatial location and represents the 2D envelope of the maximum of the ferriin concentration data [32,37]. In the resulting image (Fig. 7B), for each of the four spirals there are six spiral-shaped bands which merge into four dark spots, respectively, inside which the variations of the catalyst concentration remain significantly below those found at all outer sites of the observation area.

The structure of the upper ferriin envelope is shown in Fig. 7C by a 3D perspective display technique [36]. In this representation the core region of each spiral resembles that of a chemical "tornado". One finds that inside the core (diameter ≈ 0.7 mm) a transition takes place from the rotation center - a singular site (diameter ≤ 30 μm) at which the concentration of the catalyst remains quasi-stationary - to the surrounding area, where waves attain their full amplitude between maximum oxidation and partial reduction of the catalyst. For the given initial chemical composition the location of this singular site remains remarkably stable in time [37].

The shape of the spiral waves was determined by fitting simple mathematical functions to isointensity levels. The results of a nonlinear least square fit of an Archimedian spiral to the maximum and to the minimum level of a wave, recorded in an area of 4.5 x 4.5 mm², are presented in Fig. 8. The calculated curves are good fits to the data points of constant intensity. The distance between successive whorls of the spiral pitch is 1.2 mm. The location of the rotation center obtained by the fits precisely coincides with the location derived from the overlay technique shown in Fig. 7. It turns out that the involute of a circle fits

Fig. 7: (A) Pattern consisting of 4 counter-rotating spiral waves. (B) Digital overlay of a sequence of six subsequent images including (A), covering one revolution and showing the spiral core of each spiral as a black spot. (C) Three-dimensional perspective image of (B). The third coordinate is the measured intensity. The observer is located below the upper edge of (B). The pixel brightness of (C) as compared to that of (B) is inverted from dark to bright and vice versa

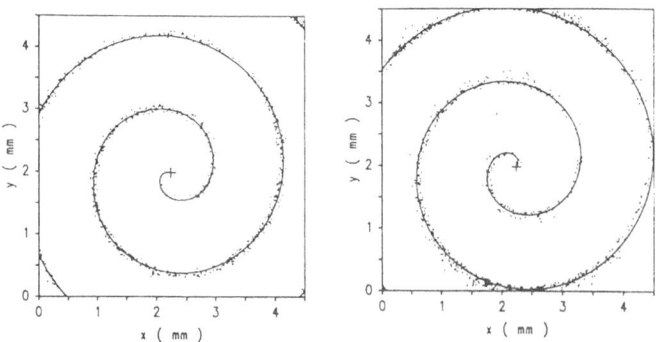

Fig. 8: Fits of an Archimedian spiral to the pixels with maximum (left) and minimum (right) grey levels in a spiral pattern observed in a 4.5 x 4.5 mm² area (from [37])

these data equally well [37]. With the available experimental techniques no decision can be made yet which spiral function is the more appropriate one. Theoretical models on the basis of realistic reaction-diffusion equations yield results close to the shape of the involute [38]. Measurements in the core region with much higher spatial resolution and detailed comparison of experimental with numerical data are in progress.

3.2. Wave Collision

As mentioned in the previous section, when two fronts of excitation travel towards each other they annihilate upon collision. No interference phenomena as expected for many other types of waves in physical systems are exhibited. Three-dimensional display techniques prove to be an efficient tool for showing details of the process of wave collision, as demonstrated in Fig. 9. The two approaching annular wave crests gradually merge into one. Subsequently, one observes the formation of a dip in the area of interaction indicating that due to the collision the crest turns into a more reduced state. The collision process implies the formation of typical cusp-like structures. The cusps are the loci of extremely high curvature of wave fronts. Therefore, as suggested in [38], the local propagation velocities should differ from those measured at low-curvature fronts.

An investigation of the interaction zone between two opposing wave fronts was performed by recording the rapid changes of the high-curvature fronts at microscopic spatial resolution on a video movie. Figure 10 presents an intermediate isoconcentration level extracted from a subsequently digitized single image of the movie with a calculated resolution of 0.7 μm per pixel. (The actual spatial resolution is somewhat smaller because of the limited bandwidth of the video recorder. This also reduces the grey level resolution.) The curvature of the cusps

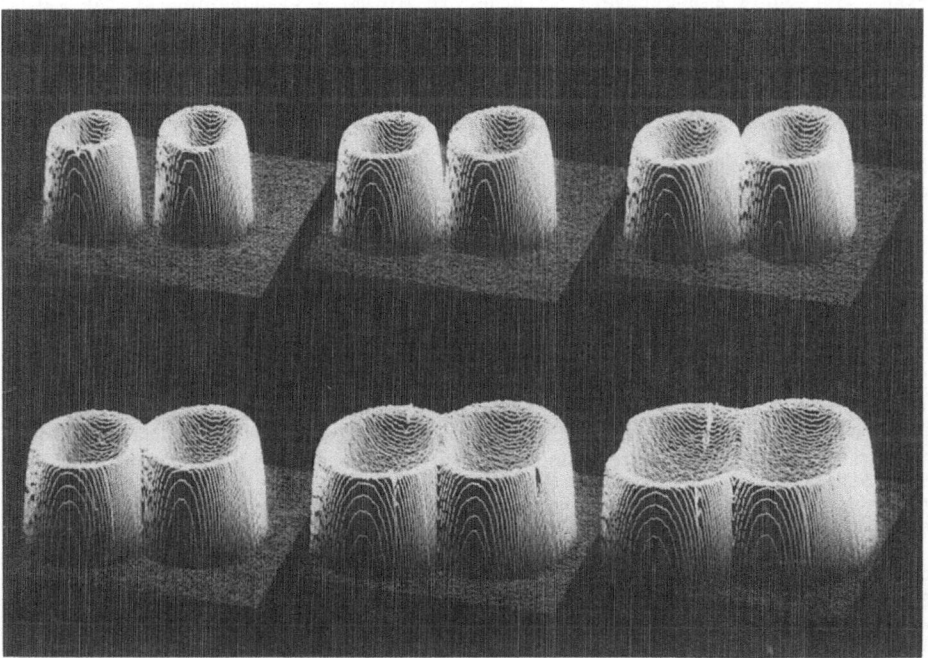

Fig. 9: Sequence of digital images in three-dimensional perspective showing the process of collision between two circular waves in an area of 11 x 11 mm²

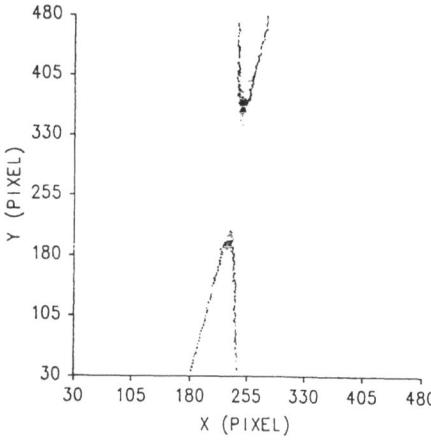

Fig. 10: Isoconcentration level in the area of wave collision, obtained from an image with microscopic resolution (0.7 μm per pixel)

was derived from fits of appropriate mathematical functions, their normal velocity was directly measured. Thus, for the first time a linear relationship between curvature and propagation velocity of a chemical wave front was determined experimentally [39]. The results are in good agreement with theoretical predictions [40,38].

4. Hydrodynamic Effects on Chemical Waves

In the experiments described in the previous section the sample dish containing the BZ reagent was covered by a glass plate (leaving an air gap of several mm between layer surface and cover) in order to minimize evaporation at the layer surface. However, if the cover is removed and the layer thickness is about 1 mm or larger, a variety of spatial effects due to evaporative cooling are readily observed. These evolve independently of travelling waves and have some influence on the geometric form of wave fronts. Evidence of such effects has been reported previously [13-17]. For instance, stationary patterns of high regularity ("mosaic" patterns) may form. There is some evidence that these patterns are correlated with a network of convection cells. These are set up in the layer due to unbalanced forces caused by temperature- or concentration-induced gradients of surface tension (Marangoni convection). They disappear when the dish is covered.

The interaction of chemical waves with convection is a highly complex matter. The sensitivity of the 2D photometric technique allows the detection of interesting structural details of the ferroin/ferriin distribution discussed in some detail in [41]. Here we only present two examples on a qualitative basis.

Figure 11 shows how the existence of a stationary mosaic pattern leads to small distortions of the wave fronts. The mosaic pattern is faintly visible inside the pair of circular waves, while no such structure appears to exist in the outside area where light absorption by ferroin is very high. However, it can be shown experimentally that the mosaic pattern extends over the whole observation territory. In the picture the intensity contrast at the front of the wave is considerably larger than that of the stationary "background" structure. Therefore, the intensity of the incident light beam had to be limited such that the camera target be not overexposed. Consequently, the pattern in the unexcited, black outer area is not resolved. It can be detected, if only a portion of the black area is imaged on the target and the incident intensity is sufficiently augmented. Then it becomes clear that the wave front distortions are strongly correlated with the network-like stationary pattern. This correlation can be demonstrated more directly if the wave amplitudes are decreased, which is achieved by appropriate choice of initial concentrations [41].

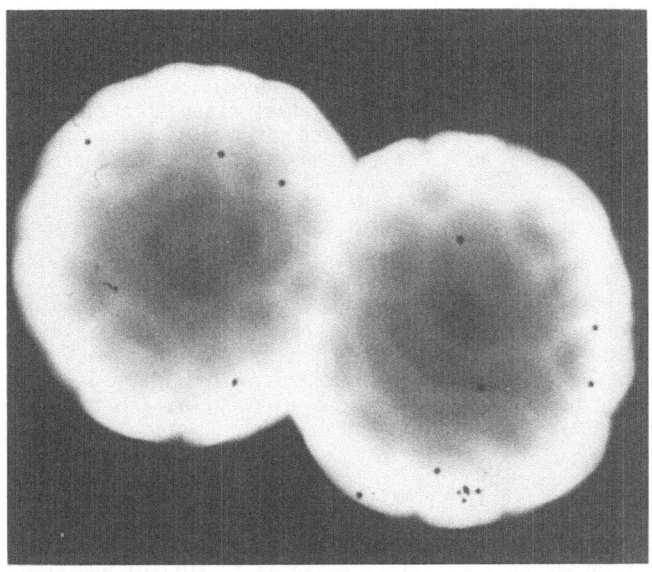

Fig. 11: Wave front with small distortions and faintly visible stationary (mosaic) pattern in an uncovered solution layer (thickness ≈ 1 mm) with concentrations similar to those in Fig. 5

The coupling of a reaction-diffusion pattern of a given wavelength with convection caused by evaporative cooling frequently results in a highly disordered structure as shown in Fig. 12. Strong distortion of wave fronts - in this case a pair of spirals - may be followed by a complete disruption of the wave. This gives rise to order-disorder transitions, which were discussed and predicted on theoretical grounds for the case of controlled Bénard-type convection [42]. Image sequences of wave decomposition and reorganization were treated by image analysis techniques with the goal of finding numerical criteria for the characterization of disorder and order in such chemical patterns [43]. In this context, the question whether spatial chaos occurs is most interesting.

Recently, the problem of hydrodynamic motion in a chemically active medium with and without wave propagation, both under covered and uncovered conditions,

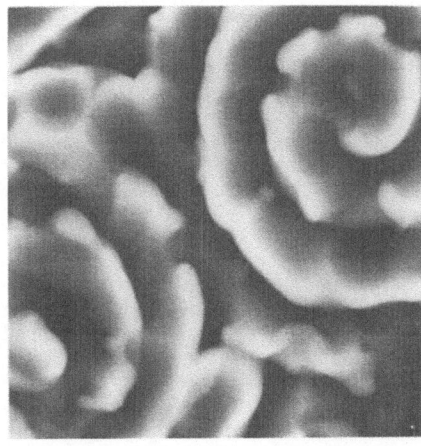

Fig. 12: Decomposition of a pair of initially regularly shaped spiral waves occurring several minutes after removing the glass cover from the dish

was investigated by light scattering and microscope video imaging techniques [44]. Using chemically inert polystyrene particles as indicators it could be shown that not only any travelling wave induces a local hydrodynamic disturbance but a periodic passage of waves can even entrain oscillatory changes in flow direction. These first results emphasize the importance of reaction-convection coupling in a system in which primarily only transport by diffusion has been taken into account.

5. Spatial Patterns in Oscillating Glycolysis

For a certain range of substrate input flux the metabolic turnover in the glycolytic pathway of yeast cells or of cytoplasm extracted from yeast cells exhibits oscillatory behaviour. For more than 20 years this system has been the most intensively investigated example of an oscillating biochemical reaction. Salient contributions have been made by HESS and coworkers for elucidating the role of control and regulation by key enzymes leading to periodic or chaotic time patterns [45-47].

In thin open layers of extracted yeast cytoplasm, to which glycolytic substrates such as trehalose or glycogen are added, one observes, at a wavelength specific for the absorption of NADH, the repetitive formation of spatial structures with a length scale of the order of 1 mm, first reported by BOITEUX and HESS [9]. While the structures tend to have predominantly rod-like shape during the first one or two phases of pattern occurrence, they break up into smaller patches at later times of an experiment.

During recent years more insight into the coupling of biochemical metabolism with transport processes was achieved by using the UV-sensitive 2D spectrophotometer combined with suitable optical techniques. By rapidly switching between transmission of a diffuse light field (λ = 380 nm) and a Schlieren-type illumination of the layer in the visible it could be shown that the inhomogeneities in NADH concentration are correlated in space with a polygonal network of convection cells [48]. Figure 13A gives direct evidence of this finding: The layer was illuminated with a parallel beam of UV light (λ = 370 nm) and slightly defocussed. The patch-like modulations of light intensity are caused by the inhomogeneous distribution of light-absorbing NADH, while the sharp bright lines reflect a focussing of light due to refractive index gradients that exist at the boundaries of the convection cells and are directly related to the distribution of temperature and biochemical composition in the liquid. This corroborates previous suggestions that in the open layer the biochemical reaction is coupled with convective flow set up by unbalanced forces at the liquid/gas interface, to be compared with the structural effects described in Section 4.

A rod-shaped NADH absorption pattern, observed with diffuse light illumination in the UV at the moment of maximum intensity contrast, can be seen in Fig. 13B as a 3D surface image. This representation shows that the NADH concentration, which is displayed along the third coordinate of the 3D coordinate system, varies smoothly in the immediate neighborhood of the thin dark lines in the center of the bright areas. The dark lines correspond to the bright ones in Fig. 13A and are, in this case of diffuse light, shadows cast by light scattering at the pronounced gradients of refractive index close to the boundaries of the convective currents. This 3D version of the NADH pattern certainly has some similarity with hydrodynamic flow patterns in the real 3D world.

The temporal correlation of glycolytic oscillations in the bulk of the layer with the time intervals of pattern formation was established by evaluating a time sequence of digital images (Fig. 14) [49]. The spatial average of the measured intensity values (Fig. 14A) reflects the oscillations. The shaded stripes indicate those time intervals during which patterns are detected by visual inspection of the images. These intervals occur twice for each oscillation and overlap with the peaks in the standard deviation of the intensity distribution (Fig. 14B),

94

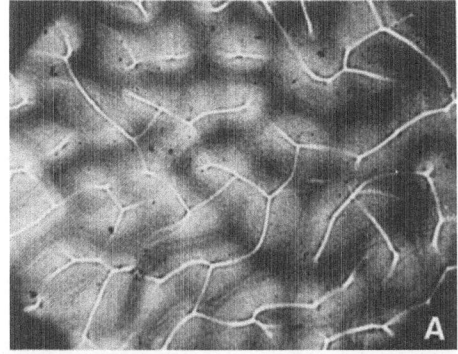

Fig. 13: Spatial patterns forming during oscillating glycolysis in a 1.8 mm layer of yeast extract.
(A) Digital image of the intensity modulation observed in parallel transmitted light (370 nm).
(B) Three-dimensional perspective of intensity modulation observed in diffuse transmitted light (370 nm). Observation area: 14 x 14 mm² (from [36])

which is a heuristic measure for structure formation. The extrema of the time derivative of the absorbance curve (Fig. 14C) fall into exactly the same time intervals. Consequently, during each glycolytic half-period, transmission patterns build up and decay just when the turnover of NADH is especially high. This cannot be explained by changes of heat production by the reaction alone, because these were measured to have the same period as the metabolic oscillations [50]. Further investigations are needed for a better understanding of the biochemical reaction-convection patterns generated by the oscillating glycolytic system.

6. Concluding Remarks

Various examples of spatial structures in chemical and biochemical solutions were presented which demonstrate the rich and complex, regular or irregular features of ordering processes in reactions under non-equilibrium conditions. The list of pattern-forming systems could be easily extended by considering, for instance, reactions at interfaces, such as the photochemically induced structures already mentioned in the Introduction [11,12] or dynamical motions at the interface between two immiscible liquids across which a reaction takes place [51,52]. Also, the spreading of a drop added to a liquid layer, when accompanied by reaction, leads to interesting structure formations, such as those studied for an enzymatic reaction in [17]. The investigations described here show that the combination of classical observation and measuring techniques with modern, sophisticated techno-

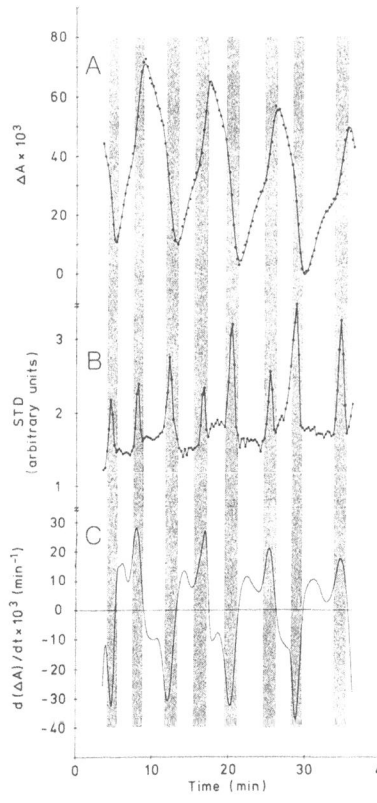

Fig. 14: Temporal relation between glycolytic oscillations and pattern formation.
Average of transmitted light (A) and its
standard deviation (B) derived from an
image sequence taken at 15s-intervals. The
peaks in (B) correspond to those time-in-
tervals during which patterns are visually
detectable on the observation screen
(shaded stripes). (C) Time derivative of
the fitted ΔA-curve in (A). (From [49])

logies significantly improves our experimental knowledge. The development of
computerized high-precision photometric methods, that are well fitted to the acquisition of concentration data of chemical patterns in two dimensions, is an important step towards a comprehensive quantitative description of the spatio-tem-
poral dynamics observed in reactive solution layers. There is no doubt that this
improvement of experimental approaches will contribute to the elucidation of the
underlying mechanisms on theoretical grounds.

Finally, it should be noted that in the context of chemical organization, biological patterns on the subcellular and cellular level are of particular interest. Ample evidence is given in this Volume for their immense variety of shapes,
forms and colors, and it is a remarkable fact that many of the structural elements can be already produced in a chemical or biochemical solution. It is therefore exciting to see to which extent the efforts to unravel the relevant mechanisms of spatial self-organization in these chemical systems will improve our
understanding of the more intricate mechanisms acting in biology.

Dedication and Acknowledgement

This article is dedicated to Professor Benno Hess on the occasion of his 65th
birthday. His scientific guidance and support through many years are acknowledged
by the author with gratitude.

References

1. A.N. Zaikin, A.M. Zhabotinskii: Nature 225, 535 (1970)
2. H.-G. Busse: J. Phys. Chem. 73, 750 (1962)
3. F.F. Runge: Der Bildungstrieb der Stoffe veranschaulicht in selbständig gewachsenen Bildern (Oranienburg, Selbstverlag 1855)
4. E.S. Liesegang: Phot. Archiv 21, 321 (1986)
5. G. Lippmann: Ann. Phys. 2nd Series 149, 544 (1873)
6. E.S. Hedges, J.E. Myers: The Problem of Physico-Chemical Periodicity (Arnold & Co., London 1926)
7. P. Glansdorff, I. Prigogine: Thermodynamic Theory of Structure, Stability and Fluctuations (Wiley Interscience, New York 1971)
8. H. Haken: Synergetics, 3rd ed. (Springer, Berlin, Heidelberg 1983)
9. A. Boiteux, B. Hess: Ber. Bunsenges. Phys. Chem. 84, 392 (1980)
10. P. Möckel: Naturwissenschaften 64, 224 (1977)
11. J.C. Micheau, M. Gimenez, P. Borckmans, G. Dewel: Nature 305, 43 (1983)
12. D. Avnir, M. Kagan: Nature 307, 717 (1984)
13. A.M. Zhabotinskii, A.N. Zaikin: J. theor. Biol. 40, 45 (1973)
14. K. Showalter: J. Chem. Phys. 73, 3735 (1980)
15. M. Orban: J. Am. Chem. Soc. 102, 4311 (1980)
16. K.I. Agladze, V.I. Krinsky, A.M. Pertsov: Nature 308, 834 (1984)
17. S.C. Müller, Th. Plesser, B. Hess: Naturwissenschaften 73, 165 (1986)
18. P.M. Wood, J. Ross: J. Chem. Phys. 82, 1924 (1985)
19. S.C. Müller, Th. Plesser, B. Hess: Anal. Biochem. 146, 125 (1985)
20. C. Vidal, A. Pagola: J. Phys. Chem. 91, 501 (1987)
21. K.H. Stern: Chem. Rev. 54, 79 (1954)
    K.H. Stern: A Bibliography of Liesegang Rings, 2nd ed. (U.S. GPO, Washington, D.C., 1967)
22. R. Fricke, O. Suwelack: Z. Phys. Chem. 124, 359 (1926)
23. H. Higuchi, R. Matuura: Mem. Fac. Sci. Kyushu Univ., Ser. C5, 33 (1962)
24. S. Kai, S.C. Müller, J. Ross: J. Chem. Phys. 76, 1392 (1982)
25. M.E. LeVan, J. Ross: J. Phys. Chem. 91, 6300 (1987)
26. W. Ostwald: Lehrbuch der Allgemeinen Chemie (Engelmann, Leipzig 1897)
27. C. Wagner: J. Colloid Sci. 5, 85 (1950)
    J.B. Keller, S.I. Rubinow: J. Chem. Phys. 74, 5000 (1982)
28. R. Lovett, P. Ortoleva, J. Ross: J. Chem. Phys. 69, 947 (1978)
    G. Venzl, J. Ross: J. Chem. Phys. 77, 1308 (1982)
29. G.T. Dee: Phys. Rev. Lett. 57, 257 (1986)
30. S.C. Müller, S. Kai, J. Ross: Science 216, 635 (1982)
31. R.J. Field, M. Burger (eds.): Oscillations and Traveling Waves in Chemical Systems (Wiley, New York 1985)
32. S.C. Müller, Th. Plesser, B. Hess: Science 230, 661 (1985)
33. J.M. Bodet, C. Vidal, J. Ross: J. Chem. Phys. 86, 4418 (1987)
34. Zs. Nagy-Ungvarai, S.C. Müller, Th. Plesser, B. Hess: Naturwissenschaften (1988), in press
35. A.T. Winfree: Science 181, 937 (1973)
36. S.C. Müller, Th. Plesser, B. Hess: Biophys. Chem. 26, 357 (1987)
37. S.C. Müller, Th. Plesser, B. Hess: Physica 24D, 71 and 87 (1987)
38. J.P. Keener, J.J. Tyson: Physica 21D, 307 (1986)
39. P. Foerster, S.C. Müller, B. Hess: In Spatial Inhomogeneities and Transient Behavior in Chemical Kinetics, ed. by G. Nicolis, P. Gray (1988), in press
40. V.S. Zykov, G.L. Morozova: Biofizika 23, 717 (1979)
41. S.C. Müller, Th. Plesser, B. Hess: In Physicochemical Hydrodynamics: Interfacial Phenomena, ed. by M.G. Velarde, B. Nichols, NATO ASI Ser. (Plenum, New York 1988)
42. D. Walgraef: In Non-Equilibrium Dynamics in Chemical System, ed. by C. Vidal, A. Pacault, Springer Ser. Syn., Vol. 27 (Springer, Berlin, Heidelberg 1984) p. 114
43. M. Markus, S.C. Müller, Th. Plesser, B. Hess: Biol. Cybern. 57, 187 (1987)
44. H. Miike, S.C. Müller, B. Hess: Chem. Phys. Lett. (1988), in press
45. B. Hess, B. Chance, A. Betz: Ber. Bunsenges. Phys. Chem. 68, 768 (1964)
46. B. Hess, A. Boiteux: Annu. Rev. Biochem. 40, 237 (1971)

47. M. Markus, B. Hess: Proc. Natl. Acad. Sci. USA $\underline{81}$, 4394 (1984)
48. S.C. Müller, Th. Plesser, A. Boiteux, B. Hess: Z. Naturforsch. $\underline{40c}$, 588 (1985)
49. S.C. Müller, Th. Plesser, B. Hess: In <u>Temporal Order</u>, ed. by L. Rensing, N.I. Jaeger, Springer Ser. Syn., Vol. 29 (Springer, Berlin, Heidelberg 1985) p. 194
50. Th. Plesser, S.C. Müller, B. Hess, I. Lamprecht, B. Schaarschmidt: FEBS Lett. $\underline{198}$, 42 (1985)
51. E. Nakache, M. Dupeyrat, M. Vignes-Adler: J. Colloid Interface Sci. $\underline{94}$, 187 (1983)
52. S. Kai, S.C. Müller: Sci. Form $\underline{1}$, 9 (1985)

Experimental Study of the Target Patterns
of the BZ Reaction Using Digital Picture Analysis

P. Hanusse, C. Vidal, and A. Pagola

Centre de Recherche Paul Pascal, CNRS, Domaine Universitaire,
F-33405 Talence Cedex, France

1. Introduction

Although the target patterns of the BZ reaction have been known for many years, the origin and nature of the leading centers are still subject to controversial discussions [1]. We give here a brief progress report of an experimental program aiming at characterizing the statistical properties of the target patterns in an oscillating medium.

In order to perform quantitative measurements of the properties of the chemical waves of the BZ reaction we have designed and used picture analysis tools. All the results presented here were obtained by studying digital pictures [2]. All experiments were carried out in a Petri dish type of cell [3]. Results presented in Figs. 1 to 5 were obtained by direct aided measurements of center position and target wavelength on computer screen [4]. Figures 6-7 report a study of structure properties near the center [5]. In the last part we briefly mention an automatic pattern recognition approach for intensive statistical target properties analysis [6].

2. Target Patterns Properties in an Oscillating Medium

In Fig. 1 we report the position of 529 target centers over 33 experiments in a 100 mm diameter cell. There are a few accumulation points, but the properties of the associated targets (frequency, wavelength) do not differ from the general behavior. In that sense, the system properties are "uniform".

We have studied the relationship between center pulsation and wave number (Fig. 2). The linear dependence supports the idea of a unique propagation speed according to the relation $\omega_1 = \nu k_1$. There seems to be no correlation between the deviations to linearity and other properties, which is evidence that they result from purely random measurement errors.

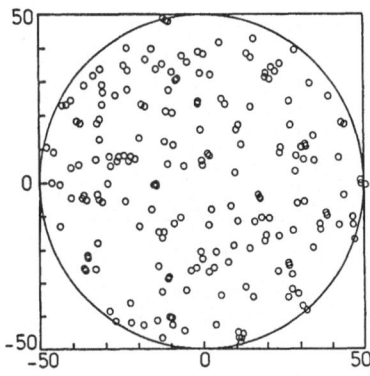

Fig. 1: Spatial distribution of target centers in the experimental cell (33 experiments, 529 targets)

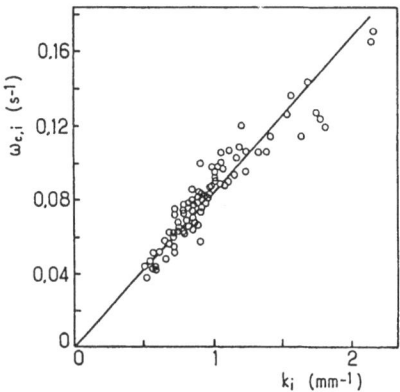

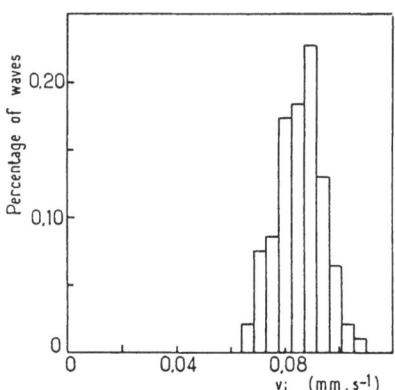

Fig. 2: Pulsation of centers versus wave number of the target pattern

Fig. 3: Speed propagation histogram

Figure 3 presents the distribution of front speeds. These measurements also support the hypothesis of a unique front propagation speed, at least within the accuracy of the present measurements. The standard deviation is of the order of 10% of the mean value which is to be expected, taking into account the unavoidable small differences in experimental conditions from one experiment to the other, and the overall precision on the speed determination.

In Fig. 4 is given the histogram of center period as a function of the reduced period (center period over bulk oscillation period). The dashed part corresponds to centers having only one ring. Their period is close to the bulk period so that they cannot expand much, one ring being removed from each target at each period of the bulk oscillation. These centers seem the most numerous, but they are also the most difficult to observe.

Finally, the wave number histogram of targets having at least two rings is given in Fig. 5. For that reason, the dashed part of Fig. 4 does not contribute here (no wavelength can be properly measured). The wavelengths vary from 3 to 10 mm. Small wavenumbers (long wavelength) correspond to low frequency centers.

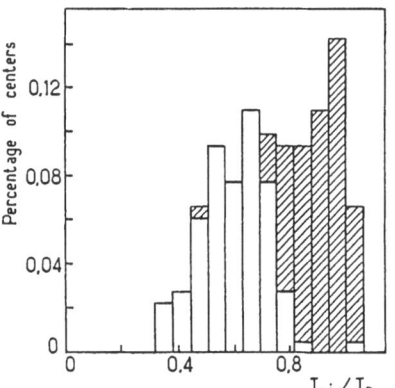

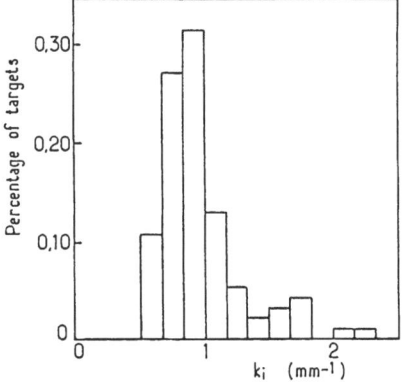

Fig. 4: Histogram of oscillation period of centers versus reduced period

Fig. 5: Wave number histogram of targets having at least two rings

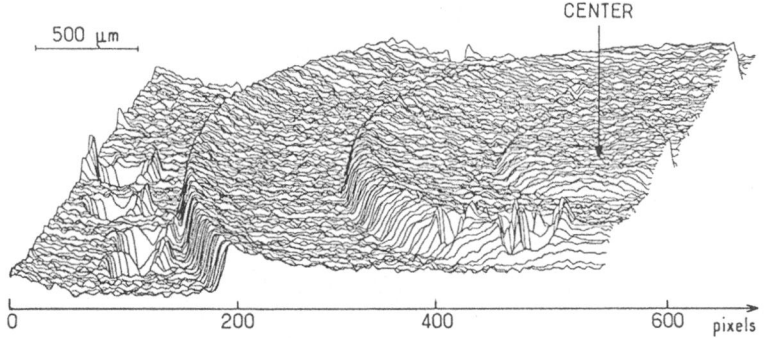

Fig. 6: 3D representation of the target pattern. The "volcanoes" are due to gaseous bubbles. Field size is 2.85 x 2.85 mm² (resolution of 6 μm). Layer depth is 0.7 mm

3. Wave Profile Near the Center of a Target

We have observed the formation of a ring at the center of a target pattern. Using our digital picture analysis system, it is possible to get down to resolution of about 1.8 μm per pixel. At this scale nothing special is observed at the center (all solutions were subjected to careful filtering). From a sequence of 20 pictures we have measured the height of the front, its width and its speed of propagation, as a function of the distance from the center of the target pattern. The region of the target center is shown in Fig. 6.

The front wave has attained its final shape at about 1.8 mm from the center (see Fig. 7). The height of the front changes the most over this distance. The

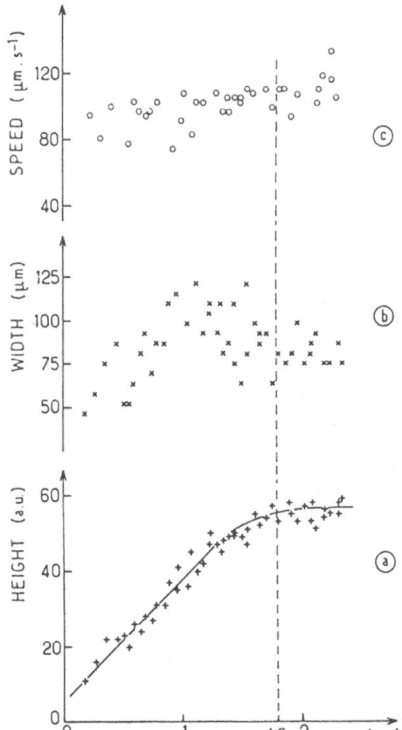

Fig. 7: Front speed, width and height versus the distance from the center

width establishes quickly at 88 $\pm$ 16 μm. So does the speed which does not seem to change much. A 10% increase can be observed until a speed value, equal to that observed far from the center, is reached ($\approx$ 100 μm/s).

4. Automatic Pattern Recognition of Targets

To perform an accurate statistical analysis of target pattern properties and to detect the correlations that may exist between them, one must face the problems resulting from the analysis of a large amount of pictures. In that case, automatic pattern recognition techniques should be used. We have designed a special software that provides automatic pattern center detection, wavelength as well as period measurements over the course of the experiment. All those morphological and dynamical properties are determined automatically [6] from a video tape of the experiment. These measurements allow us to reconstruct the global picture at any moment of the evolution and study its transformations (Fig. 8). We plan to perform intensive statistical analysis using such an approach to extend the results obtained so far manually, although on digital pictures.

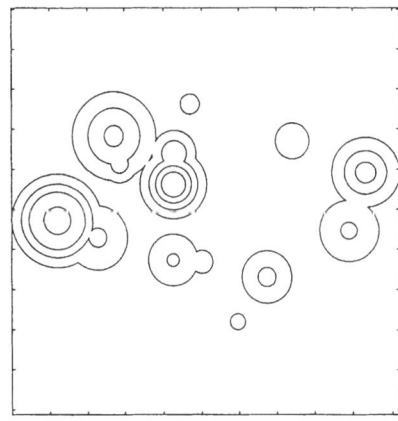

Fig. 8: Reconstructed picture of a target pattern system of the BZ reaction from the data obtained automatically from the video tape of the experiment

References

1. C. Vidal, P. Hanusse: Int. Rev. Phys. Chem. <u>6</u>, 1 (1986)
2. We use a PERICOLOR 2000 digital picture system of NUMELEC company. Pictures have 512 x 512 x 8 bits format
3. Cell size 10 cm. Typical initial composition: $[H_2SO_4]$ = 0.22 M, [MA]=0.08 M, $[NaBrO_3]$ = 0.31 M, [ferroin] = 4.10^{-3} M (see [4] for details)
4. C. Vidal, A. Pagola, J.M. Bodet, P. Hanusse, E. Bastardie: J. Phys. <u>47</u>, 1999 (1986)
5. A. Pagola, C. Vidal: J. Phys. Chem. <u>91</u>, 502 (1987)
6. P. Hanusse, E. Bastardie, C. Vidal: to be published

Dispersion Curves and Pulse Wave Propagation in Excitable Systems

M. Marek[1] *and H. Ševčíková*

Hošťálkova 73, CSSR-16900 Prague 6, Czechoslovakia

1. Introduction

The excitation and wave propagation phenomena are common in systems of varying nature such as plasmas, excitable tissues in living organisms, heterogeneous catalytic
reactions, combustion, homogeneous autocatalytic reactions, populations of living
organisms and so on /1,2/. Results of empirical observations, planned experiments
and mathematical modelling have accumulated in recent years but comparisons and ge-
neralizations valid for common phenomena in different systems are scarce. Here we
attempt to briefly compare generation and propagation of action potential in nerve
tissue and initiation and propagation of concentration waves in chemical reaction-
diffusion systems. We use experimental facts well known from the studies of receptors and propagation of the action potential in nerve fibers /3,4,5/ and modelling
results for the processes of nerve conduction /6,7,8,9/ on one hand and the results
of experimental and modelling studies of pulse wave initiation and propagation in
spatially quasi one- and two- dimensional reaction-diffusion systems on the other
hand. Experimental systems quoted here are represented by a thin layer of reacting
mixture of the Belousov-Zhabotinski (B-Z) type /10/; either kinetic model of the
B-Z reaction (Oregonator /11,12,13/) or kinetic SH model /14,15/ were considered.

For the reaction-diffusion system we present for the first time the "dispersion
relation", an experimentally determined dependence of the velocity of the pulse wave
propagation on the period of wave initiation and discuss its use in the interpretation of complex wave patterns, as it is common in modelling of nerve conduction.
Even though the wave-like variables in chemical and neurophysiological systems are
often different - the species concentration in the chemical system and transmembrane
electric potential and current in the neurophysiological system - the underlying
mechanism and dynamics in the phase space are similar and the comparison might be
useful. The overall process of the wave initiation and propagation consists of several stages which are schematically shown in Fig.1.

In neuronal systems /4/ the stimulus can be either some form of energy /heat,
pressure, light/ acting on terminals of sensory units or the presynaptic potentials
in the synapses of neuron cells. In both cases the chemical transformation processes
together with transport processes evoked by stimulus generate a generator potential
/receptor or postsynaptic potential, respectively/. Both forms of potentials are
transmembrane electric potentials arising in general at the electrically nonexcitable parts of a cell membrane. The longitudinal electric current in the arising potential difference causes the change of the transmembrane potential in the adjacent
electrically excitable parts of the membrane. The potential change when exceeding
certain - threshold - value evokes a rapid growth of a transmembrane potential -
a spike of an action potential. Hence the longitudinal potential gradient is reesta-
blished, the longitudinal current excites adjacent non-excited parts of the system
and a spike of the action potential propagates along the nerve fiber.

In the chemical excitable medium of the B-Z type, e.g., a local heating of the
reaction solution /16,17/ or a local change of Br⁻ ion concentration caused by
controlled dissolution of Ag electrode /18,19/ can serve as an initiating stimulus.

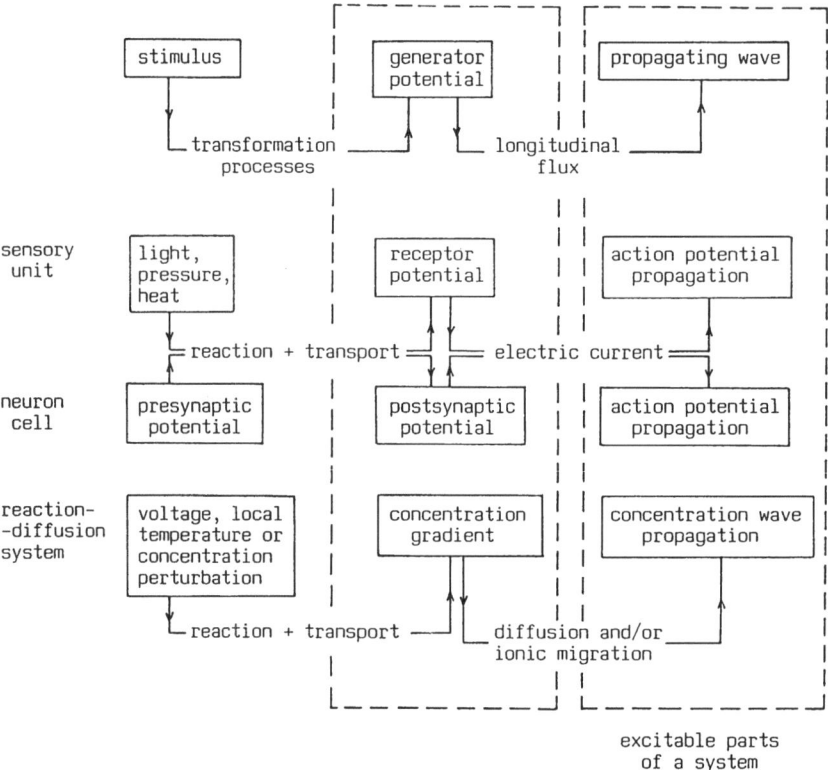

Fig. 1: Schematic picture of the wave generation and propagation in chemical and
neurophysiological systems

In other cases where the wave generation on inhomogeneities in the solution or at
liquid-solid-air interfaces was observed the nature of the stimulus and mechanism
of initiating processes are less obvious. However, the effects evoked by the stimulus generate concentration gradients (i.e. "generator potential") of Br^- ions and
other components which start autocatalytic reaction steps producing abrupt changes
in species concentrations. The longitudinal diffusion of species in established
concentration gradients transfers the conditions suitable for the autocatalytic activity to the adjacent parts of the medium and the repetition of the whole process
is displayed as a propagating wave of concentration disturbance /20/.

Different stages of the propagation often occur in dynamically different parts
of the system. For example, in the nerve tissue the formation of the generator potential is localized in electrically nonexcitable parts of the membrane and its
size increases with the magnitude and length of the stimulus and with the size and
numbers of synapses or sensitive endings of the receptor exposed to the stimulus -
graded response. On the other hand in the excitable parts of the membrane (forming
a nerve axon) the increase of the stimulus strength causes the change of the originally stable stationary state of the system to the periodic state (repeated waves
of action potential) with the amplitude (i.e., the magnitude of the action potential)
practically independent on the stimulus character - all-or-none response.

Mathematical models of pulse initiation and propagation are usually based on
empirical observations and enable both to classify types of behaviour in the phase
space in dependence on the chosen set of parameters (reflecting studied experimental situation) and to study individual processes of pulse initiation and propagation in detail.

Excitable one-dimensional systems are most often described by the set of partial differential equations which are generally in the form

$$X_t = D X_{zz} + F(X),$$ (1)

where X and F are vectors of system variables and their functions, respectively, D is a diagonal matrix and the indices represent first and second derivatives with respect to time and spatial coordinate, respectively.

In reaction-diffusion systems /14/, the X represents species concentrations, D their diffusion coefficients and F(X) consists of nonlinear rate relations related to the kinetic mechanism of chemical reaction.

In the Hodgkin-Huxley (HH) model /21/ of nerve conduction, the individual terms of a vector X are transmembrane potential and phenomenological coefficients describing a degree of activation and deactivation of individual ionic channels.The diagonal matrix D then usually contains only one non-zero term for the transmembrane potential. This term consists of electric properties of a nerve membrane and the product of this term and the second derivative of transmembrane potential describes the longitudinal electric current. The term F(X) describes processes related to the ionic current density and to conductance of individual ionic channels.

FITZHUGH /22/ and NAGUMO /23/ simplified the above model and obtained a two-variable system (FHN model) where all three phenomenological coefficients are summarized and give one new variable - the recovery current.

2. Wave Propagation

2.1 Excitability, dispersion relation

Wave propagation itself occurs at excitable parts of a system when an appropriate perturbation of system variables is applied. The excitability of a system is illustrated in the phase portrait in Fig.2 /14,24/ (other phase portraits may also correspond to excitability). The considered system has three stationary solutions, one stable state of a focus type and two unstable ones - a saddle and a focus. The stable solution describes the initial rest state of the system. When the perturbation is applied, the system can either diminish the initial perturbation restoring its original state or the perturbation can rapidly grow before it drops back to the original state. Drawn trajectories demonstrate that to evoke a large excitation loop the perturbation has to fall on the right hand side of the trajectory a . Hence the trajectory a consists of threshold values of perturbations and we shall call it a critical trajectory.

Figure 2 describes a two-dimensional case - a lumped parameter system with two independent variables. The actual phase space of the system under consideration has an infinite dimension as it is spatially distributed and its mathematical description is most often provided by a set of partial differential equations. In the infinite dimensional phase space every fixed point represents a time independent spatial profile of system variables and the critical trajectory a is a manifold of at least codimension one /25,26/. The perturbation, i.e., the change of spatial profiles of system variables, has to cross the critical manifold to bring the system to the excitation.

From the physical point of view, the perturbation of system variables occurs locally and when it has a superthreshold value the local excitation takes place building up the longitudinal gradient of system variables. A flow in this gradient (diffusion or ionic migration of species in chemical case or an electric current in neurophysiological case) evokes superthreshold changes in adjacent parts of the system and, successively, a new excitation.

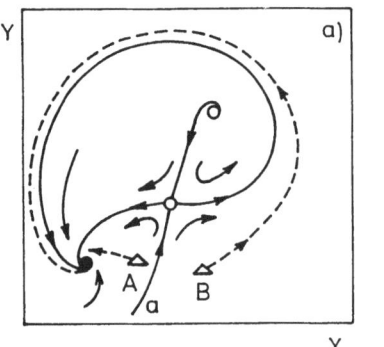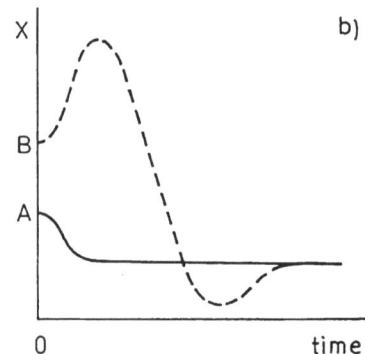

Fig. 2: a) Phase portrait of an excitable system, ● - stable and ○ - unstable
solutions, A - subthreshold and B - superthreshold perturbation,
a - critical trajectory
b) Response to the subthreshold (A) and superthreshold perturbation (B)

Repetitively or permanently evoked perturbation of a superthreshold value gives
rise to repetitive wave trains of excitation. The resulting wave trains are characterized by the velocity of wave propagation and by its period, i.e., the time interval between two successive waves. The characteristic values of velocities and
periods are affected by actual conditions - e.g., by the diameter of a nerve fiber,
the thickness of a membrane, the distance between nodes in nerve fibers /3,4/ or
by the initial composition or temperature in the B-Z reaction /27,28/. Moreover, it
was found both experimentally and theoretically that even an excitable medium at the
same conditions can support a variety of periodic wave trains differing both in their
periods and velocities /4,9,11,14,28-33/. It was also found that the velocity of an
individual wave (v) and its time spacing (T) from the preceding one can change
during the propagation /4,8,9,14,34,35/.

A variety of possible wave patterns was theoretically studied for several models
of nerve conduction assuming constant shape, velocity and spacing in the wave pattern propagating on an infinite length interval or on a circle. One-periodic wave
trains both with uniform velocity and interval between the waves were calculated
for the HH model /9/ and FHN model /35/, the existence conditions for double periodic wave patterns with two different alternating intervals but with the same velocity of both waves were found for the FHN model /30/ and multiple-periodic and chaotic wave trains were found for the FitzHugh model /31/. The wave trains of corresponding characters were also obtained as numerical solution of reaction-diffusion
model with the SH kinetics /14/.

Obtained velocity and period data can be summarized in the form of the dispersion
relation - the dependence of the wave propagation velocity on the time interval between the successive waves. A variety of shapes of dispersion relations were obtained
depending on the used model and chosen values of parameters. Several examples are
shown in Fig.3. The monotonous dispersion relations were also found in the model
reaction-diffusion system with the SH kinetics /14/ and the Oregonator kinetic model of the B-Z reaction /13/.

However, all dispersion relations have common features:
i) The finite minimal velocity and time interval T_A allowed for successive waves
to be generated. During the time interval $t \in \langle 0, T_A \rangle$ the medium is in a state of
"absolute refracterity", it restores its original state and is unable to undergo
another excitation. This period is followed by a "relative refractory" period during which the medium further restores its initial state but its wave supportive
properties are already recovered.
ii) When the time interval between successive waves is sufficiently long the waves

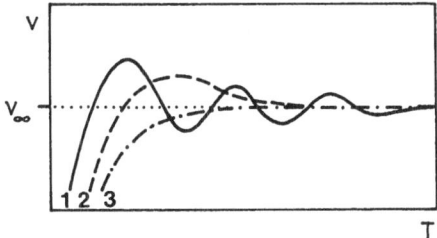

Fig. 3: Dispersion curves, 1,2 - FitzHugh-Nagumo model /35/, 2,3 - Hodgkin-Huxley
model /9/, 2 - experimental measurements on frog optic nerve /34/,
3 - models of reaction-diffusion systems /13,14/

move with the same velocity v_∞ which is equal to the velocity of a wave travelling
through the undisturbed medium.

The dispersion relation, in general, gives simple condensed information about
wave propagation. On one hand it relates admissible velocities and periods of wave
trains steadily propagating through the medium (i.e., waves preserving their shape,
velocity and period in the course of propagation) and on the other hand it can be
used to describe the development of nonsteadily propagating wave patterns. The kinematic method of calculations of time and space evolution of periods and velocities
of individual waves in the pattern from the known initial distribution of periods
and known dispersion relation was discussed for example by CHRAMOV /36/ and by RINZEL
et al. /9,32/. The basic idea is as follows: Let us consider the wave moving in
the undisturbed medium with the velocity $v_1 = v_\infty$ followed by the second wave in
such a time interval that the corresponding velocity $v_2 < v_1$. As both waves move
with unequal velocities, the time interval between them changes (it grows in this
case) and the velocity of the second wave changes so as to agree with the actual
time interval. The shape of the dispersion relation then determines whether the second wave will slow down or accelerate. A comparison of kinematic results with the
results of numerical simulation made by RINZEL /9,32/ confirmed that the wave velocity and period satisfy the dispersion relation in the course of propagation.

The dispersion relation can be considered to be a basic characteristic of wave
supporting excitable media. We performed measurements of the pulse wave velocities
in the reaction medium of the B-Z type in dependence on the period of their initiation and the results are presented in the next section.

2.2 Experimental

Measurements were performed in a thin layer (depth ≈ 0.9 mm) of reaction solution
in a thermostated Petri dish at the temperature 298 K. Experimental procedure and
composition of reaction mixture were the same as in previously reported measurements
/14/ except the bromic acid concentration (0.225 M $HBrO_3$ in these measurements).
Circular wave patterns were generated by an immersed Ag wire and frequency of wave
initiation was controlled by the intervals between successive immersions of the wire. The wire stayed immersed for the time necessary to initiate the wave. Positions
of individual waves depending on time were measured to determine propagation velocity. The propagation of waves in two directions from the center was followed.

Several experiments were performed using various periods of the wave generation.
In each wave pattern the first wave had a highest velocity; its averaged value calculated from all experimental runs is $\overline{v}_1 = 3.4 \pm 0.2$ mm/min. The velocities of consecutive waves were the same or lower than the velocity of the first wave depending
on the period of initiation in the given experiment. After several waves were initiated and left the center the pattern reached a stationary state where consecutive
waves propagated with the same velocity /v_T/ and the same time interval /T/ between

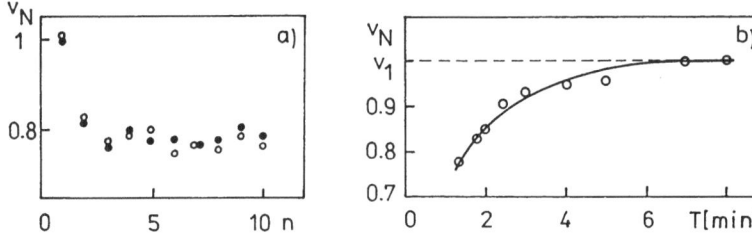

Fig. 4: a) Dependence of the normalized wave velocity on the order of waves in the pattern; experimental conditions see text, period of initiation $T_{in} = 80s$; ● , ○ - waves propagating to the left and to the right from the center, respectively;

b) Dependence of the wave velocity on the period of initiation

them /cf. Fig.4a/. In measured cases the resulting value of T was equal to the period of wave initiation T_{in} . To compare different experiments we normalized the actual value of the wave velocity (v_T) by the velocity of the first wave in the given pattern: $v_N = v_T/v_1$. The dependence of the normalized wave velocity on the period of generation is given in Fig.4b. The dependence is monotonous and displays general features of dispersion relations. The minimal period of initiation determining the absolute refractivity of the medium was measured to be about 80s. Corresponding wave velocity was approximately equal to 75% of the value of the velocity of the first wave. For T > 7 min the propagation velocity of the individual wave is not affected by the preceding one and the medium behaves like an undisturbed one.

The measured dependence was compared with the dispersion relation calculated for the model Oregonator /13/ and results are shown in Fig.5. The dimensionless velocity c and period T_P were calculated from our v and T data according to relations /13/

$$c = v / \sqrt{k_3 A D} \quad ; \quad T_P = T k_5 B, \tag{2}$$

where the values of kinetic constants k_3 , k_5 and diffusion coefficient D were taken from /13/ and are equal to: $k_3 = 40 \times [H^+] M^{-2}s^{-1}$, $k_5 = 0.4 M^{-1}s^{-1}$ and $D = 1.5 \times 10^{-5} cm^2s^{-1}$. Concentration of reaction species $A = [BrO_3^-]$, $B = [MA]$ + + [BrMA] and $[H^+]$ were chosen to be equal to their initial concentration in the experimental solution ($A = 0.0225$, $B = 0.05$, $[H^+] = 0.225$). However, the reaction mixture in our case is not a classical B-Z reaction medium and this could cause differences in measured and predicted dispersion curves.

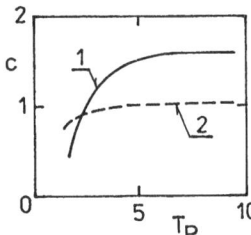

Fig. 5: Distribution curves, 1 - calculated for model Oregonator /13/ and 2 - measured for the reaction medium of the B-Z type

3. Formation of Wave Patterns

As discussed above the dispersion relation determines properties of all possible wave patterns which may arise in the given excitable medium. Type of the wave train

108

which will appear in the medium depends on initial distribution of periods, i.e., on the time intervals in which the waves start their propagation from the site of their initiation (axon hillock in nerve cells, pacemakers in reaction-diffusion systems). On one side these intervals are limited by the course of the entire sequence of processes transforming the initial stimulus via the generator potential to the superthreshold changes of variables in the excitable parts of the system. On the other side the time intervals reflect the actual processes of wave formation after the superthreshold change occurred and the refractory conditions of the medium, i.e., the "preparedness" of the medium to support wave propagation.

Two model situations are usually considered in studies of the effects of local changes of system variables - permanent or periodic perturbations of system variables.

3.1. Permanent perturbation

Here belong experimental studies of stimulation of nerve axons by a constant value of electric current (space-clamped experiments) /3,4,37/ and simulations of the corresponding situations by the Hodgkin-Huxley /38/ or FitzHugh-Nagumo models /29/. Such studies are often used to determine the threshold value of current necessary to evoke a repetitive firing of a nerve axon under different physical conditions.

In the chemical case the similar situation was studied numerically on the reaction-diffusion model of reacting medium with the SH kinetics using local perturbation of species concentration /14,15/. The propagating wave trains were investigated in dependence on the magnitude of the local perturbation and several qualitatively different types of behaviour were distinguished - cf. Fig.6.

In the first case, when the perturbation has a subthreshold value, no wave was formed (cf. Fig.6b). Maintained level of the perturbation caused diffusion-like annihilation of disturbances decaying to the stationary value of concentration in a finite distance from the site of the perturbation. Thus the state of the system is spatially nonhomogeneous but independent of time. The state of the system in an infinite dimensional phase space is represented by a fixed point.

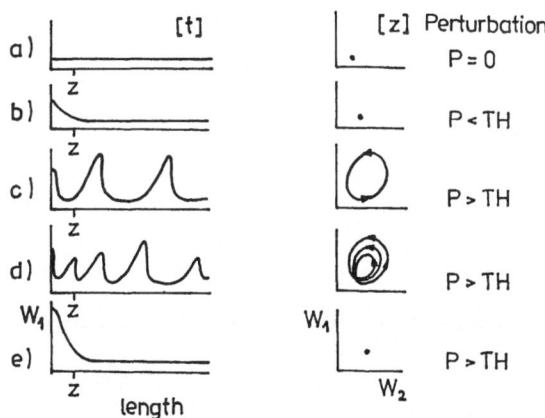

Fig. 6: Behaviour of the excitable medium under the permanent local perturbation;
first column - spatial profiles of system variable W_1 at the fixed time
t , second column - local phase portraits taken at the position z ,
a) initial spatially uniform stationary state,
b) and e) spatially nonuniform stationary states,
c) one-periodic wave train,
d) multiple-periodic wave train, TH - threshold magnitude of the perturbation P

When a superthreshold perturbation is applied, the train of waves propagating along the system is evoked. Each wave essentially preserves its shape and velocity and these characteristics as well as the period of the wave pattern depend on the amplitude of the perturbation.

The wave pattern can be one-periodic (cf. Fig.6c) when every single wave has the same shape and velocity and the periods between the successive waves remain constant. In another words the spatial functions of system variables are periodic in time and the behaviour of the system is represented by a limit cycle in an infinite dimensional phase space. If we observe the local phase portrait taken at an arbitrary point along the system the single loop appears corresponding to the unique wave repetitively travelling in the system.

Multiple-periodic trains were also observed (cf. Fig.6d), i.e., sequence of waves which differ in their shapes, velocities and mutual periods but form a regularly repeating wave packet; functions describing the spatial profiles of dependent variables are in this case again periodic in time. In an infinite dimensional phase space the solution is also of a limit cycle type, however, the limit cycle exhibits several loops in certain projections. The local phase portrait also consists of several loops each of them belonging to one wave in the packet. Performed numerical calculations point to the existence of alternating intervals of one- and multiple- periodic solutions in dependence on the magnitude of the perturbation (cf. Fig.7) and this can be connected with the existence of chaotic wave trains.

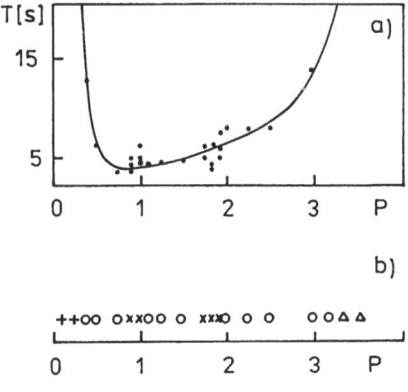

Fig. 7: Dependence of
a) the period of wave patterns and
b) the character of wave train solutions on the magnitude of the applied perturbation ,
+ - no propagating waves - stationary nonhomogeneous solution (cf. Fig.6b),
o - one-periodic solution (cf. Fig.6c),
x - multiple-periodic solution (cf. Fig.6d),
△ - single propagating wave as a transition to nonhomogeneous stationary state (cf. Fig.6e)

An application of a strong local perturbation may lead to a cessation of the repetitive activity of the excitable medium (so called "nerve block" in neurophysiological systems /29/). The numerical simulations of reaction-diffusion system with the SH kinetics illustrated in Fig.6e showed only one wave passing through the system before the spatially nonuniform time-independent state was established in the system.

3.2. Periodic application of local perturbation

When the system is perturbed repetitively with the period T_{in} then for superthreshold values of the perturbation the periodic wave trains of the period $T = T_{in}$ appear if T_{in} is greater than the absolute refracterity of the medium (entrainment by periodic perturbation). From the simulation of this phenomenon by a reaction-diffusion system with the SH kinetics /14/ it follows that the value of the threshold perturbation slightly depends on the period of the perturbation or, vice versa, the absolute refractory period slightly changes with the threshold value of the perturbation. The same behaviour is known from the experiments on nerve axons in the form of the recovery curve /4/.

The dependence of the threshold value on the period of the perturbation can be explained on the situation shown in Fig.2. When the system is recovering from the first excitation, its state (value of X and Y for the case shown in Fig.2 or spatial functions of system variables in a general spatially distributed case) is given by the elapsed time. The actual state of the system affects the minimal value of the perturbation necessary for the crossing of the critical manifold and excitation of a new excitation loop.

Another situation occurs when a periodic stimulation is a subthreshold one or has a period less than the absolute refractory period of the medium $T_{in} < T_A$. In both cases the repetitive wave trains may still appear but their period differs from the period of the perturbation. The numerical simulations with the reaction-diffusion model with the SH kinetics /14/ showed in this case the existence of one-periodic, multiple-periodic and chaotic wave trains. To describe the evolution of dynamic behaviour according to the control parameter value the firing number was defined /19, 24/ as a ratio of the number of elicited waves to the number of applied perturbations. The dependence of the firing number on the value of the perturbation or on the period of its application was studied in the B-Z reaction mixture /19/, in the simplified reaction-diffusion model with the SH kinetics /24/ and on the difference equation describing the repetitive activity of a nerve fiber /39/. In all cases the devil's-staircase-like structure of firing numbers was found.

The devil's-staircase-like structure of wave patterns follows from the refractory properties of the medium, from the dependence of the threshold perturbation on the time elapsed and from the processes responsible for the wave formation after a superthreshold perturbation of system variables. The situation where a superthreshold perturbation with a period $T_{in} < T_A$ is applied is studied by numerical simulations for the reaction-diffusion model with the SH kinetics in /14/.

Every superthreshold perturbation evokes a wave, i.e., the excitation at the site of its application. If the medium close to it is in the state of absolute refracterity the wave cannot start to propagate from the site of the initiation and is localized there until the recovery of the neighbouring medium from the absolute refractory period is completed. However, during the time of localization the excitation proceeds along the excitation loop and wave height decreases (cf. Fig.2). The time of localization is given by the difference between the absolute refractory period and by the time interval in which the perturbation is applied after the preceding wave started to propagate. When this time is not too large, the little progress along the excitation loop is done, the wave height is sufficient and the wave can start to propagate. When the localization time gets longer the excitation loop approaches the stationary state and the wave ceases to exist. Thus the perturbation fails to excite a propagating wave. This can occur periodically or chaotically.

4. Concluding Remarks

In the following we shall briefly discuss several general aspects of the applicability of the dispersion relation which we consider to be the basic experimentally obtainable quantitative characteristics of excitable media. As it was already discussed, the dispersion relation defines wave velocities and periods in wave patterns which are permissible in the studied excitable medium and determines refractory properties of the medium, particularly the period of absolute refracterity. The absolute refracterity affects the initial period of wave propagation from the point of initiation and in quasi two-dimensional and three-dimensional media it also determines the dimension of an "obstacle" behind which a spiral wave may be formed. The dimension must be such that the time necessary for the wave to travel around the obstacle is higher than the period of absolute refracterity. If the spirals arise at boundaries of two media with differing refracterities the difference between the refracterities and the wave velocities in the media again determine a critical dimension of the boundary necessary for formation of spiral waves. These considerations are interesting with respect to generation of arrythmias and fibrillations connected with the formation of spiral waves in heart tissue /6,40/. The spiral waves rotate around the center /17/ and the speed of the rotation depends also on the wave curvature. ZYKOV /6/ and TYSON and KEENER /13/ have shown that the dependence of the wave velocity on curvature and the dispersion curve determine the velocity and period of spiral wave rotation and the diameter of the spiral center. Attempts to measure the dependence of the wave velocity on the curvature are under way in our research group.

The character of formed wave patterns (i.e., periodicity or chaoticity) is determined by the absolute refracterity, magnitude and character of the perturbation and by excitable properties of the medium (i.e., dependence of the critical perturbation magnitude on the time elapsed from the previous perturbation). Until now mostly situations, where the perturbation amplitude is constant and the perturbation is either periodic or permanent, were studied. Situations with various forms of the time dependence of the amplitude and form of the perturbation (common in biological systems) will be certainly a subject of interest in chemical and modelling studies in the future. In biological systems the particular form of the perturbation follows from the action of the generator potential on the excitable parts of the system; the generator potential itself reflects both the properties of initial stimulus and the character of transformation processes of the system and its properties are affected by integration over time and space (graded response).

The action potential propagating along the nerve fiber causes a formation of an electric field which affects the excitability in the neighbouring neurons. The effects of the electric field on the refracterity, generation frequency and velocity of propagated waves were also studied both experimentally and on mathematical models in reaction-diffusion systems /14,19,28/.

Simple reaction-diffusion systems (e.g., the quoted B-Z system) enable to compare experimental results both on the generation and propagation of waves more directly with the results of mathematical modelling than it is in the case of biological systems. Methodology of such studies can be thus worked out and, hopefully, help in the interpretation of corresponding biological phenomena. Fast development of molecular biology and, particularly, development of methods of genetic manipulations in principle enables, for example, to produce in larger quantities such enzymes as acetyl cholinesterase and thus reconstruct common neurophysiological excitable system choline - acetylcholine - acetyl cholinesterase (in solution or in an immobilized form), hence, to study in vitro its properties with the above methodology.

References

1. V.I. Krinski, V.G. Jachno eds: Autowave Processes in Systems with Diffusion (IPF AN SSSR, Gorkij 1981)
2. V.I. Krinski ed.: Self-Organization. Autowaves and Structures Far from Equilibrium (Springer-Verlag, Berlin, Heidelberg 1984)
3. I. Cooke, M. Lipkin, Jr.,eds: Cellular Neurophysiology (Holt, Rinehart and Winston, Inc., New York 1972)
4. J. Field, H.W. Magoun, V.E. Hall, eds.: Handbook of Physiology. Neurophysiology, Vol. 1 (American Physiological Society, Washington, D.C. 1959)
5. A.C. Scott: Rev. of Mod. Physics 47, 487 (1975)
6. V.S. Zykov: Modelling of Wave Processes in Excitable Media (Nauka, Moscow 1984)
7. V.S. Markin, V.F. Pastushenko, Ju.A. Chizmadzkev: Theory of Excitable Media (Nauka, Moscow 1981)
8. J. Rinzel: Lecture Notes in the Life Sci. 8, 125 (1976)
9. R.N. Miller, J. Rinzel: Biophys. J. 34, 227 (1981)
10. R.J. Field, M. Burger,eds.: Oscillations and Travelling Waves in Chemical Systems (Wiley, New York 1985)

11. J.J. Tyson, P.C. Fife: J. Chem. Phys. 73, 2224 (1980)
12. E.J. Reusser, R.J. Field: J. Am. Chem. Soc. 101, 1063 (1979)
13. J.P. Keener, J.J. Tyson: Physica 21D, 307 (1986)
14. H. Ševčíková, M. Marek: Physica 21D, 61 (1986)
15. H. Ševčíková, M. Kubíček, M. Marek: In Math. Modelling in Sci. and Technol., ed. by X.J.R. Avula, R.E. Kalman, A.I. Ljapis, E.Y. Rodin (Pergamon Press, New York 1984) p. 477
16. M. Marek, J. Juda: Sci. Papers of the Prague Institute of Chemical Technology K13, 129 (1978)
17. S.C. Müller, T. Plesser, B. Hess: Physica 24D, 71 (1987)
18. K. Showalter, R.M. Noyes, H. Turner: J. Am. Chem. Soc. 7463 (1979)
19. M. Marek, I. Schreiber, L. Vroblová: In Proc. from MIDIT 1986 Workshop: Structure, Coherence and Chaos, ed. by R.D. Parmentier, P.L. Christiansen (Manchester Univ. Press 1987)
20. R.J. Field, R.M. Noyes: Nature 237, 390 (1972)
21. A.L. Hodgkin, A.F. Huxley: J. Physiol. (London) 117, 500 (1952)
22. R. FitzHugh: Biophys. J.: 1, 445 (1961)
23. J.S. Nagumo, S. Arimoto, S. Yoshizawa: Proc. IRE. 50, 2061 (1962)
24. M. Marek, I. Schreiber: In Bifurcation: Analysis, Algorithms, Applications, ed. by T. Küpper, R. Seydel, H. Tröger (Birkhäuser, Basel 1987) p. 201
25. M. Kubíček, M. Marek: Computational Methods in Bifurcation Theory and Dissipative Structures (Springer-Verlag, New York 1983)
26. N. Kopell, L. Howard: Advances in Math. 18, 306 (1975)
27. R.J. Field, R.M. Noyes: J. Am. Chem. Soc. 96, 2001 (1974)
28. H. Ševčíková, M. Marek: Physica 9D, 140 (1983)
29. J. Rinzel, J.P. Keener: SIAM J. Appl. Math. 43, 907 (1983)
30. J.W. Evans, N. Fenichel, J.A. Feroe: SIAM J. Appl. Math. 42, 219 (1982)
31. A. Lahiri, D.K. Goswami, U. Basu, B. Dasgupta: Phys. Letts. 111A, 246 (1985)
32. S.P. Hastings: SIAM J. Appl. Math. 42, 247 (1982)
33. J.A. Feroe: SIAM J. Appl. Math. 42, 235 (1982)
34. S.A. George: Biol. Cyber. 26, 209 (1977)
35. J. Rinzel, K. Maginu: In Nonequilibrium Dynamics in Chemical Systems, ed. by A. Pacault, C. Vidal (Springer-Verlag, Berlin 1984) p. 107
36. R.N. Chramov: In /1/ p. 99
37. S. Rotshenker, Y. Palti: J. Theor. Biol. 41, 401 (1973)
38. E.N. Best: Biophys. J. 27, 87 (1979)
39. J. Nagumo, S. Sato: Kybernetik 10, 155 (1972)
40. V.I. Krinski: Pharmac. Ther. B. 3, 539 (1978)

Chemical Structures and Convection

P. Borckmans and G. Dewel

Faculté des Sciences, Université Libre de Bruxelles,
Campus Plaine, C.P. 231, Bd. du Triomphe, B-1050 Brussels, Belgium

1. Introduction

Whereas cellular patterns in driven hydrodynamic systems have been known for nearly a century /1/, the observation of organized behavior (periodic oscillations,
waves, ...) in complex chemical systems is much more recent. It goes back to the
work of Belousov and Zhabotinsky and came as a surprise to most chemists. They were
indeed at first considered as rather exotic phenomena (although they had been
shown, by Glansdorff and Prigogine /2/, not to be in contradiction with basic thermodynamic principles), but it is now apparent that an increasing number of complex
chemical networks may exhibit this type of behavior when functioning sufficiently
far from thermal equilibrium /3/. Much effort has lately gone into the precise determination of the characteristics of the various kind of waves (fronts /4/, target patterns /5/ and spiral waves /6,7/) in order to understand the mechanism of
their formation.

However, besides chemical waves, chemical structure formation should also occur
in the form of stationary space periodic concentration patterns as proposed by Turing as of 1952 /8/. In this case however, in contrast to most hydrodynamical patterns, the characteristic wavelength is intrinsic, as it is determined solely by
the interplay between chemical reaction and diffusion and not by the boundary conditions.

But, although the analysis of theoretical models (Brusselator /9/, Oregonator
/10/, ...) corroborates the possibility of existence of such structures, few if any
real stationary patterns have been experimentally reported. Besides that in yeast
extracts /11/, the examples arise in systems related to the Belousov-Zhabotinsky
reaction /12-14/.

Interpretation of the origin of these "mosaic patterns" is unfortunately ambig-
uous because of the possible presence of convective and interfacial effects.

Some causes behind these experimental difficulties are that:

- The analyses of the theoretical models imply that the diffusion coefficients of the activator and inhibitor substances should differ sufficiently from
one another /9,10/. In this respect it is worth noting that pseudo-Turing structures have been observed, and their stability studied, in networks of mutually connected continuous well-stirred reaction cells where mass transfer coefficients may
easily be controlled /15/. Alas for the typical species playing a role in experimental systems presenting organized chemical behavior, diffusion coefficients are
usually of the same order . However, non-uniform steady patterns bifurcating from
homogeneous unstable steady states have been predicted to exist /16,17/ even in
systems with equal diffusion coefficients. /18/

- Evidence of chemical structures should be found in unstirred batch reactors
(no feeding) in order to get rid of the interferences of hydrodynamic fluxes.
Therefore, because of the consumption of reagents, the system drifts in parameter
space and the structures may at best appear as transients, which makes their detailed study very difficult. The study of waves is plagued by the same problems. On
the other hand, gel media permit feeding but give rise to non-uniform basic states.

- The avoidance (or the control) of natural convective currents may prove
difficult in fluid phase multicomponent reactive systems, as we shall discuss.

2. Concentration Patterns Generated at Interfaces

A rich variety of processes have been shown to produce spatial structures at liquid interfaces. In a pioneering work, Möckel /19/ observed in 1977 the appearance
of inhomogeneous concentration stripes while irradiating the system KI/CCl$_4$/starch
in water. Thereafter the generality of this phenomenon has been proved by Kagan et
al. /20/ and Gimenez and Micheau /21/ who reported the formation of photochemical
structures for a large number of photochromic (reversible) and chromogenic (irreversible) compounds. As a result, because of the diversity, the nature of the chemical reaction does not seem to play an important role in the onset mechanism. Moreover photochemistry is not even an essential factor: patterns are also formed
when gases diffusing through a liquid interface react with the solute or when the
reagents in the same solvent are separated by a dialysis membrane /22/. In all these cases a thin layer of product first appears in the vicinity of the interface and
then breaks down into inhomogeneous zones. The patterns do not form when convection
motion is prevented from appearing; for instance, when the reactions are carried
out in a gel or in a very thin layer.

The interpretation of this class of spatial phenomena has led to a controversy
about whether the convective motions are generated by the chemical reactions or
whether the chemical reactions merely serve to visualize pre-existing hydrodynamic
currents. It now appears that there is probably not a unique mechanism of patterning and that one must clearly distinguish experiments conducted in open air situations from those performed in a Petri dish sealed with a glass cover.

2.1. Pre-existing convection (Experiments in open air conditions)

The mechanism of photolysis of the halogen compounds first studied by Möckel and
Avnir was complex and largely unknown. This led Micheau and Gimenez to study simpler photochemical reactions with known mechanisms /21/: first or second order
reactions. These experiments reveal the existence of patterns whether the irradiation is performed from above or from below. The following facts strongly suggest a
hydrodynamic origin for these patterns /23/:
  - The wavelength of the structure increases almost linearly with the thickness of the layer.
  - The structure disappears when the Petri dish is covered and it reappears after removal of the plate.
  - Photographs of striations have been obtained before irradiation (prepatterns) using the Schlieren technique. Moreover the patterns revealed under illumination coincide with the prepatterns.

Consequently, in these experiments chemistry only serves to trace pre-existing
convective structures driven by the mechanism of evaporative cooling. The evaporation of the solvent effectively cools the liquid surface, thereby inducing a potentially unstable density gradient in the layer and also initiating surface tension
variations. Both effects contribute to the destabilization of the conduction state.
Surface tension effects are seen to dominate in thin layers, while for thicker
layers convection is driven mainly by the buoyancy effects. Concurrently a modification in the morphology of the pattern is observed. At small d, surface tension
forces tend to favor hexagons (cellular structures) whereas rolls (vermiculated
structures) characteristic of buoyancy forces become more and more the rule as d is
increased. Both structures may coexist giving rise to quite irregular patterns. The
signature of this transition appears as a discontinuity, with hysteresis effects,
in the plot of the yield of the photochemical reaction versus the layer thickness.

Deformations of the free surface accompany these convective motions and modify
the threshold conditions. The profile is concave when d is small and convex when
the layer is deep. These undulations can lead via Beer-Lambert's law to a spatial
variation of the light intensity transmitted through a solution. For instance,
Müller and Plesser /24/ have reported spatial modulations in the absorbance of an
aqueous solution of reduced nicotinamide adenine dinucleotide (NADH) which is an

important intermediate of glycolysis. These observations help to understand the results obtained when a light beam with a wavelength specific for the absorption of NADH passes through an open-to-air layer of yeast extract exhibiting glycolytic oscillations. Structures form and disappear at intervals of several minutes /11,24/. They result from a strong coupling between the chemical oscillations and the network of convective cells induced by the Benard-Marangoni instability. The convective patterns can be detected whenever the NADH concentration passes through a maximum.

These studies point out that considerable care must be taken to ensure that all pre-existing hydrodynamic motions are properly taken into account in the interpretation of the symmetry-breaking instabilities observed in chemical systems.

2.2. Convective motion induced by chemical reactions at interfaces

Growing experimental evidence suggests that chemical reactions can also promote hydrodynamic movements /22/; a well-documented example is provided by photochemical reactions in closed Petri dishes. A large number of possible mechanisms have been proposed to explain the onset of these structures. Among them, the case of surface catalyzed reactions leading to adverse density gradients has been analyzed using the methods developed for the study of the classical Rayleigh-Benard problem /25/.

Double-diffusion has long been recognized as the major mechanism leading to convection in multicomponent systems /26/. Because these patterns can form even in the presence of a hydrostatically stable density gradient we think that this instability provides the "engine" of many chemically driven convective motions in isothermal systems. Inhomogeneous chemical reactions are indeed supposed to create only weak density gradients via the expansion coefficients. For the sake of simplicity we consider a photochemical reaction taking place in the absorbing layer of a shallow solution:

$$A + h\nu \longrightarrow B \quad . \tag{1}$$

Furthermore we assume that $D_A > D_B$ where D_i is the molecular diffusion constant of the species i. This reaction induces in the layer opposite gradients of reactant and product which can then generate various instabilities /27/.

2.2.1. Fingers /28/
When the system is irradiated from above, fingers perpendicular to the interface develop (tending to increase their length and width) when

$$\alpha_B D_A > \alpha_A D_B \tag{2}$$

(where α are the expansion coefficients). These fingers can thus appear when the reactant diffuses more rapidly than the product. In this network, the upgoing fingers lose reactant to the downward moving fingers making the former less dense. Indeed very little product (B) is transferred between the fingers because of the lower diffusivity of B. It is this density difference that triggers the motion even though the mean density gradient over the fingers is hydrostatically stable, as is the case when the expansion coefficient of the reagent is larger than that of the product: $\alpha_A - \alpha_B > 0$. This situation seems to occur in the case of the photoreduction of ferric ion (irradiation from above) by oxalic acid followed by complexation with potassium ferricyanide to produce the soluble blue dye known as Turnbull's blue. Avnir and Kagan have shown that spectacular patterns can form in a dish covered with a plate to avoid evaporation /29/. In that case the diffusion coefficient of Turnbull's blue (the product), $D_B = 2.5 \times 10^{-7}$ cm^2/s, is much smaller than that of the reactant, $D_A = 10^{-5}$ cm^2/s, and the expansion coefficient of the blue dye is also lower than that of the starting species ($\alpha_B \ll \alpha_A$); therefore the condition (2) can easily be satisfied in this experiment.

The presence of a stabilizing temperature gradient resulting for instance from the absorption of the radiation by the product (a process which fixes the initial position of the absorbing layer) does not modify these conclusions. Indeed since the thermal diffusion coefficient is larger than the solute diffusion constant this gradient reinforces the destabilizing mechanism leading to the formation of the fingers /26/.

2.2.2. Oscillatory convection /26/

When the system is irradiated from below, oscillatory convection is possible when

$$\alpha_A > \alpha_B \qquad \text{or} \qquad v_B > v_A \tag{3}$$

(where v are the specific volumes), i.e. when the volume increases during the reaction. Again these conditions are satisfied in the case of the photoreduction of ferric ion by oxalic acid when the Petri.dish is illuminated from below /29/.

Despite a flurry of works the nature of the periodic structures which appear beyond the overstability threshold is still the subject of many experimental and theoretical investigations /30/. In a laterally infinite system, standing waves are unstable to travelling waves, and states consisting of convective rolls which move laterally have indeed been observed. The effects of finite geometry play an important role in this competition between travelling and standing waves and they can lead to interesting spatial structures, including multistability of confined travelling waves /31/.

In the photochemical experiments, time-dependent convection has not been observed yet. For some range of the parameters the instability corresponding to the threshold can also lead to finite-amplitude nearly steady state convection /30/. Oscillatory convection can thus easily be missed by stepping too quickly through the threshold. It would be interesting to perform further experimental works on photochemical systems satisfying the condition (3) because they provide one of the few systems where dissipative waves can emerge at a primary bifurcation point.

Because uncontrolled convection is so widespread, it seems interesting to try to observe chemical periodic structures either in the presence of well-controlled flows (which also play a role in the feeding mechanism) or in media such as gels or fritted glass to prevent convection altogether. In the latter case the system is only constrained at the boundaries. However, the flows or concentration gradients then introduce specific directions and one has to tackle a problem of pattern selection in anisotropic systems. These situations are discussed in the following paragraphs./32/

3. Chemical Instabilities and Convection

We consider a physicochemical system confined between two plates distant from 2d and discuss the effect of the simple flows, such as shear, Poiseuille or time-independent periodic flow (Rayleigh-Benard configuration) on the Turing instability.

These convective motions greatly enhance the mass transfer in the direction of the flow. Since the work of Taylor /33/, it has well been known that this enhancement can often be described by an effective diffusion coefficient K which takes the general form

$$K = D + C/D , \tag{4}$$

where $C \sim (dV_m)^2$ and V_m is the maximum velocity in the flow. For time-independent periodic flows this last expression is valid at low Peclet numbers

$$Pe = V_m d/D < L/d , \tag{5}$$

where L is the longitudinal characteristic length. Because ordered flows can be well controlled only in cells of small aspect ratio, it is experimentally difficult to reach values of Pe smaller than 10 in the case of periodic flows. For higher values of Pe the effective diffusion coefficient in this case takes the form /34,35/

$$K = k \, (DV_m d)^{1/2} \, ,$$ (6)

where k is a numerical constant. This expression is related to the flux across the narrow boundary layers that develop between adjacent convective rolls.

When chemical reactions are taken into account, the parameters become numerous and a complete analysis has not been done even for idealized models. In the vicinity of a chemical instability one generally distinguishes the long relaxation time of the marginal mode t_c and the characteristic times t_r associated to the other rapid relaxation processes. When the following conditions are satisfied:

$$t_c \gg t_r \gg d^2/D \, ,$$ (7)

the concentrations of the species are essentially constant over a section of the channel on the chemical time scales. In this regime and in a coordinate system moving with the mean velocity, the system is described by a reaction-diffusion model including the anisotropy resulting from the renormalization of the diffusion coefficients in the direction of the flow.
The reaction-diffusion equations then read

$$\frac{\partial X}{\partial t} = F(X) + D \nabla^2 X + D^A \frac{\partial^2 X}{\partial x^2} \, ,$$ (8)

where the flow is in the x-direction and $D^A = K-D$ is a measure of the anisotropy induced by this flow which will select an orientation for the critical wave vector of the Turing structure.

Calculations /36/ on the Brusselator model show that if the system already presents a Turing instability in absence of flow then this flow will align the axis of the one-dimensional concentration pattern in a direction parallel to that of the velocity, without modifying either the threshold or the wave vector. In the alternate case when a Turing instability is not possible, flows at small Peclet numbers can induce a concentration pattern, the axis of which is perpendicular to the flow with a modified wavenumber.

An experiment devised to illustrate this latter effect would take place in a flow-through channel connected with a well-mixed flow reactor. In the reactor the system is in a steady state (in this zone the mass transfer coefficients are all equal to the turbulent diffusion coefficient). When flowing through the channel the mass transfer coefficients are renormalized differently and could then induce a Turing instability. The corresponding chemical pattern would move with the mean velocity of the fluid. By magnifying the differences between the diffusion coefficients, laminar flows could provide favorable conditions for the onset of spatial chemical structures through the Taylor diffusion phenomenon. However, if the system brings more than two determining species into play, the instability could lead immediately to travelling waves. Experiments of this type, exhibiting spatial structure formation have been reported by Marek and Svobodova /37/.
When condition (7) is replaced by

$$t_c \gg d^2/D \gtrsim t_r$$ (9)

the renormalized diffusion coefficient then contains mixed chemo-diffusive contributions as shown for a shear flow /38/. In this case the flow is also able, in particular circumstances, to produce a diffusive instability leading to a chemical pattern, even when no such pattern is obtainable in absence of convection.

118

4. Localized Steady State Dissipative Structures

In this section we consider a long rectangular slab filled with a gel in order to eliminate the interference with spurious convective motion. The chemicals are fed uniformly along the length of the device ($\parallel$ to the Ox axis) with, for instance, the inhibitor along one side and the activator along the other one. (This set-up bears some resemblance to the experimental cell recently used by Noszticzius et al./39/.) The autocatalytic reaction takes place inside the reactor in the zone where the corresponding diffusive fronts meet and where the reagents are present in significant amounts. Both effects, chemical reactions and diffusion, thus generate characteristic concentration profiles (spatial rampings).

If the conditions for the onset of a Turing instability are satisfied, a localized steady chemical pattern can appear in the region where the local bifurcation parameter exceeds its critical value; this region is confined between two isoconcentration lines parallel to the length of the system. Such localized dissipative structures have, for instance, been obtained numerically by solving the kinetic equations of the Brusselator in the presence of a spatial profile of species A /40/. They have also been constructed analytically in the particular limit where $D_B=0$, D_X/D_A, $D_Y/D_A \longrightarrow 0$ /41,42/.

Because these authors only considered one-dimensional systems, the wave vector of the pattern was necessarily parallel to the imposed gradient. However, a related problem (as far as pattern selection in anisotropic media is concerned), the study of the onset of cellular convection in a shallow two-dimensional container heated non-uniformly from below, has shown that convective rolls can appear with their wave vector either parallel (longitudinal mode) or perpendicular (transverse mode) to the ramping direction /43/.

It is therefore important to consider the orientational effect of the gradients of inhibitor and/or activator on Turing structures. Various situations must be considered according to the relative values of three characteristic lengths of the problem, the critical wavelength (q_c^{-1}), the characteristic length of the spatial ramping (l_r) and the width of the slab (d)/44/.

In the simple case where $q_c^{-1} \gg d$, the profile is mainly generated by the autocatalytic reaction. A longitudinal structure could only be confined in such a small box by reducing the wavelength of the pattern, but this would shift the corresponding threshold to higher values of the bifurcation parameter as most of the chemical systems presenting a Turing instability have a neutral stability curve with a minimum at $q = q_c$, the intrinsic critical wavenumber. Hence the transverse mode will be selected with this wavelength. Boissonade /45/ has recently obtained such a structure for the Brusselator model in a two-dimensional system.

In the opposite limit when $q_c^{-1} \ll d$, the slow spatial modulation on a scale intermediate between q_c^{-1} and d can be described by an amplitude equation when the local value of the bifurcation parameter remains close to the critical value in the zone of width $d'=d'(l_r)$ where the basic nonuniform state becomes unstable. These equations can be derived by using the standard techniques of multiple scale analysis /46/ which have been introduced to study the weakly nonlinear regime near the onset of pattern formation.

If one assumes a profile with a local maximum in the slab, it may be shown that the amplitude is largest when $q \perp O_x$ and the mode selected on this basis is thus the transverse mode. Moreover very near threshold a longitudinal pattern is prevented from appearing. When the bifurcation parameter is increased, the instability zone fills the whole box (d' $\longrightarrow$ d). If furthermore one studies the influence of a slow spatial ramping on a well-developed structure one reaches the same conclusions. If no further instability takes place, this mode will remain even in the fully nonlinear regime.

However, in systems with two active species, when the diffusion coefficients are of the same order of magnitude a Hopf bifurcation usually occurs before the pattern-forming instability. In such a case the effect of a spatial ramping will induce frequency and/or phase variations eventually able to trigger chemical waves.

On the other hand, if the system brings more than two species into play the first instability could immediately lead to travelling waves. The orientational effect of the gradients on such waves would be of the same nature and select patterns propagating perpendicularly to those gradients. Further theoretical and experimental analysis is necessary to determine to which class of phenomena the waves obtained in /39/ belong.

5. Conclusion

Despite their prediction thirty-five years ago, the so-called Turing structures have not yet been observed experimentally.

Because of the necessity of imposing fluxes to feed the chemical system to keep it far from equilibrium on the one hand and to control convective motion on the other hand, Turing structures stand a better chance of being characterized in anisotropic systems.

However, we now have the tools at hand that not only permit the mapping of the velocity field but furthermore the quantification of moving or stationary chemical gradients in reaction-diffusion-convection systems (see /4,5,47/).

Acknowledgments

We thank H. Swinney and J. Boissonade for making references 18, 39 and 45 available to us before publication.
The authors are Research Associates at the National Fund for Scientific Research (Belgium).

References

01. H.L. Swinney and J.P. Gollub (Eds.), Hydrodynamic Instabilities and the Transition to Turbulence (Springer, Berlin 1985)
02. P. Glansdorff and I. Prigogine, Thermodynamic Theory of Structures, Stability and Fluctuations (Wiley, NY 1971)
03. R.J. Field and M. Burger, Eds. Oscillations and Traveling Waves in Chemical Systems (Wiley, NY 1984)
04. P. Wood and J. Ross, J. Chem. Phys. 82 1924 (1985)
    J.M. Bodet, J. Ross and C. Vidal, J. Chem. Phys. 86 4418 (1987)
05. C. Vidal, J. Stat. Phys. 48 1017 (1987)
06. K.I. Agladze and V.I. Krinsky, Nature 296 424 (1982)
07. S.C. Müller, T. Plesser, B. Hess, Physica 24D 71, 87 (1987)
08. A.M. Turing, Philos. Trans. Roy. Soc. London B237 37 (1952)
09. G. Nicolis and I. Prigogine, Self Organization in Nonequilibrium Systems (Wiley, NY 1977)
    D. Walgraef, G. Dewel, P. Borckmans, Adv. Chem. Phys. 49 311 (1982)
10. P.K. Becker and R. Field, J. Phys. Chem. 89 118 (1985)
11. A. Boiteux and B. Hess, Ber. Bunsenges. Phys. Chem. 84 392 (1980)
12. A.M. Zhabothinsky and A.N. Zaikin, J. Theor. Biol. 40 45 (1973)
13. K. Showalter, J. Chem. Phys. 73 3735 (1980)
14. M. Orban, J. Am. Chem. Soc. 102 4311 (1980)
15. M. Marek in Modelling of Patterns in Space and Time. Lecture Notes in Biomathematics Vol.55 (Ed. W. Jäger and J.D. Murray) (Springer, Berlin 1984) page 214
    K. Bar-Eli and W. Geiseler, J. Phys. Chem. 85 3461 (1981)
16. D. Bedeaux, P. Mazur, R.A. Pasmanter, Physica 86A 355 (1977)
17. D. Hefer and M. Sheintuch, Chem. Eng. Sci. 41 2285 (1986)
18. J.A. Vastano, J.E. Pearson, W. Horsthemke, H.L. Swinney, Phys. Lett. A124 320 (1987)
19. P. Möckel, Naturwissenschaften 64 224 (1977)
20. M. Kagan, A. Levi, D. Avnir, ibid. 69 548 (1982), 70 144 (1983)
21. M. Gimenez and J.-C. Micheau, ibid. 70 90 (1983)

22. D. Avnir and M. Kagan, Nature 307 717 (1984)
23. J.-C. Micheau, M. Gimenez, P. Borckmans, G. Dewel, Nature 305 43 (1983)
24. S.C. Müller and Th. Plesser, in ref.15 page 246
    S.C. Müller, Th. Plesser, B. Hess in Temporal Order, Synergetics N°29.
    L.Rensing and N. Jaeger Eds. (Springer, Berlin 1985)
25. J. Bdzil and H. Frisch, Phys. Fluids 14 476, 1077 (1971)
26. J. Platten and J.-C. Legros, Convection in Liquids (Springer, Berlin 1984)
    Part D.
27. G. Dewel, D. Walgraef, P. Borckmans, Proc. Natl. Acad. Sci. U.S.A. 80 6429
    (1983)
28. T.J. McDougall, J. Fluid Mech. 126 379 (1983)
29. M. Kagan and D. Avnir in Interfacial Phenomena : Proc. NATO ASI and EPS
    Summerschool and Conference on Physicochemical Hydrodynamics.Spain 1986.
    M.G. Velarde and B. Nichols Eds. (Plenum Press, NY to be published)
30. V.Steinberg and E. Moses in Patterns, Defects and Microstructures in Nonequili-
    brium Systems: NATO ASI (Series E: Applied Sciences N° 121)
    D.Walgraef Ed. (Martinus Nijhoff, Dordrecht 1987)
    C.M. Surko, P. Kolodner,A. Passner, R.W. Walden, Physica D23 220 (1986)
31. E. Moses, J. Fineberg, V. Steinberg, Phys. Rev. A35 2757 (1987)
    R. Heinrichs, G. Ahlers, D.S. Cannell, Ibid. 2761 (1987)
32. G. Dewel, P. Borckmans, D. Walgraef in Chemical Instabilities: NATO ASI
    (Series C: Mathematical and Physical Sciences N° 120).G. Nicolis and F.
    Baras Eds. (Reidel, Dordrecht 1984)
33. G.I. Taylor, Proc. Roy. Soc. London A219 186 (1953), A225 473 (1954)
34. J.P. Gollub and T.H. Solomon in Chaos Related Nonlinear Phenomena
    I.Procaccia Ed. (Plenum Press, New York 1987)
35. B.I. Shraiman, Phys. Rev. A36 261 (1987)
36. P. Borckmans, G. Dewel, D. Walgraef, Y. Katayama, J. Stat. Phys. 48 1031
    (1987)
37. M. Marek and E. Svobodova, Biophys. Chem. 3 263 (1975)
38. E.A. Spiegel and S. Zalesky, Phys. Lett. 106A 335 (1984)
39. Z. Nosziczius, W. Horsthemke, W.D. McCormick, H.L. Swinney, W.Y. Tam, Nature
    329 619 (1987)
40 M. Herschkowitz-Kaufman and G. Nicolis, J. Chem. Phys. 56 1890 (1972)
41. J.F.G. Auchmuty and G. Nicolis, Bull. Math. Biol. 35 323 (1975)
42. J. Boa and D.S. Cohen, SIAM J. Appl. Math. 30 123 (1976)
43. I.C. Walton, J. Fluid Mech. 131 455 (1983)
44. G. Dewel, D. Walgraef, P. Borckmans, J. Chim. Phys. 84 1335 (1987)
45. J. Boissonade, J. Physique (France) 49 541 (1988)
46. A. Newell and J.A. Whitehead, J. Fluid Mech. 38 279 (1969)
47. S.C. Müller, Th. Plesser, B. Hess, Biophys. Chem. 26 357 (1987) and in ref.29

The Path to Hydrodynamic Instability During Reactions at Liquid Interfaces: Comparison of Experimental Image Analysis Results and Simulations

M.L. Kagan, R. Kosloff, and D. Avnir

Department of Organic Chemistry and the F. Haber Research Center for Molecular Dynamics, The Hebrew University of Jersualem, Jerusalem 91904, Israel

Abstract

The photoreduction of Fe^{+3} to Fe^{+2} which leads to patterns of photoproducts and to convections, was analysed by *computerized image analysis* of the concentration gradients along the vertical gravity axis, (perpendicular to the horizontal, covered, layer product). The observed time-evolution of the concentration profiles was analysed by modeling it with the actual photochemical reaction kinetics coupled to the diffusion of the components. Good agreement between experiments and simulations was obtained (see e.g., Fig. 4). The time evolution of the Rayleigh number of the reacting system was obtained by numerical solution of the appropriate one-dimensional reaction/diffusion equations, employing a recently developed Fast Fourier Transform algorithm. The calculated time for reaching the Rayleigh critical value of about 660 (two free boundaries) was in close agreement with the experimental value for the time of onset of convections (around 310 sec for both), and in agreement with the experimental system of a rigid cover (top) and a free liquid interface (bottom). It was also found that such a chemical reaction accelerates the evolution time of the hydrodynamic instability, compared to a double-diffusion system without a reaction.

1. Introduction

Of the many reactions that were shown to drive horizontal covered liquid interfaces into hydrodynamic instability as revealed by the evolution of product-patterns and convections [1-4], we concentrated our efforts on the photoreduction of Fe^{+3} to Fe^{+2}, visualized by complexation with $Fe(CN)_6^{-3}$ to give Turnbull's Blue [5]:

$$2Fe^{+3} + (COOH)_2 \rightarrow 2Fe^{+2} + 2H^+ + CO_2 \tag{1}$$

$$Fe^{+2} + K_3\,Fe(CN)_6 \rightarrow K\,Fe^{II}Fe^{III}(CN)_6 + 2K^+ \ . \tag{2}$$

Utilizing high-sensitivity computerized image analysis under a microscope we have shown recently (for 1 mm thin films) the existence of a bifurcation point for the onset of convective patterns in this system, and have shown the sensitivity of that point to the concentration of the starting material [5]. Dependence of the bifurcation point on light intensity and depth was also revealed recently, and is reported separately [6].

So far we have confined ourselves to a phenomenological exploration of this very wide and very complex phenomenon. However, the accumulated experimental observations have now led us to the second stage of this project, i.e., to the ability of suggesting feasible mechanisms for the phenomenon, and to their testing by model simulations.

Here we wish to report on a very good agreement between two experimental aspects of the above reaction (in thick layers of depths greater than 10 mm) and model simulations of the diffusion/reaction process:

a) The general characteristics of the time evolution of the concentration gradients up to the unstable bifurcation point.
b) The time it takes the system to reach the unstable bifurcation point.

The experiments involved sideview scanning (vertical to the horizontal reacting interface) by computerized image-analysis densitometry. The simulations involved the calculation of the time evolution of the Rayleigh number by numerical solutions of one-dimensional reaction/diffusion equations, utilizing a recently developed Fast Fourier Transform algorithm [7]. Details are given below.

2. Experimental Details

2.1 The Reaction

Of the many reactions and types of reactions that gave rise to patterning we chose the one described above for an indepth study for the reasons that: the absorption spectrum has a maximum at 360 nm (and not at 256 nm as in most other cases), thus allowing for the use of Pyrex containers; the colour contrast from pale yellow to intense blue is ideal for imaging and He/Ne laser absorption; the solution is aqueous and easily prepared; the patterns form rapidly under a variety of conditions. Potassium ferroferricyanide is known as soluble Turnbull's Blue [8-10] and is similar in molecular structure to Prussian Blue [11]. It can also be produced thermally according to the reaction:

$$Fe(NH_4)_2(SO_4)_2 + K_3Fe(CN)_6 \rightarrow KFeFe(CN)_6 + K_2(NH_4)_2(SO_4)_2 \qquad \{3\}$$

We found that Turnbull's Blue is perfectly soluble in water below concentrations of 1×10^{-2} M. Standard solution concentrations were: 3×10^{-2} M oxalic acid (Merck), 1.4×10^{-3} M $K_3Fe(CN)_6$ (BDH) and 1.4×10^{-3} M $FeCl_3$ (60 w/v, Merck). The oxalic acid is in large excess and therefore only $FeCl_3$ enters the kinetic equations.

2.2 Measurement of Physical Parameters

Diffusion rates of Turnbull's Blue and ferric chloride were measured by single point light absorption of the diffusing species in a shear cell. (For more details see [12] and [13,14]). Density coefficients (coefficients of volume expansion) were measured in a PAAR thermostated densitometer with six decimal places accuracy. Extinction coefficients at 360 nm were computed from absorption data measured on a Cary spectrophotometer. Results are summarized in Table 1.

TABLE 1: Physical Parameters of the Turnbull's Blue Reaction

Solute		Extinction $[cm^2]$	Diffusion $[cm^2.sec^{-1}]$	Starting Conc.[M]	Density Coeff. $[M^{-1}]$
$FeCl_3$	[A]	900	1×10^{-5}	1.4×10^{-3}	0.145
$FeCl_2$	[B]	-	7×10^{-6}	-	-
$K_3Fe(CN)_6$	[C]	320	9.4×10^{-6}	1.4×10^{-3}	0.14
$KFeFe(CN)_6$	[E]	100	9.3×10^{-7}	-	0.395

2.3 Computerized Image Analysis

Image analysis [15,16] was used for the recording and analysis of the development of concentration profiles of the product before and up to the critical break-up time.

A Grinnel digitizer converted the Vidicon camera image into 512×512 picture elements (pixels), each with 256 greylevels. A VAX computer processed and stored the digitized images. Time sequences of the build-up of the product were obtained by scanning, at fixed time intervals, a predefined window in the vertical side of the reaction vessel, of typical dimensions 40 pixels wide by 150 pixels long (0.235cm×0.882cm). The minimum time step could be set at 2 sec. The greylevel of each scanned line was averaged and stored in an accumulative data file. Non-random inhomogeneities due to lighting, marks on the cell wall, etc., were reduced by subtraction of the first frame, line by line, from all the subsequent frames. Calibration of the greylevel values to product concentrations was made by passing a laser beam horizontally through the layer and scanning for maximum absorption.

3. Simulation Details

The photochemical formation of Turnbull's Blue in a quiescent solution irradiated from above can be reduced to the following simple sequence:

$$A \rightarrow B \tag{4}$$

$$B + C \rightarrow E \tag{5}$$

in which (see {1} and {2}) A is Fe^{+3}, B is Fe^{+2}, C is $K_3Fe(CN)_6$ and E is Turnbull's Blue. In this simplified scheme, which, as shown in Section 4, agrees with experimental observations, it is taken into account that the excitation stage $A \rightarrow A^*$ is very fast and it is assumed that the rate of formation of B is determined by the flux rate of the photons, and that the steady state concentration of B is small to a degree that it does not contribute to light absorption:

$$\frac{\partial A}{\partial t} = -I \varepsilon A + D_A \nabla^2 A \tag{6}$$

$$\frac{\partial B}{\partial t} = I \varepsilon A - kBC + D_B \nabla^2 B \tag{7}$$

$$\frac{\partial C}{\partial t} = -kBC + D_c \nabla^2 C \tag{8}$$

$$\frac{\partial E}{\partial t} = kBC + D_E \nabla^2 E \tag{9}$$

$$I = I_0 e^{-\int_0^z (\varepsilon_A A + \varepsilon_C C + \varepsilon_E E) \, dz} \tag{10}$$

$$\rho = \rho_0 (1 + \alpha_A A + \alpha_C C + \alpha_E E) \tag{11}$$

where A, B, C, E are the concentrations of these components, A_0, C_0 are the initial concentrations, the D's are the diffusion constants (Table 1), I is the light intensity at depth z as defined in {10} (Beer-Lambert law), I_0 is the lamp intensity (photons $cm^{-2} sec^{-1}$), the ε's are the extinction coefficients (at

124

360 mm, Table 1), the α's are the expansion coefficients (Table 1), k is taken as 10^3 sec/M, ρ is the density as defined in {11}, ρ_0 is the initial density

$$\rho_0 = \rho_{H_2O} \, (1 + \alpha_A \, A_0 + \alpha_c \, C_0) \qquad \{12\}$$

where ρ_{H_2O} is the density of pure water and $\nabla^2 = \partial^2/\partial z^2$ where z is the depth measured along gravity axis.

The strong non-linear coupling between the equations has so far prevented us from reaching an analytical solution even at steady states. We have therefore solved the equations by numerical analysis using a powerful pseudo-spectral algorithm developed by Kosloff and Kosloff (for details see [17]). The method has been used extensively for numerical solutions of the Schrodinger equation [18] and only recently applied to the diffusion equation [19,20]. The algorithm was used here to produce an exact time-stepped solution with boundary conditions of {6-9}, which was then used for calculating the time/space evolution of ρ {11}. With this in hand one can calculate the time evolution of the Rayleigh equation for deep solutions [21,22]:

$$R = \frac{g \, \dfrac{\Delta\rho}{\Delta h} \, (h)^4}{\nu \, \rho_0 \, D_m} \qquad \{13\}$$

where g is the gravitational acceleration, $\Delta\rho/\Delta h$ is the negative density gradient, h is the thickness of the layer covered by the negative density gradient (see Fig. 6 below), ν is the kinematic viscosity, D_m is the average of the diffusion rates, and R is the dimensionless Rayleigh stability number. Convections will ensue if $R > R_c$, a critical number dependent upon the boundary conditions of the system. For aqueous solutions R_c equals 0, 660, 1100, 1770 for horizontal gradients, vertical gradients between two free boundaries, one free and one rigid boundary, two rigid boundaries, respectively [23].

4. Results and Interpretation

4.1 The Time Evolution of Concentration Gradients:
Comparison of Image Analysis and Simulations

Figure 1a shows a typical set of experimental profiles for the build-up of the product layer of Turnbull's Blue from 20-340 sec in increments of average 20 sec. The product is seen to grow with right-skewed bell-shaped distribution with the maximum a short distance below the interface and spreading downwards. The build-up of the product continues until reaching a maximum value equivalent to the complete depletion of the local starting material. The last curve appears to be more irregular than the others with a sudden drop of the product front downwards. This occurs as a result of the break-up of the layer and the start of the formation of horizontal patterns and vertical fingers. The insert to Fig. 1a shows the profiles after the critical time with the descent of fingers from the highest concentration clearly illustrated. The critical time occurs between 310 and 330 sec.

Figure 1b shows concentration curves from computerized simulation ({6}-{10}) of the reaction/diffusion set of equations {4,5}. The reaction was run from 0-500 equivalent seconds (5000 steps with a 0.1 time step and with time increment between each curve of 25 sec. The product grows beneath the interface reaching a maximum value almost equivalent to that of the starting materials while simultaneously descending. Simulations in which the incident light was "turned-off" before break-up confirmed the experimental result that the concentration front descended because of light

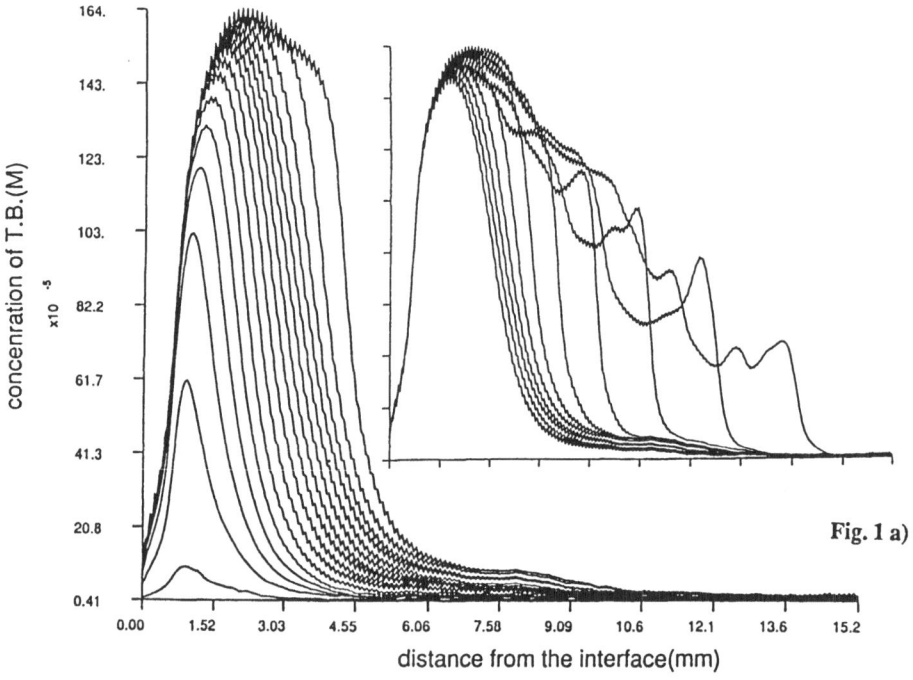

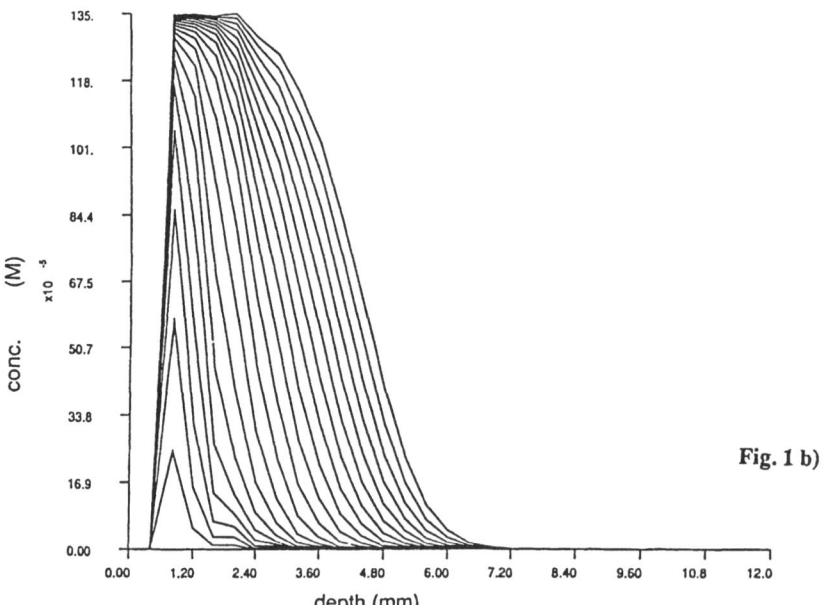

Figure 1a)- The time evolution of the concentration profile of Turnbull's Blue, taken every 20 sec from 20 to 320 sec, when convections start. **Insert:** Irregular profiles indicating fingering and convections taken from 240 to 460 sec, every 20 sec. b) The calculated time evolution of the concentration profile of the product E, according to the model-simulation (25 to 500 sec, every 25 sec). c) The calculated profiles of light penetration at 360 nm due to depletion of starting materials A,C

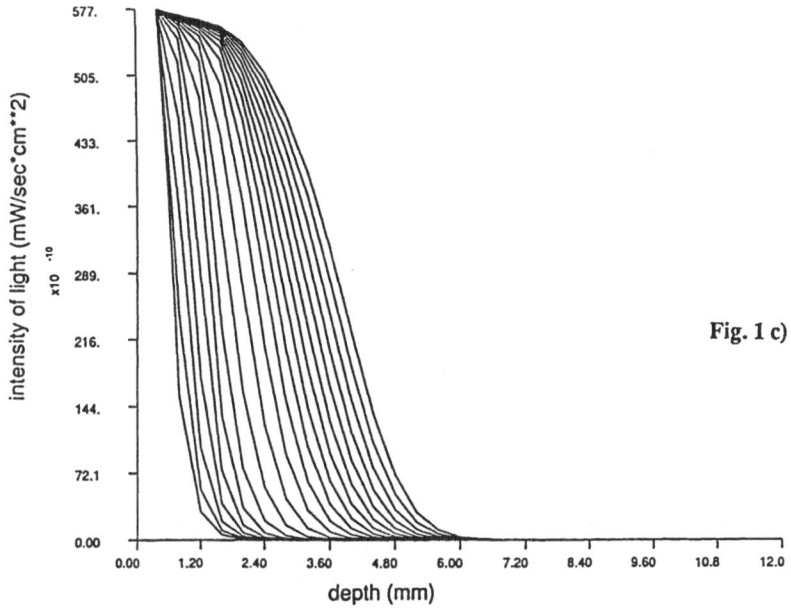

Fig. 1 c)

penetration (Fig. 1c) due to the consumption of starting materials and not primarily because of diffusion. Notice the general similarity between parts a,b,c of Fig. 1.

Figure 2a compares the rate of descent of the concentration gradient fronts for the experimental (open circles) and the simulated (closed circles) results. The cross-section was taken at 5×10^{-4} M for both examples in Figs 1a and 1b. The sudden change in the gradient of the experimental curve at 310 sec indicates the onset of the convective mode and the collapse of the layer. The disorder after about 400 sec is due to the descent of the fingers. (It should be remembered that the simulation results are only for the non-convective, reaction/diffusion mode of the phenomenon and do not include hydrodynamic equations).

Figure 2b shows the position of the front as a function of $t^{0.5}$. Least means square fitting for the pre-convective section of the curve reveals a straight line (correlation coeff. of 0.999) for both the experimental and simulated results indicating normal Fickian diffusion. The slopes in both cases are identical.

Figure 3 shows the increase in the total concentration of the product integration of the curves in Figs. 1a and 1b as a function of time. Again the curves have matching slopes and the experimental curve shows the gradient change at 310 sec as in Fig. 2.

Close correlation between experiment and simulation is revealed in Fig. 4 for the rate of increase in the maximum of the concentration profiles. The onset of heavy fingering occurs at ~400 sec.

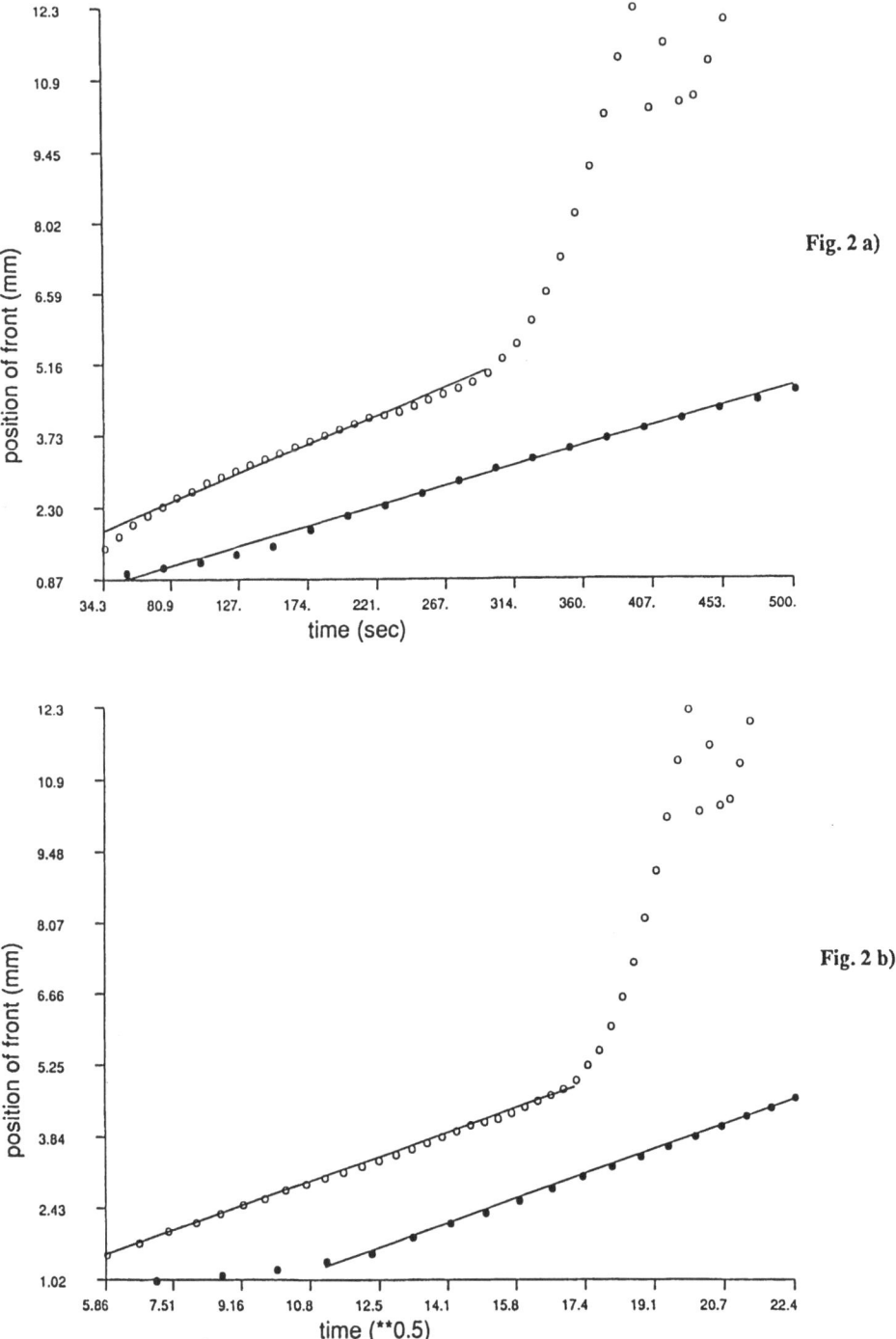

Figure 2. - Experimental (open circles) and calculated (closed circles) position of product front as a function of reaction time (a) and as a function of $t^{0.5}$ (b). Convections start $t \sim 310$ sec

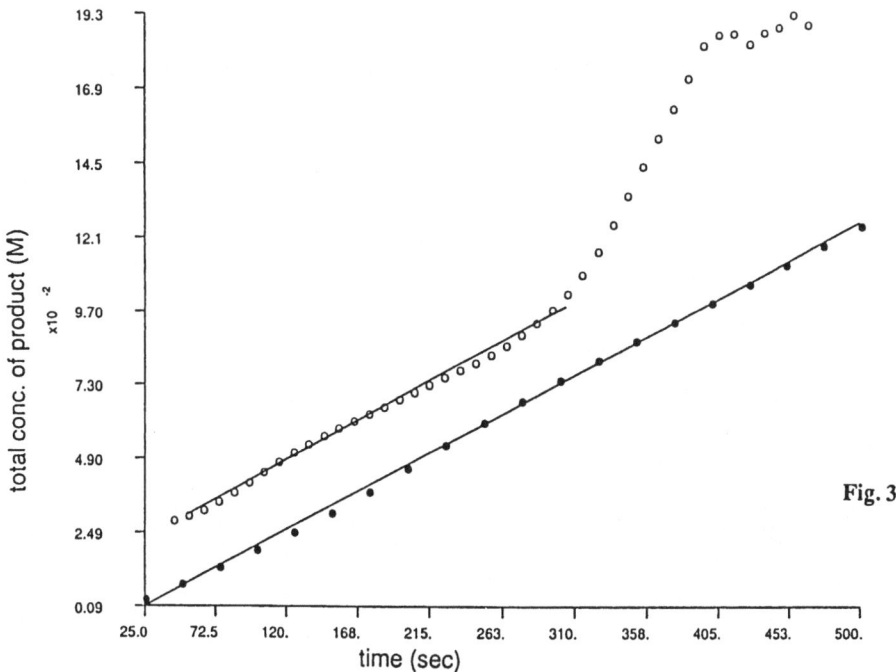

Figure 3.- Experimental (open circles) and calculated (closed circles) reaction rate, shown as total concentration of product as a function of time

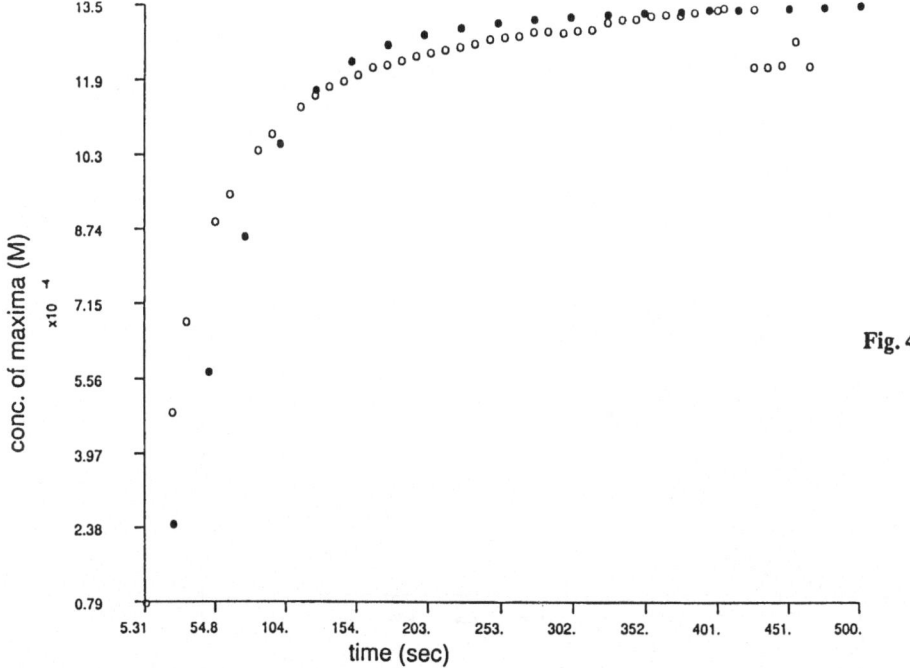

Figure 4.- Experimental (open) and calculated (closed) time-evolution of the maximum values of product concentration

4.2 Density Gradients, Stability Curve and Bifurcation Time

For the calculation of the time-evolution of the density profile of the reacting solution according to {12}, the concentration changes of A and C are required, in addition to the concentration profile changes of E shown above in Fig. 1b. These are shown in Figs. 5a and 5b, respectively. The result of

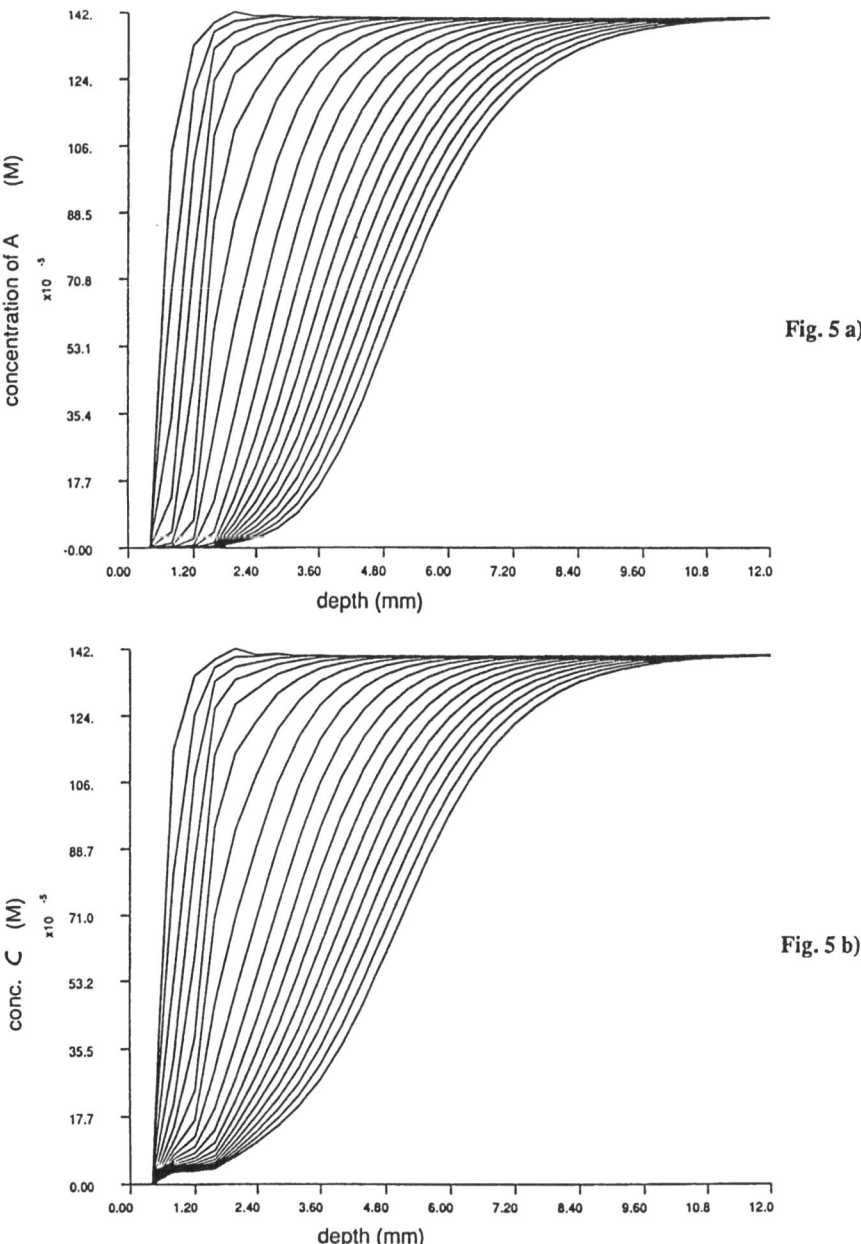

Figure 5. - Calculated time-evolution of the concentration profiles of the starting materials A (a) and C (b) at the same time intervals as Fig. 1b

applying {12} is then shown in Fig. 6. Since $\alpha_E > \alpha_A + \alpha_C$, a negative (top heavy) gradient forms from the very beginning of the reaction. The increase in the negative density gradient is not only due to the development of the product layer but also to the manifestation of a double diffusion mechanism [24,25,26] that operates between the solutes because of the differences in their diffusion rates (Table 1). (The starting materials are diffusing up into the reaction zone faster than the product can diffuse out). Obviously, however, the negative density gradient does not imply an immediate hydrodynamic instability; the latter will be determined by the time evolution of the Rayleigh number ({13}), calculated from the gradients of Fig. 6 and shown in Fig. 7. Also shown in Fig. 7 is the R = 660 and 1100 levels, which, as mentioned above, correspond to two free boundaries and one free and one rigid boundary, respectively. The crossing time of the R = 660 value, around 310 sec, is in close agreement with the experimental time of onset of convections (the right-most curve in Fig. 1a), also around 310 sec. We can conclude from this that the system behaves like one with two free boundaries. This is probably because the destabilizing gradient as seen in Fig. 6 is removed by a small distance from actual contact with the upper boundary.

Finally, it should be noted (Fig. 8) that R changes with time as t^3. For double-diffusion without a chemical reaction, R is a function of $t^{3/2}$ [25]. Thus, in this particular case, the reaction increases the rate of evolution of the instability of the system.

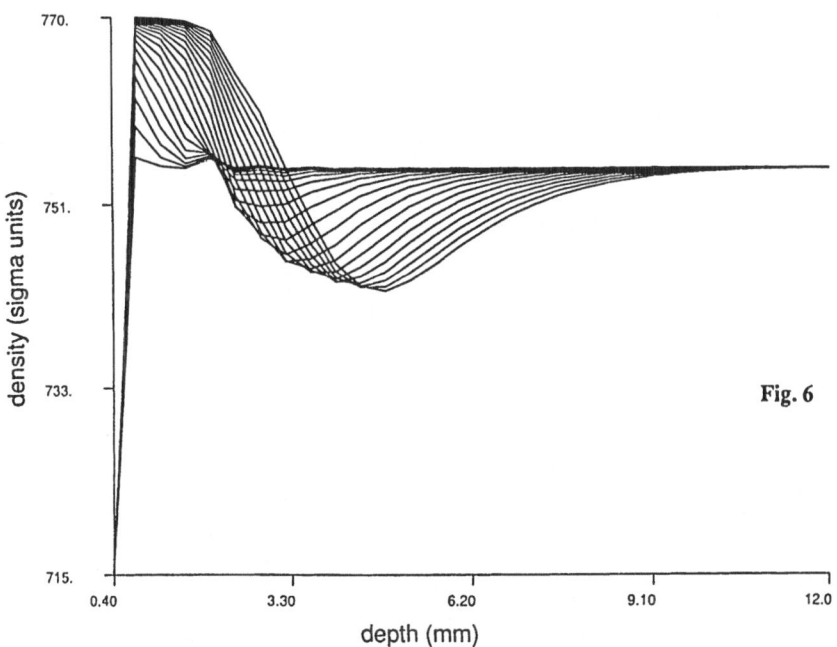

Figure 6. - The calculated time evolution of the density profiles. The layer thickness, h, is taken as the distance from the left hand maximum to the right hand minimum, for each curve. Notice the evolution of the top-heavy unstable gradient. (Sigma density units: $(\rho - 0.999) \times 10^6$)

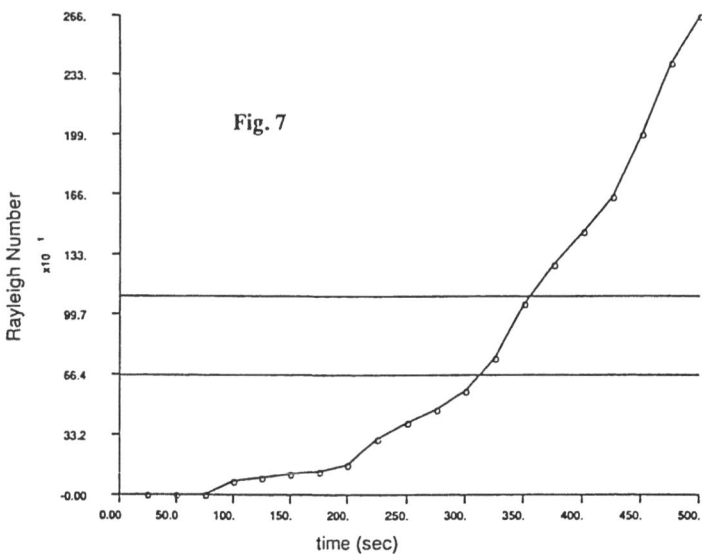

Figure 7. - The calculated time evolution of the Rayleigh Number. The crossing line with the critical values of 660 and 1100 are shown

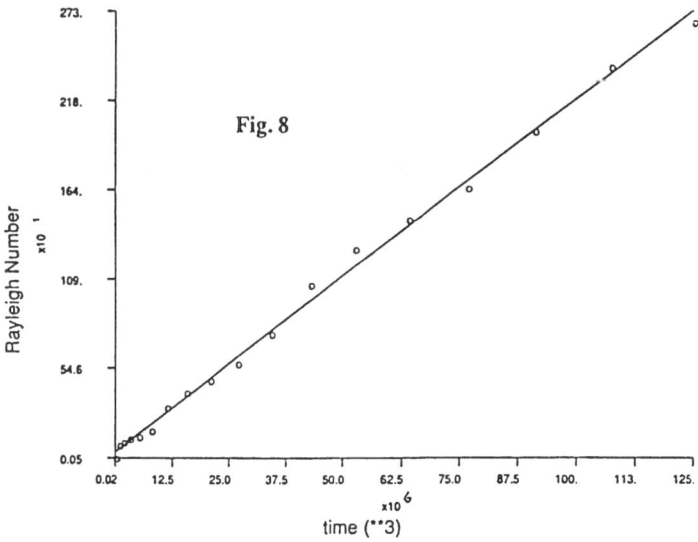

Figure 8. - The relation between the Rayleigh number and t^3

Acknowledgements
Sponsored by the Volkswagen Foundation (under Synergetics) and by the Israel Space Agency through the National Council for R&D.

References

1. M. Kagan, A. Levi and D. Avnir: *Naturwiss.* **69**, 548 (1982)
2. D. Avnir, M. Kagan and A. Levi: *ibid.* **70**, 141 (1983)

3. D. Avnir and M. Kagan: *ibid.* **70**, 361 (1983)

4. D. Avnir and M. Kagan: *Nature* **307**, 717 (1984)

5. D. Avnir, M.L. Kagan and W. Ross: *Chem. Phys. Lett.* **135**, 177 (1987)

6. To appear in: *The Proceedings of Spatial Inhomogeneities and Transient Behaviour in Chemical Kinetics,* eds. G. Nicolis and P. Gray, Brusselles, 1987

7. D. Kosloff and R. Kosloff: *Comput. Phys.* **52**, 35 (1983)

8. K.V. Krishnamurty and G.M. Harris: *Chem. Rev.* **61**, 213 (1961)

9. A.H.I. Ben-Bassat: *Israel J. Chem.* **6**, 91 (1968)

10. G.G. Rav, G. Aravamudan and N.C. Venkatamma: *Z. Anal. Chem.* **146**, 161 (1955)

11. A.Ito, M.Suenaga and K.Ono: *J. Chem. Phys.* **48**, 3597 (1968)

12. M.L. Kagan: *Ph.D Thesis*, The Hebrew University of Jerusalem, 1987

13. L.G. Longsworth: *Ann. N.Y. Acad. Sci.* **46**, 211 (1945)

14. L-O. Sundelof: *Anal. Biochem.* **127**, 282 (1982)

15. M. Kagan, E. Meisels, S. Peleg and D. Avnir: *Lecture Notes in Biomathematics* **55**, 146 (1984), and M.L. Kagan, S. Peleg, A. Tchiprout and D. Avnir: in *Non-Equilibrium Dynamics in Chemical Systems*, eds. C. Vidal and A. Pacault, (Springer, Berlin, Heidelberg 1984) p.223

16. M.L. Kagan, S. Peleg and D. Avnir: *in preparation.*

17. R.H. Bisseling: *Ph.D. Thesis*, The Hebrew University of Jerusalem, 1987

18. R.B. Gerber, R. Kosloff and M. Berman: *Computer Physics Reports* **5**, 59 (1986)

19. N. Agmon and R. Kosloff: *in press*

20. D. Avnir, M.L.Kagan, R. Kosloff and S. Peleg: in *Non-Equilibrium Dynamics in Chemical Systems*, eds. C. Vidal and A. Pacault, (Springer, Berlin, Heidelberg 1984) p.118

21. P.M. Gresho and R.L. Sani: *Int. J. Heat Mass Transfer* **14**, 207 (1977)

22. E.M. Sparrow, R.J. Goldstein and V.K. Jonsson: *J. Fluid Mech.* **18**, 513 (1964)

23. S. Chandrasekhar: *Hydrodynamics and Hydromagnetic Stability*, Dover Edition, N.Y. 1981

24. J.S. Turner: *Ann. Rev. Fluid Mech.* **17**, 11 (1985)

25. H.E. Huppert and J.S. Turner: *J. Fluid Mech.* **106**, 299 (1981)

26. M.L. Kagan, R. Kosloff and D. Avnir: *in preparation*

Cellular Automata Simulating the Evolution of Structure Through the Synchronization of Oscillators

A.W.M. Dress, M. Gerhardt, and H. Schuster

Department of Mathematics, University of Bielefeld,
Postfach 8640, D-4800 Bielefeld, Fed. Rep. of Germany

Abstract: It is documented that the evolution of spatial-temporal organization in catalytic oxidation processes on metal surfaces through the synchronization of oscillators can be simulated in an at least qualitatively satisfying way by means of an "infection" model expressed in form of a cellular automaton.

Since several years, by now, cellular automata have become an important tool for the study of complex systems [1-5]. In this note we want to report on some results which were obtained while studying the complex behaviour of catalytic oxidation processes on metal surfaces by means of cellular automata. Let us recall that behavioural patterns indicating reaction schemes of rather intricate and highly nonlinear dynamics have been observed in many experiments (cf.[6-22]) ever since the landmark paper [6] by E. WICKE et al. was published in 1972 (cf. Fig. 1).

In close cooperation with N. Jaeger, P. Plath and their collaborators in Bremen our own investigations started with a mathematical analysis of the following problem: can oscillations be generated by a simple feedback mechanism by which the "reactivity" $A = A(t)$ of a system, which keeps a certain amount $B = B(t)$ of reactive substance in store and gets refilled continuously by a constant amount C of reactive substance, is coupled to the amount $A(t) \cdot B(t)$ of actually reacting substance? Using discrete time steps for the sake of convenience and simplicity the "reactivity" $A = A(t)$ is defined implicitly as a number between 0 and 1 by the equation

$$B(t+1) = B(t) + C - A(t) \cdot B(t) \quad , \tag{1}$$

while its dependence on $A(t) \cdot B(t)$ is expressed by the equation

$$A(t+1) = F(A(t) \cdot B(t)) \quad , \tag{2}$$

where F is assumed to be a monotonously increasing function, defined on the positive real axis with values in the open unit interval $(0,1)$ (cf. Fig. 2). The analysis (cf. [23]) of the discrete dynamical system, defined by (1) and (2), and its unique steady state

$$A_0 = F(C) \quad , \quad B_0 = C/F(C) \tag{3}$$

showed that the system can indeed exhibit oscillatory behaviour.

More precisely, for standard choices of F and with increasing values of the control parameter C, the system will undergo two (discrete) Hopf bifurcations, first from a low-level stable equilibrium to an intermediate periodic attractor and then back to another stable equilibrium, now of high reactivity.

Since this coincided well enough - at least on the purely phenomelogical level pursued so far - with what was known from experiments, we continued our investigations by trying to establish a realistic model of at least one particu-

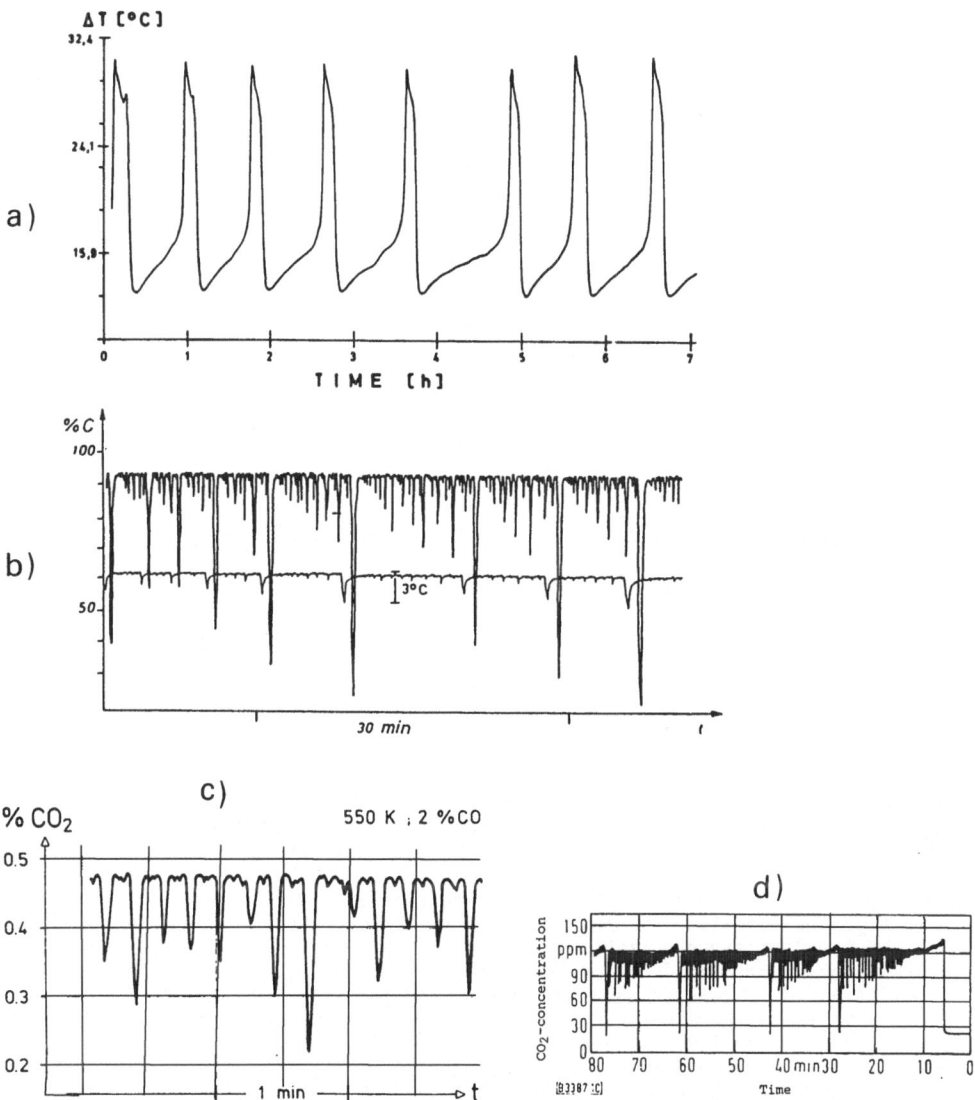

Fig. 1: Experimentally observed oscillation patterns: a) temperature oscillations during the oxidation of methanol on a palladium supported catalyst [9,11,13]; b) oscillations of the CO_2 production (U: conversion of CO to CO_2), resp. the temperature, during the oxidation of CO on a palladium loaded zeolite [10, 17,18,21]; c) and d) oscillations of the CO_2-production during the oxidation of CO on platinum [12,6]

lar catalytic process. For this purpose we chose the catalytic decomposition and oxidation of methanol on a Pd-supported catalyst, studied by A. HABERDITZL et al. [9,11,13] in Bremen. We based our model on the hypothesis proposed by the Bremen group that by some feedback mechanism the reactivity and the storage capacity of the catalytic system is coupled to changes in the solid state phase of the catalysts themselves. The results of these investigations (cf. [24]) were satisfying enough (cf. Fig. 3) to set out to use the model developed so far for

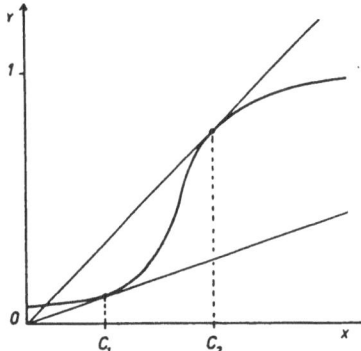

Fig. 2: Graph of a function F of interest in the model of an ideal storage and tangents from the origin. For values of C (rate of influx) lower than C_1 a stable steady state occurs. The increase of C leads to oscillations at $C = C_1$ and to a new stable steady state at $C = C_2$

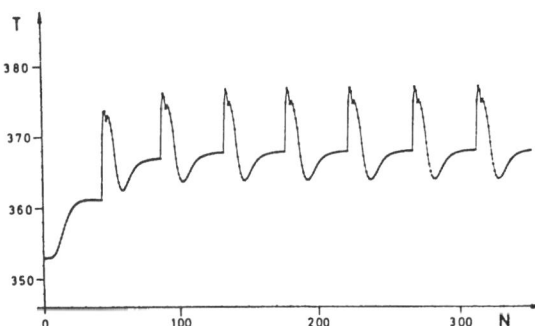

Fig. 3: Time evolution of a dynamical system describing typical temperature oscillations of the methanol oxidation (comp. Fig. 1a)

tackling the most pressing problem in this context: can local interaction, induced e.g. by heat conductivity, be responsible for the synchronization of myriads of individually oscillating, catalytically active crystallites?

To answer this question we had to be aware that at least some of the observations, e.g. the oxidation of carbon monoxide on Pd-crystallites, showed rather intricate forms of synchronization leading to time series of "blocking events" exhibiting even some mild forms of "self similarity" (cf. Fig. 1b) and suggesting the evolution of a complex spatial-temporal organization inside the catalytic system, - a hypothesis which was corroborated soon after we started our work by E.E. WOLF et al. (cf.[19,20]) and by R.A. SCHMITZ et al. (cf. [15]). Hence, we considered the following intriguing observation (cf. [25]) as a first hint of how to approach this problem: if for any integer n=0,1,2,... we count the number $K(n)$ of integers k between 0 and n for which the binomial coefficient $\binom{n}{k}$ is odd, then the resulting function $K(n)$ depicted in Fig. 4 looks surprisingly similar to the time series depicted in Fig. 1b). It is well known and follows immediately

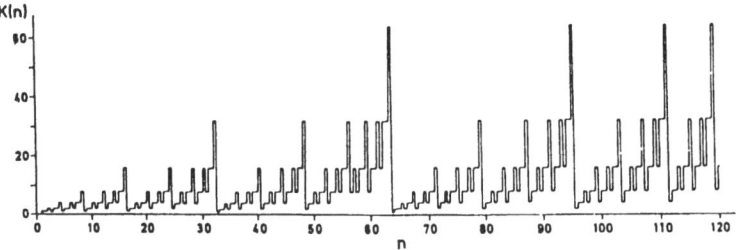

Fig. 4: Representation of the number sequence defined by K(n)

from the scheme, by which binomial coefficients may be computed in Pascal's triangle, that the number $K(n)$ counts the number of black fields in a column of black and white fields at time n (cf. Fig. 5), once we start with such a column at time zero where just the lowest field at height 0 is coloured black while all others are white and continue according to the following recursive "infection principle": if at time n-1 the field at height k is coloured black and the field above at height k+1 is still white, the field "k+1" gets "infected" by the field "k", i.e. it becomes black at time n, while the field "k" itself remains black unless the field "k-1" directly below was "infected" (i.e. black) already at time n-1, in which case "k" becomes white again. In addition, if neither "k" nor "k+1" were infected at time n-1, "k+1" remains uninfected at time n. We concluded from the surprising coincidence between the curves in Fig. 1b and in Fig. 4 that the spatial-temporal organization of the catalytic systems in question could perhaps be modeled in an at least qualitatively satisfying way in terms of an "infection" model expressed in form of a cellular automaton.

Hence, before reporting on our findings using this approach, it may be worthwhile to shortly introduce the abstract mathematical concept of cellular automata. According to some fundamental principles in theoretical physics, the behaviour of many spatially inhomogeneous physical systems can be described most conveniently by associating to any such system a so-called "fibre bundle" $f: E \to B$, that is a mathematical construct, consisting of the "base space" B, representing (in general) the physical space, and the "total space" E, representing all possible "local" events one may find at any place in B, together with the map $f: E \to B$ which associates to any "local" state $e \in E$ the place $f(e) \in B$ where this local state is supposed to occur. Hence, for any place $b \in B$ the fibre $F_b := f^{-1}(b) := \{e \in E \mid f(e)=b\}$ represents the local state space of our system at b. The global states of our system are given by the set of (admissible) "sections" $s: B \to E$ (i.e. maps from B back into E such that $f(s(b))=b$ for all b), representing at each $b \in B$ its local state $s(b)$, and the dynamics of the system is described in terms of "local" rules by which for any (admissible) section $s = s_t$, representing the global state at time t, the sections $s' = s_{t'}$, representing the global state at times $t' > t$, can be computed. Here such a rule is called local if, for some appropriate concept of *neighbourhood*, the local state $s'(b)$ the global state $s' = s_{t'}$, assumed at some $b \in B$, depends only on how the original global state s_t looks in some neighbourhood around b, at least as long as t' is not much larger than t.

In general, to define admissibility of sections and to state the local rules, describing the dynamics of the system, the spaces B and E and the map $f: E \to B$

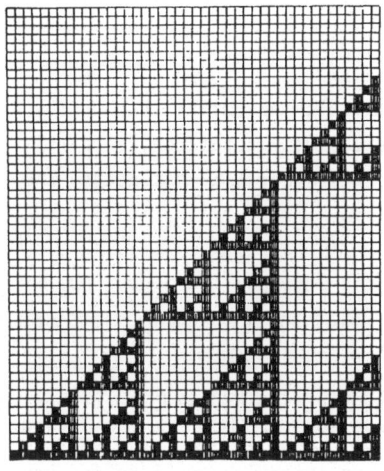

Fig. 5: Time evolution of a one-dimensional cellular automaton corresponding to the function $K(n)$ depicted in Fig. 4. The hatched areas mark the "infected" cells for each time $t = n$

⟶ n

are assumed to carry additional structure, mostly of a continuous, non-discrete nature, so that one may use concepts from differential topology and related fields and the deep mathematical theorems proved in these fields to analyse such systems. Mathematically, much less is known, if B is assumed to be a discrete space. Hence in this case it may be justified to start investigations under the (in this case not even very restrictive) assumption that the given fibre bundle $f: E \to B$ is "trivial", i.e. that all fibres F_b ($b \in B$) are alike, just some set F, that E is the product space $F \times B$ and that $\hat{f}: E \to B$ is just the projection map $F \times B \to B: (x,b) \mapsto b$. In this case the sections $s: B \to F \times B$ can be identified with the maps $s: B \to F$; associating to each $b \in B$ its individual state $s(b) \in F$. To define a concept of neighbourhood as well as local rules representing the dynamics one may choose a "model" neighbourhood N, i.e. some arbitrary, generally finite set. Then for each $b \in B$ one may choose a "chart" $\varphi_b: N \to B$, i.e. a map from the model neighbourhood N into B, "coordinatizing" the relevant neighbourhood $\varphi_b(N) \subseteq B$ around b. Finally one may specify the local rule in terms of a map ψ from the set F^N of all "global" states of the model neighbourhood N, i.e. of mappings $r: N \to F$, into the set F of local states, associating to any such map $r: N \to F$ some local state $\psi(r) \in F$, representing the local state at the next time step at the imagined "center" of the model neighbourhood. Then the resulting discrete dynamical system is defined on the global state space F^B of all maps from B into F in terms of a transition map

$$T: F^B \to F^B : s \mapsto s'$$

where $s'(b)$ is defined by

$$s'(b): = \psi(s \circ \varphi_b) ,$$

i.e. by evaluating ψ on the "restriction" $s \circ \varphi_b: N \xrightarrow{\varphi_b} B \xrightarrow{s} F$ of $s: B \to F$ onto the model neighbourhood N via the "coordinatization" $\varphi_b: N \to B$ of the neighbourhood around b.

Hence one may define an arbitrary transition map $T: F^B \to F^B$ to represent a cellular automaton on B with values in F, if there exists some set N, some family of maps $\varphi_b: N \to B$ ($b \in B$) and some $\psi: F^N \to F$, such that $s' = T(s)$ coincides at every $b \in B$ with $\psi(s \circ \varphi_b)$ for all $s \in F^N$. It is easy to see that indeed any cellular automaton considered so far in the literature fits into the above (admitted rather abstract) framework and that in addition this approach allows to deduce some interesting invariants of cellular automata as well as some formal consequences such as the following

Lemma: If $T_1 : F^B \to F^B$ and $T_2 : F^B \to F^B$ represent cellular automata, then $T_1 \circ T_2 : F^B \to F^B : s \mapsto T_1(T_2(s))$ represents also a cellular automaton, in particular $T_1^n : F^B \to F^B : s \mapsto \underbrace{T_1(T_1(...T_1(s)...))}_{n \text{ times}}$ represents a cellular automaton.

We have taken the liberty to propose this abstract framework for the study of cellular automata since it allows to clarify its relations with more classical approaches towards modeling spatially inhomogeneous physical systems.

We now come back to our original purpose, the study of synchronization processes between spatially distributed and locally coupled oscillators. Simulating such processes by means of cellular automata, it was hoped to get some insight into the various possible forms of spatial-temporal organization evolving in such systems which we believed to be responsible for the observed synchronization phenomena. In each simulation a two-dimensional array of oscillators, represented by the cells of the automaton, was subjected to various forms of local interaction between neighbouring oscillators. Two different approaches were pursued. The first one was closely related to our former studies of methanol oxidation, mentioned above. Since the original model appeared to be too complicated to be executed in parallel at 16 or even at 64 different places for several hundreds or even thousands of time steps in an acceptable time span by the available computing machinery, a simplified version was developed whose local state space F was defined to be $F = R^3 \times \{0,1\}$. Uncoupled, each cell oscillated with the same frequency, while the phases of the cells were distributed randomly. The amazing result (cf. [26,27]) after introducing local interaction in form of heat conductivity was that essentially only three different types of behaviour evolved: either the whole system synchronized eventually or the system partitioned itself into precisely two, not necessarily spatially connected blocks B_1 and B_2 which oscillated either with a phase shift of around 180° in case both were of approximately the same size (and hence misleadingly suggesting a period doubling phenomenon in the global output), while in case $\#B_1 \ll \#B_2$ the larger block B_2 followed the smaller block B_1 with a phase shift of approximately 60° (cf. Fig. 6). Even though a satisfactory mathematical explanation for this rather high degree of regularity evolving in our system is still missing, the findings not only appear to be relevant for any further theoretical considerations concerning the spatial-temporal organization of oscillating chemical processes, but they suggest also the existence of a rather wide and promising, while still almost untouched field for further mathematical investigations.

The second approach was motivated by the one-dimensional "infection" model, based on Pascal's triangle, mentioned above. Taking into account some ideas con-

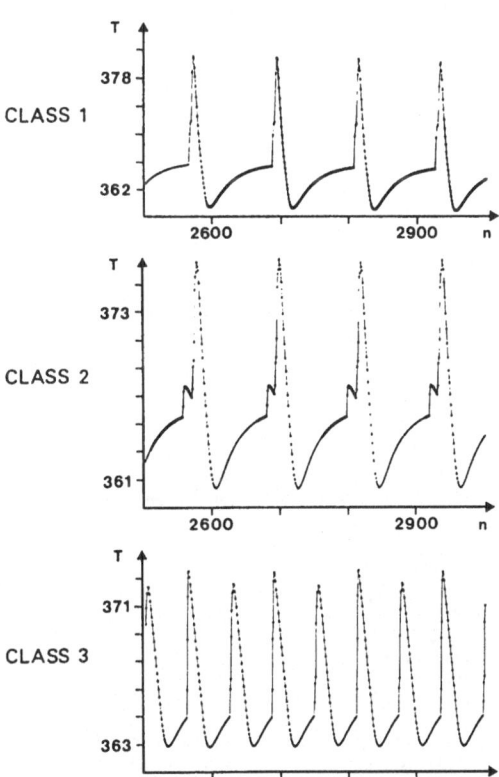

Fig. 6: Evolution of the average temperature of sixteen coupled dynamical systems (describing the methanol oxidation) in the three different classes of attractors

cerning the chemical mechanisms responsible for the experimentally observed intricate temporal patterns of catalytic CO-oxidation on Pd-crystallites mentioned already above, and trying to simplify these ideas to the utmost, the following model was studied: each cell in a quadratic array of altogether 400 up to 22500 cells was allowed to show various degrees of infection, measured by a certain natural number, varying between 0 - the "healthy" state - and 100 - the state of "definite illness". A healthy cell remains healthy unless the number of infected and ill cells in its neighbourhood exceeds a certain threshold value, the first control parameter of the system. At each time step the degree of infection of an infected, but not already ill cell, first gets enlarged by a certain amount, a second control parameter of the system - later on called g in Fig. 7 - and then averaged to a certain degree by an exchange process between neighbouring infected cells, by which less infected cells get more infected on the expense of its already more infected neighbours. Finally, ill cells are supposed to always get healthy at the next time step.

Simulation studies of this model were executed for a large variety of values of the control parameters and initial states and running through many time steps

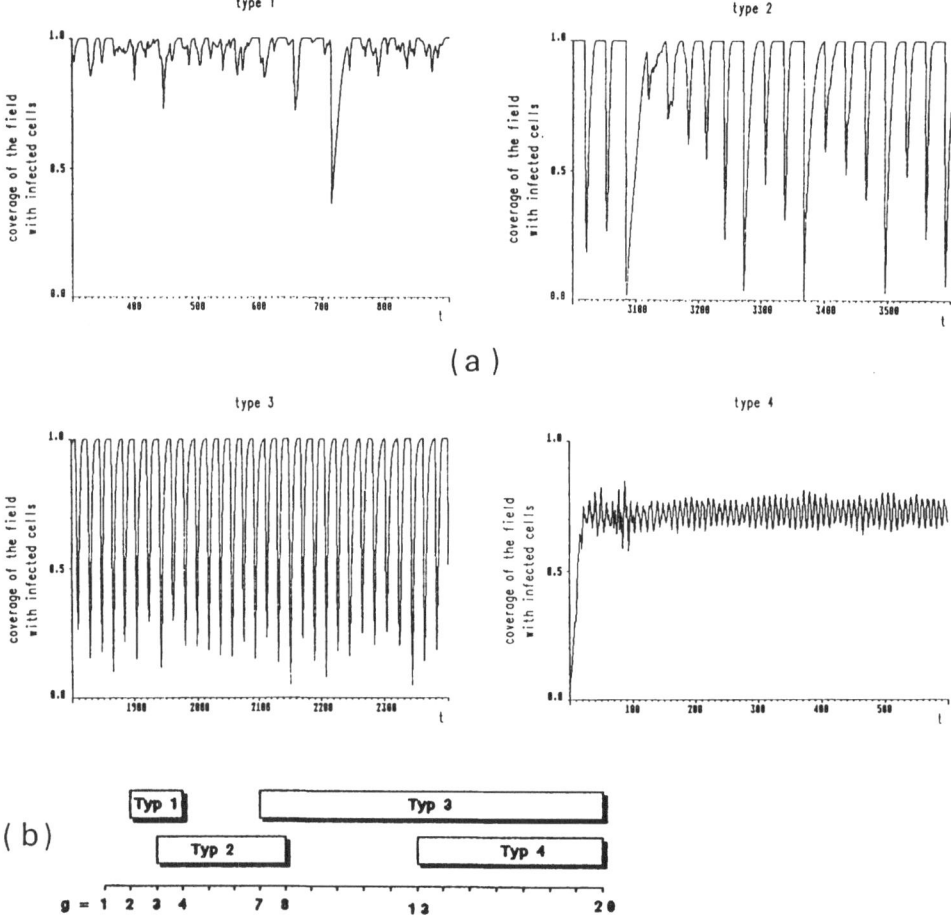

Fig. 7: The four different types of behaviour occurring in the time evolution of the cellular automaton describing the CO-oxidation (a) and the ranges of the control parameter g where the various types have been observed (b)

- generally 10,000 steps. Similarly as above, the amazing result (cf. [28,29]) was that again only very few types, in this case just four types of essentially different forms of behaviour evolved. They are described in Fig. 7. Even more amazing was the discovery that not only systems could switch from one type to the other one and back again, indicating some form of bistability (cf. Fig. 8), but that for all four types fascinating spatial-temporal structures like those observed by SCHMITZ et al. (cf. Fig. 9) and even meandering spiral formed waves evolved from random inputs and kept running for many periods (cf. Fig. 10). Similar spiral waves could experimentally be observed e.g. in the Belousov-Zhabotinskii-reaction (cf. [30,31]), and have given rise so far to different kinds of mathematical models for their theoretical description (cf. [32-35]).

Unfortunately, it is not possible to discuss all these findings and their chemical and mathematical implications and correlations any further in detail in this note. Let us close instead by stating that we believe to have demonstrated that simulating non-linear and spatially inhomogeneous processes like synchronization processes of spatially distributed and locally coupled oscillators by simple cellular automata may reveal the evolution of sometimes surprisingly definite forms of spatial-temporal organization during such processes as well as the parameter dependence of the various observed forms of such organization. Thus such simulations may lead at least to new interesting hypotheses concerning such processes which then can be tested by new experiments. In addition, the often surprising results of such simulations represent a considerable challenge for theoretical mathematics to develop a conceptual framework and machinery which could be used to explain the observed phenomena in purely mathematical terms - a task which may keep mathematicians busy for several decades.

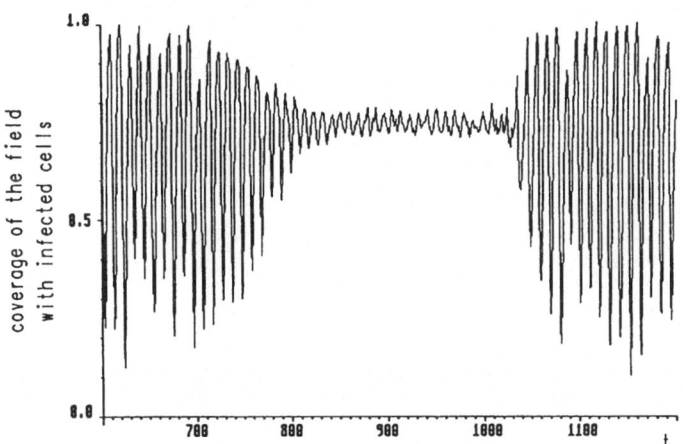

Fig. 8: Time evolution of a cellular automaton in which the system switches from a type - 3 - to a type - 4 - behaviour and back again

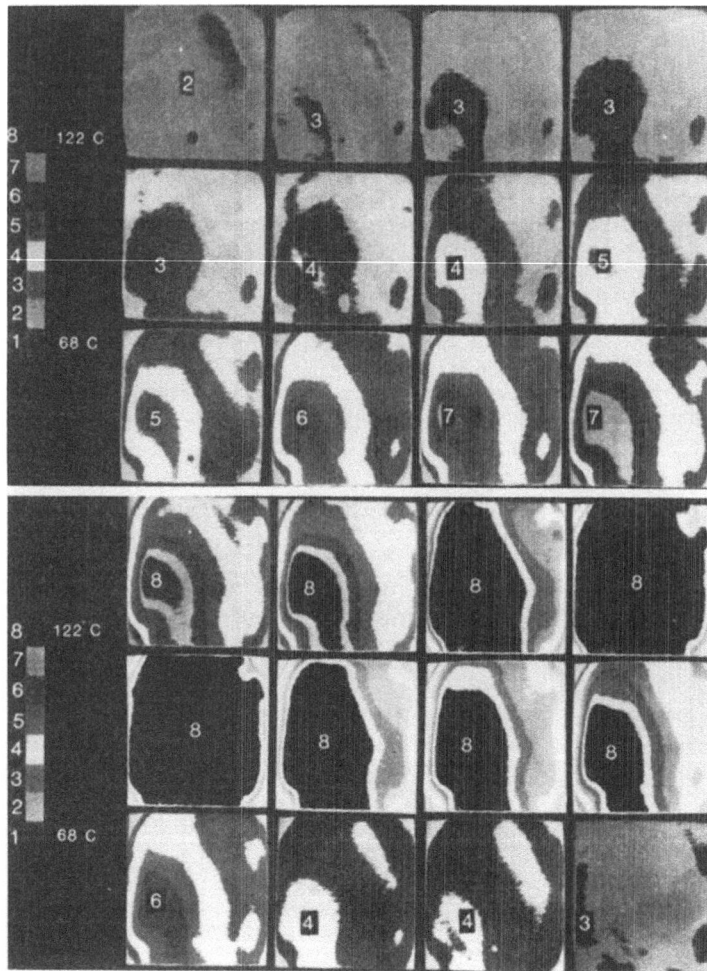

Fig. 9: Thermograms of a platinum foil during an oscillatory oxidation of hydrogen observed by R.A. SCHMITZ et al. [15]

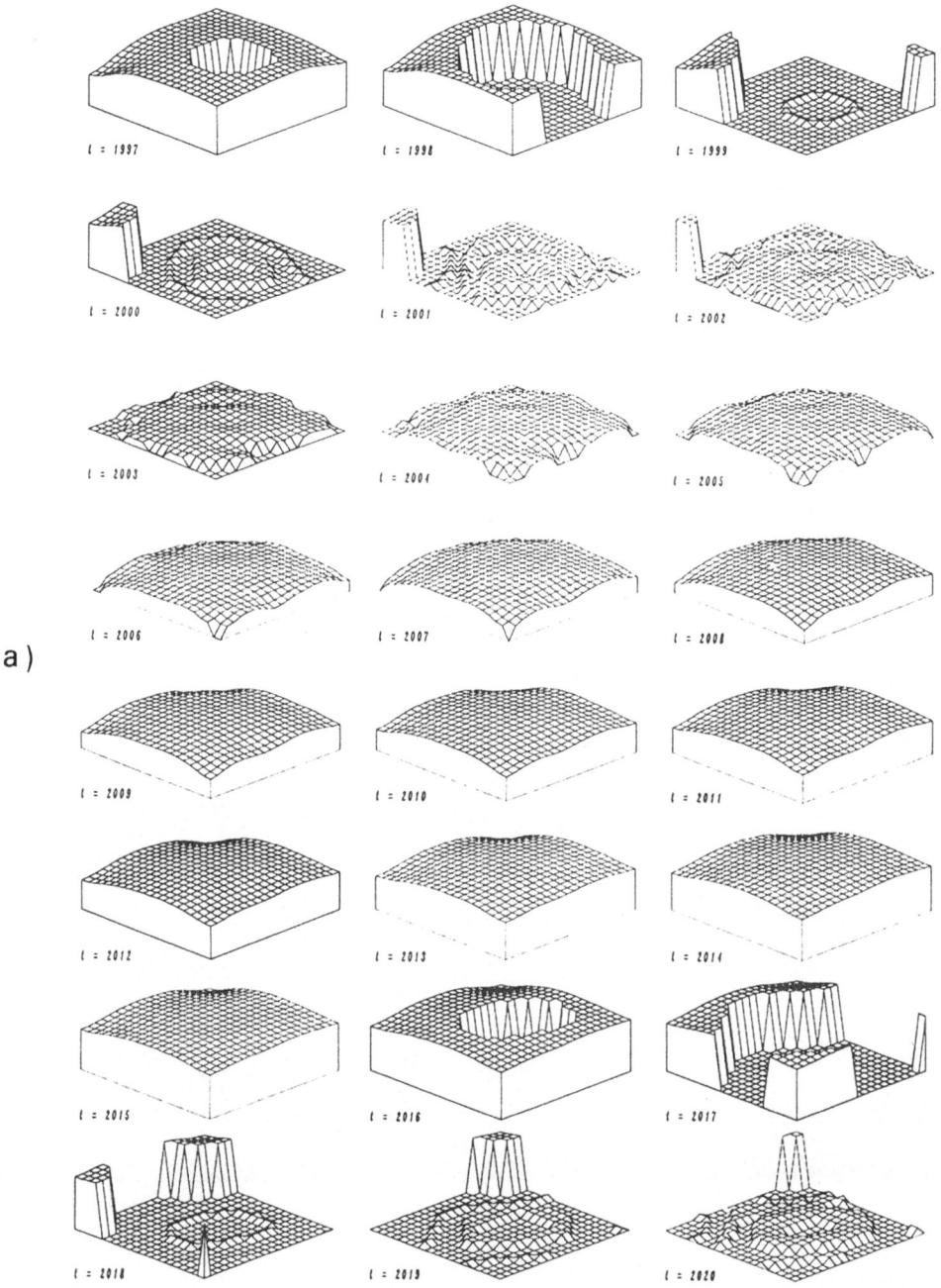

Fig. 10: a) A typical time evolution of the infection states of the cells corresponding to the behaviour of type 3. b) A typical time evolution of the infection states of the cells corresponding to the behaviour of type 4. c) and d) Typical infection states of the cellular automaton with a 50 x 50-lattice of cells corresponding to the behaviour of type 3 (c) and type 4 (d). (Different infection states of a cell are represented by different shadings.)

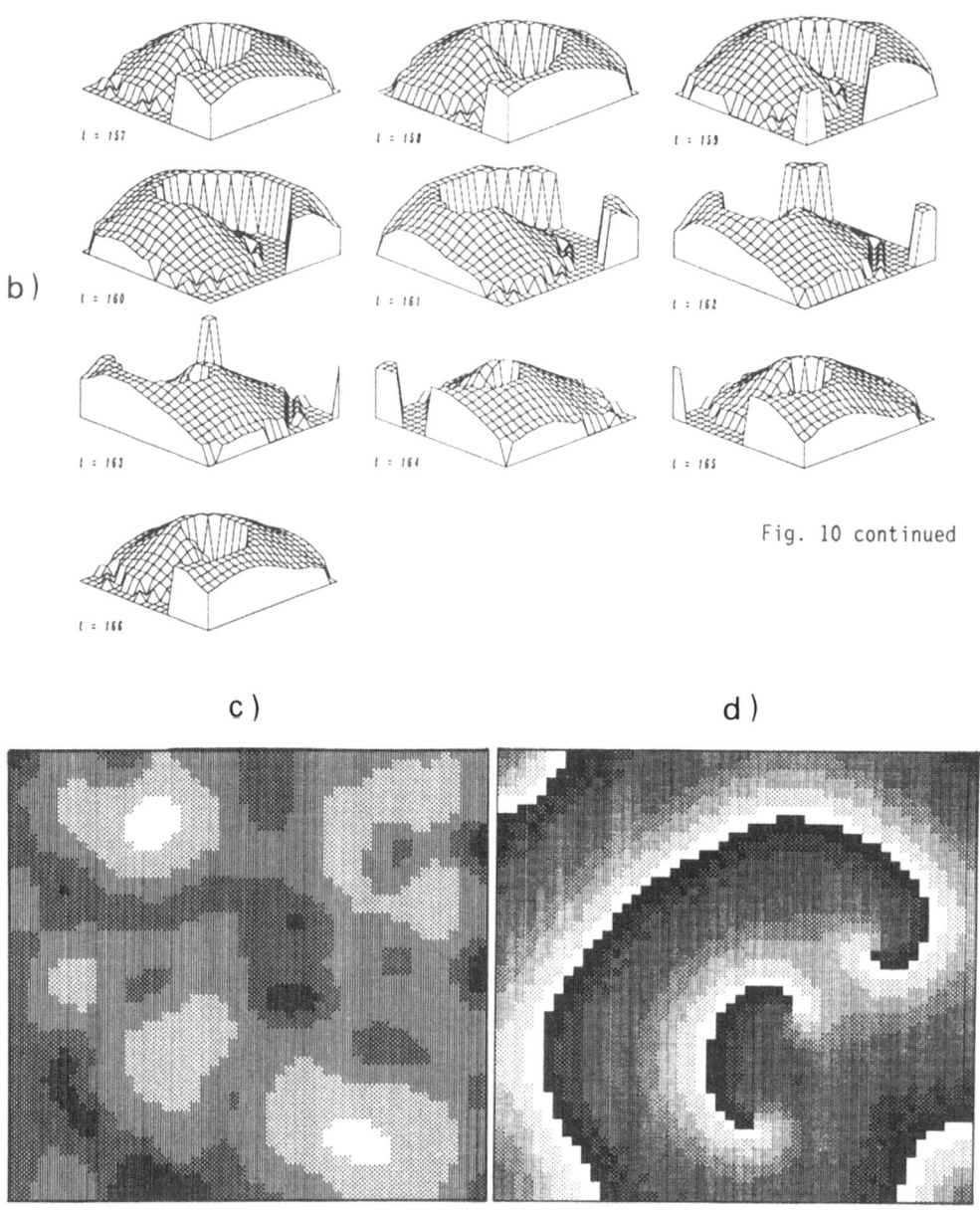

$l = 157$ $l = 158$ $l = 159$

b) $l = 160$ $l = 161$ $l = 162$

$l = 163$ $l = 164$ $l = 165$

Fig. 10 continued

$l = 166$

c) d)

Fig. 10 continued

References

Cellular Automata:

1. J. von Neumann: <u>Theory of Self-Reproducing Automata</u>, ed. by A.W. Burks (University of Illinois, Urbana 1966)
2. E.F. Codd: <u>Cellular Automata</u> (Academic Press, New York 1968)
3. <u>Cellular Automata: Proceedings of an Interdisciplinary Workshop</u>, ed. by S. Wolfram, J. Doyne Farmer, T. Toffoli (Physica D, Vol. 10D, Nos. 1 and 2, 1984)

4. Dynamical Systems and Cellular Automata, ed. by J. Demongeot, E. Golès, M. Tchuente (Academic Press, New York 1985)
5. S. Wolfram: Theory and Application of Cellular Automata, Advanced series on complex systems, Vol. 1 (World Scientific Publishing Co. Pte. Ltd, Singapore 1986)

Experimental Results:

6. H. Beusch, P. Fieguth, E. Wicke: Chem. Ing. Tech. 44, 445 (1972)
7. M. Sheintuch, R.A. Schmitz: Catal. Rev.-Sci. Eng. 15(1), 107 (1977)
8. W. Keil, E. Wicke: Ber. Bunsenges. Phys. Chem. 84, 377 (1980)
9. N.I. Jaeger, P.J. Plath, E. van Raaij: Ber. Bunsenges. Phys. Chem. 84, 417 (1980)
10. N.I. Jaeger, K. Möller, P.J. Plath: Z. Naturforsch. 36a, 1012 (1981)
11. N.I. Jaeger, P.J. Plath, E. van Raaij: Z. Naturforsch. 36a, 395 (1981)
12. D. Böcker: Thesis (Münster 1984)
13. A.Th. Haberditzl, N.I. Jaeger, P.J. Plath: Z. Phys. Chemie Leizpig 265, 449 (1984)
14. D. Böcker, E. Wicke: In Temporal Order, ed. by L. Rensing, N.I.Jaeger, Springer Series in Synergetics, Vol. 29 (Springer, Berlin, Heidelberg 1985) p. 75
15. J.R. Brown, G.A. D'Netto, R.A. Schmitz: In Temporal Order, ed. by L. Rensing, N.I. Jaeger, Springer Series in Synergetics, Vol. 29 (Springer, Berlin, Heidelberg 1985) p. 86
16. M.P. Cox, G. Ertl, R. Imbihl: Phys. Rev. Lett. 54, 1725 (1985)
17. N.I. Jaeger, K. Möller, P.J. Plath: Ber. Bunsenges. Phys. Chem. 89, 633 (1985)
18. N.I. Jaeger, K. Möller, P.J. Plath: Temporal Order, ed. by L. Rensing and N.I.Jaeger, Springer Series in Synergetics, Vol. 29 (Springer, Berlin, Heidelberg 1985) p. 96
19. D.J. Kaul, E.E. Wolf: J. Catalysis 91, 216 (1985)
20. D.J. Kaul, E.E. Wolf: J. Catalysis 93, 321 (1985)
21. N.I. Jaeger, K. Möller, P.J. Plath: J. Chem. Soc., Faraday Trans. 82, 3315 (1986)
22. N.I. Jaeger, R. Ottensmeyer, P.J. Plath: Ber. Bunsenges. Phys. Chem. 90, 1075 (1986)

Theoretical Results:

23. A. Dress, N.I. Jaeger, P.J. Plath: Theoret. Chim. Acta (Berl.) 61, 437 (1982)
24. M. Gerhardt, H. Schuster: Diplomarbeit (Bielefeld 1984)
25. A. Dress, M. Gerhardt, N.I. Jaeger, P.J. Plath, H. Schuster: In Temporal Order, ed. by L. Rensing, N.I. Jaeger, Springer Series in Synergetics, Vol. 29 (Springer, Berlin, Heidelberg 1985) p. 67
26. M. Gerhardt, P.J. Plath, H. Schuster: Ber. Bunsenges. Phys. Chem. 90, 1040 (1986)
27. H. Schuster: Thesis (Bremen 1987)
28. M. Gerhardt: Thesis (Bremen 1987)
29. M. Gerhardt, H. Schuster: in preparation

Spiral Waves:

30. A.T. Winfree: Science 175, 634 (1972)
31. S.C. Müller, Th. Plesser, B. Hess: Science 230, 661 (1985)
32. B.F. Madore, W.L. Freedman: Science 222, 615 (1983)
33. J.M. Greenberg, B.D. Hassard, S.P. Hastings: Bull. Am. Math. Soc. 84, 1296 (1978)
34. O.E. Rössler, C. Kahlert: Z. Naturforsch. 34a, 565 (1979)

Part III

Biochemical Organization

Potential Functions and Molecular Evolution

P. Schuster

Institut für theoretische Chemie und Strahlenchemie, Universität Wien, Währingerstraße 17, A-1090 Wien, Austria

Abstract

Selection and molecular evolution are often considered as processes on potential surfaces which are characterized as "fitness landscapes". Two classes of potentials are particularly useful: "selection potential" for the selection process within a given population and "value landscapes" for evolutionary adaption. Among the various types of selection dynamics we distinguish rare and frequent mutation scenarios as well as different mechanisms for replication. The selection potential for independently replicating entities is a linear function of their concentrations. Evolutionary optimization leads to "corner equilibria" which represent pure states in the rare mutation scenario, or "quasispecies" if mutations occur frequently. The existence of a selection potential is a direct proof for the absence of complicated dynamics and dissipative structures. Dynamical systems for which no potential functions can be found are interesting in their turn because they may lead to oscillations and chaotic dynamics. Value landscapes provide direct insight into the course of evolutionary optimization. They are, however, very hard to determine even for the most simple examples which deal with "test-tube evolution". We can discuss here only the results of a computer model which is thought to be representative also for real systems.

1. Evolution as a "Hillclimbing" Process

Since the development of population genetics by the three famous scholars FISHER, HALDANE and WRIGHT [1-3] the process of evolutionary optimization is commonly viewed as a kind of "hillclimbing" leading to local or global fitness optima of populations. The notion of "fitness", which was introduced soon after the earliest editions of CHARLES DARWIN's "Origin of Species" in the famous phrase of the "survival of the fittest" [4], turned out to be exceedingly useful for the illustration of selection and evolution. According to a commonly held view biological species are assumed to reside in local maxima of "fitness landscapes" which often were identified with ecological niches. Real fitness landscapes, however, are exceedingly complicated objects. Fitness is not simply a property of species or types in isolation, since all other individuals living within the same ecosystem contribute to it. Mutual dependence obscures the simple picture and complicates every attempt to analyse evolutionary phenomena. Harsh critics sometimes commented nastily: "The problem with the attractive concept of fitness landscapes is that nobody knows what is plotted on the ordinate and on the abscissa axis". Indeed, the enormous complexity of biological systems is prohibitive when it comes to derive quantitative measures of fitness and distance between organisms. Detailed knowledge of the molecular processes going on during reproduction of organisms is required if concepts like evolutionary distance or fitness are to be put on more solid grounds.

In conventional population theory fitness is expressed as the mean number of descendants reaching the age of reproduction. Fitness is a property to be assigned to a type of individuals commonly called the "phenotype". Selection chooses between phenotypes thereby optimizing the mean fitness of the population.

Mutations leading to new types are assumed to be rare events. Most of them are deleterious in the sense that the new variants have lower fitness and hence they will be eliminated soon after formation. Eventually a mutation results in a type of higher fitness and then it may replace the previously selected phenotype, provided it succeeds to survive a stochastic phase of multiplication. Evolutionary optimization is visualized as a succession of phenotypes with increasing fitness. This simple scenario is substantially complicated by recombination through sexual reproduction, by population-dependent fitness and by frequent mutation. We shall not deal with the first case here which is even more sophisticated, because the unfolding of the phenotype of a multicellular organism involves cell differentiation and morphogenesis. But we shall consider here the frequent mutation case and examples of population-dependent fitness factors.

The discovery of DNA being the seat of genetic information provided a natural measure of distance between two organisms which is the Hamming distance of their polynucleotide sequences - this is the number of positions at which the two sequences have different bases. With respect to fitness the situation is much more involved. At present it is not conceivable to estimate the fitness of an organism seriously, not even in the simplest cases. Tremendous reduction in complexity is inevitable. On the other hand a phenomenological kinetic theory of molecular evolution of simple system has been developed [5-8] and tested on virus RNA replication *in vitro* [9-11]. The results obtained justify the use of the fairly simple expression derived from mass action kinetics for multiplication of polynucleotide molecules under idealized environmental conditions as found in flow reactors. Actually most of the results hold also under much more general conditions of non-stationary, growing systems [6].

Recent data on populations of RNA viruses have shown that mutations cannot be considered as rare events. The accuracies of virus specific RNA polymerases are so low that errors in the replication of the entire virus genomes occur frequently. To give an example: the replicase of the RNA bacteriophage $Q\beta$ replicates with an error rate of 1/3000 per digit [12]. Therefore the newly synthesized RNA sequences, which are 4220 bases long, contain one error on the average. Most virus populations studied so far were found to be "heterogeneous" [13-16]: they contain a great variety of different polynucleotide sequences. Bacterial populations are less well known in detail, but the analysis of experiments on bacterial growth in chemostats [17] indicates that mutations are frequent events there, as well. The accuracy of the DNA replicating machinery is much higher than that of RNA replicases, but most bacterial genomes exceed viral genomes in length by factors between 10^2 to 10^3. Therefore, a frequent mutation scenario seems to be more realistic than a model based on rare mutations, at least for prokaryots.

2. Two Classes of Potentials for Evolutionary Processes

First we consider a somewhat idealized scenario of the selection process which is based on the following assumptions:

(1) Individuals replicate producing thereby correct replicas and mutants according to physically determined probability distributions.

(2) Individuals have finite lifetimes. They are either destroyed by degradation or diluted out of the system.

(3) Individuals are grouped into "types". If an individual is assigned to a type, all its properties are determined. The population contains n types: I_k; $k = 1,\ldots,n$.

(4) The system is limited with respect to formation of new types. We have n different types - n eventually being very large - and we are interested in the stationary distribution of types and in the dynamics of the approach towards it.

(5) The environment of the system is constant and therefore the expressions for the rates of replication, mutation and degradation as well as the dilution fluxes do not change during the selection process.

If selection occurs in a system as described above, it leads to an asymptotically stable stationary state. Commonly this approach is monotonous, no oscillations occur. Selection, in addition, implies that the final state is a pure state which, in general, is approached from a mixed initial state. There exist Lyapunov functions for selection processes, but can these Lyapunov functions be interpreted as "selection potentials"?

Let us assume that the dynamics of our system is described within the frame of conventional chemical kinetics by the differential equation

$$\frac{dx_k}{dt} = F_k(\vec{x}); \quad k = 1, \ldots, n. \tag{1}$$

The variables are the concentrations of individual types which we denote by $x_k(t) = [I_k]$. The system and its evolution is thus described by the vector

$$\vec{x}(t) = \Big(x_1(t), \ldots, x_n(t)\Big).$$

The existence of a selection potential implies that the differential equation (1) can be formulated as a gradient system

$$\frac{dx_k}{dt} = \frac{\partial V(\vec{x})}{\partial x_k}; \quad k = 1, \ldots, n \quad \text{or} \quad \frac{d\vec{x}}{dt} = \text{grad}\, V(\vec{x}) \tag{2}$$

with $F_k(\vec{x}) = \partial V(\vec{x})/\partial x_k$.

Gradient systems have several general properties which are particularly useful in the analysis of their dynamics:

(1) The trajectories of gradient systems intersect with the constant level sets of the potential $V(\vec{x})$ at right angles.

(2) Gradient systems approach an extremum or an optimum of the potential function $V(\vec{x})$.

(3) No oscillations and/or other more complicated dynamical phenomena like deterministic chaos occur in gradient systems.

(4) No spatial dissipative structures are formed in gradient systems under no flux boundary conditions.

The nonexistence of a selection potential, if it can be proven, is also of interest since it makes the system a candidate for the search for complex dynamics. In section 3 we shall discuss several dynamical systems with selection potentials.

The scenario of selection dynamics developed above does not meet the common view of evolution properly. The numbers of types that can exist in principle exceed all imagination. In terms of polynucleotides we have 4^ν different sequences of chain length ν. Apparently it is impossible to visualize numbers like $4^{1000} \approx 10^{600}$! And a chainlength of $\nu = 1000$ is smaller than the lengths of the smallest viral genomes. Hence, any realistic population covers only a tiny fraction of all possible types. It is more appropriate to consider populations open with respect to the formation of types: new types are formed and sometimes grow to macroscopic concentrations during evolution, old types are eventually lost. Thus we may visualize populations migrating through a huge abstract space of potentially existing types. Since we shall follow the molecular view from now on, this abstract space is a space of polynucleotide sequences and we shall call it therefore the "sequence space" [18].

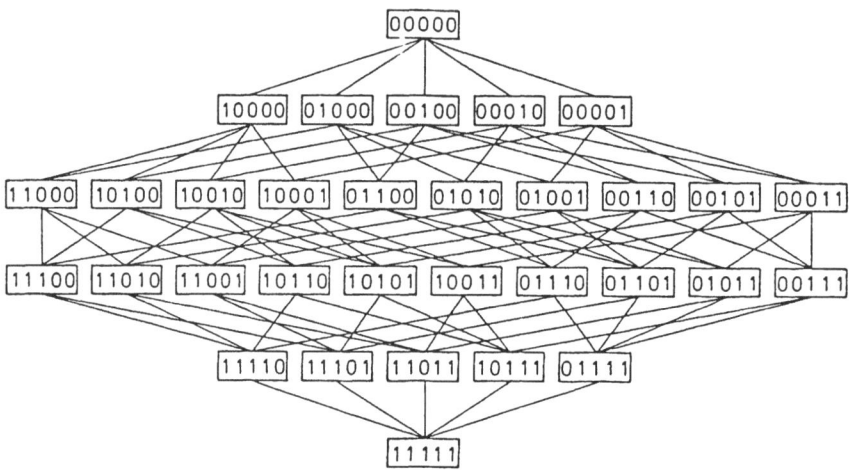

Fig. 1: The sequence space of all binary sequences with chain length ν = 5. All pairs of sequences which differ only in a single symbol - these are the sequences with Hamming distance d = 1 - are connected by straight lines. The graph obtained therby is a hypercube of dimension ν

Using the concept of Hamming distance we can order sequences and introduce a topology into the sequence space (Fig. 1). Instead of natural polyribonucleotide sequences which are based on the four-letter alphabet (G,A,C,U) we use binary sequences derived from only two symbols (0,1) in Fig. 1. This is no loss in generality, since we can represent every (G,A,C,U)-sequence of chain length ν by a binary sequence of length 2ν. The sequence space of natural four-letter sequences is a rather sophisticated object and therefore, binary sequences are much better suited for qualitative studies of general principles.

Let us first consider the evaluation of polynucleotide sequences through replication and selection. This evaluation can be formalized by means of a "value landscape" which assigns a selective value to every point of the sequence space. In all realistic systems the evaluation of a sequence is done in two steps. In the first step the sequence, commonly denoted as "genotype" in the language of genetics, is unfolded to yield the "phenotype". Fitness, as we mentioned above, is a property of the phenotypes. Mutations change the genotype; their effect on fitness is evaluated only after the phenotype had been developed. In Table 1 we show several examples of genotype-phenotype relations taken from *in vivo* as well as from *in vitro* evolution. All these examples are by far too complex in order to allow serious estimates of the corresponding value landscapes with our present knowledge. In order to illustrate the problem we make use of a rather simple computer model of a value landscape (section 4).

It is essential for the evolutionary process that the relation between genotype and phenotype is "nontrivial". This means that there is no easy way to predict phenotypic properties from the known polynucleotide sequence of the genotype. Properties and features of the phenotype are strongly coupled and hence clear-cut one-to-one relations between genes and phenotypic features are very rare exceptions.

Let us now consider a population in sequence space. The area it covers depends on population size and mutation rates. It drifts through sequence space by formation of new and loss of old, mostly less efficient variants. To account for this scenario a diffusion model of evolution has been developed [19]. The sequence space is replaced by a continuum of genotypes and population dynamics is described by a diffusion equation. In general, analytical solutions are available

Table 1: The mapping of genotypes in value landscapes

System	Genotype		Phenotype		Selective Value
		Folding			
Molecular	*RNA − Sequence*	$\Longrightarrow$	*Tert.Structure* $\Longrightarrow$		*Net Replication Rate*
		Cell Cycle			
Unicellular	*DNA − Sequence*	$\Longrightarrow$	*Cell*	$\Longrightarrow$	*Cell Division Rate*
		Development			
Multicellular	*Zygote*	$\Longrightarrow$	*Organism*	$\Longrightarrow$	*Fitness*

only for the degenerate case of selectively neutral variants. This case corresponds to the trivial value landscape of a multidimensional horizontal plane. Then the population diffuses freely on this hyperplane according to the laws of random drift. Free diffusion, in essence, is an appropriate model of "neutral evolution". Of course, it is unable to account for evolution properly, which has to include optimization and adaptation, and thus requires complicated value landscapes as discussed in section 4.

3. Selection Potentials

As an example of selection dynamics we consider competition of replicating RNA molecules. We make use of the available data on the mechanism of RNA replication and derive a simple but general reaction network which describes replication, mutation and degradation:

$$A + I_k \xrightarrow{A_k \cdot Q_{kk}} 2I_k \quad k = 1,\ldots,n \tag{3}$$

$$A + I_k \xrightarrow{A_k \cdot Q_{jk}} I_j + I_k \quad j,k = 1,\ldots,n \tag{4}$$

$$I_k \xrightarrow{D_k} B \quad k = 1,\ldots,n \ . \tag{5}$$

Individual RNA molecules represent the types and we denote them by I_k, k = 1,...,n. As we mentioned in section 2 the numbers of different molecular species, n, may be "hyperastronomically" large. Low molecular weight material which is consumed during RNA synthesis is symbolized by A. Similarly we use B as a symbol for the degradation products.

The rates of replication and mutation depend on the concentration of template $[I_k]$ = x_k, on the concentration of low molecular weight material $[A]$ = a, on the mutation probability $-Q_{jk}$ for the process $I_k \rightarrow I_j$ - and on a "rate function" $A_k(\vec{x})$. Since we consider here only systems in constant environment the dependence on low molecular weight material is just a constant factor which can be incorporated into the rate function A_k. Degradation usually is a much simpler process than replication and hence the rate function $D_k(\vec{x})$ is described well by a constant: $D_k(\vec{x})$ = d_k.

The probabilities of error-free replication and mutation are described by means of a stochastic matrix

$$Q = \{Q_{jk}\}: \quad \sum_{j=1}^{n} Q_{jk} = 1 \ . \tag{6}$$

In the limit of error-free replication the mutation matrix Q converges to the unit matrix: $Q_{jk} \to \delta_{jk}$.

In order to compensate for molecules produced in excess an unspecific dilution flux $\Phi(t)$ is introduced whose contribution to the instantaneous change in the concentration of type I_k is given by

$$-x_k.\Phi(t)/\sum_{j=1}^{n} x_j.$$

Following the principles of mass action kinetics we expand the rate function A_k as a polynomial in the concentrations x_k:

$$A_k(\overrightarrow{x}) = a_k + \sum_{j=1}^{n} a_{kj}x_j + \dots ; \quad k = 1,\dots,n . \tag{7}$$

Here we shall deal with two cases only which are characterized by

(1) constant replication rates, $A_k = a_k$ and

(2) replication rates which depend linearly on polynucleotide concentrations,

$$A_k(\overrightarrow{x}) = \sum_{j=1}^{n} a_{kj}x_j.$$

Case (1) describes independently replicating and mutating polynucleotide molecules. It is representative for SPIEGELMAN's test tube evolution experiments [20]. The physical properties of "case(1)-systems" are well understood by now. The mathematical analysis is straightforward, since the problem can be transformed into a linear differential equation, which is then studied by conventional techniques. We shall characterize case(1)-systems as "quasilinear replication mutation systems". A recent summary can be found in [21].

Case (2) is more involved, since it requires twofold action of RNA molecules: template action as in case (1) and, in addition, catalytic action on RNA synthesis. RNA catalysis on RNA processing has been found in nature [22]. Reaction networks which include catalytic steps are of actual interest therefore. But little is known on the kinetic details of RNA catalysis at present and we study here only some general examples of "second order autocatalysis".

The differential equation corresponding to "case(2)-systems" plays an important role also in several other fields like population genetics or theoretical ecology [23]. The mathematical analysis of case(2)-systems, nevertheless, is rather sophisticated. Despite general interest, only systems without mutation terms and only special cases with certain restrictions to the rate constants a_{jk} were analysed so far.

3.1. The Selection Potential of the Quasilinear System

The kinetic equation of the reaction mechanism (3-5) for independent replication, $A_k = a_k$, is of the form

$$\frac{dx_k}{dt} = \sum_{j=1}^{n} W_{kj}x_j - x_k.\Phi(t)\bigg/\sum_{j=1}^{n} x_j ; \quad k = 1,\dots,n . \tag{8}$$

Here a "value matrix" W is introduced. Its elements

$$W_{kj} = a_j.Q_{kj} - d_k.\delta_{kj} \tag{9}$$

determine the dynamics of the system. We called the diagonal elements of the value matrix, $W_{kk} = a_k \cdot Q_{kk} - d_k$, the "selective values", since they represent the most important contribution to the fitness of the individual types I_k.

The dilution flux $\Phi(t)$ controls the total concentration of polynucleotides and compensates the total excess or net production of the system. We define a mean net production

$$\bar{F}(t) = \Phi(t) \Big/ \sum_{k=1}^{n} x_k = \sum_{k=1}^{n} f_k x_k \Big/ \sum_{k=1}^{n} x_k \; , \tag{10}$$

which is the expectation value of the net productions $f_k = a_k - d_k$ of individual sequences. Without losing generality we redefine the variables x_k as normalized or relative concentrations:

$$x_k \longrightarrow x_k \Big/ \sum_{k=1}^{n} x_k \quad \text{and hence} \quad \sum_{k=1}^{n} x_k = 1 \; . \tag{11}$$

Let us now first consider error-free copying which implies $Q_{jk} = \delta_{jk}$. The corresponding differential equation

$$\frac{dx_k}{dt} = \Big(W_{kk} - \bar{F}(t)\Big)x_k = \Big(f_k - \bar{F}(t)\Big)x_k \; ; \quad k = 1, \ldots, n \tag{8a}$$

can be solved analytically. It is verified straight away that the mean net production $\bar{F}(t)$, or the dilution flux $\Phi(t)$, are non-decreasing functions of time.

Although $\bar{F}(t)$ is a Lyapunov function of (8a), this differential equation is no gradient system in ordinary Euclidean space: the trajectories do not cross the constant level sets of $\bar{F}(t)$ at right angles. It can be formulated, nevertheless, as a generalized gradient, if the metric underlying the definition of the "grad" operation is changed [24-27]: for the "Shahshahani gradient" denoted by "Grad" the Euclidean metric based on the conventional scalar product

$$\langle \vec{x} \, | \, \vec{y} \rangle = \sum_{k=1}^{n} x_k \cdot y_k$$

is replaced by a Riemannian metric using the scalar product

$$[\vec{x}, \vec{y}]_{\vec{z}} = \sum_{k=1}^{n} \frac{1}{z_k} x_k \cdot y_k \; .$$

This metric changes the definition of the angle in such a way that trajectories of the differential equation

$$\frac{d\vec{x}}{dt} = \text{Grad} \, V(\vec{x})$$

are perpendicular to the constant level sets of the potential function

$$V(\vec{x}) = \Phi = \sum_{k=1}^{n} f_k \cdot x_k \; . \tag{12}$$

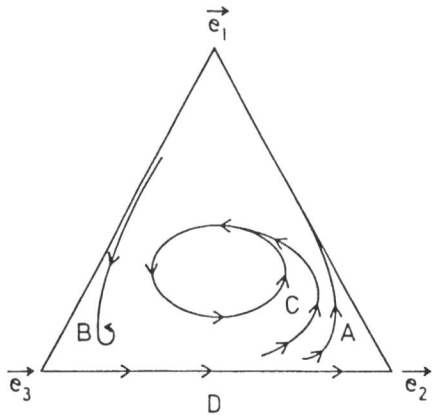

Fig. 2: Some properties of the solutions of replicator equations on
the simplex S_3. Either the trajectories approach the boundary (A) or
they stay in the interior having
equilibrium points (B) or closed
orbits (C) as ω-limits. Trajectories
starting on the boundary remain
there (D)

Equation (8a) is a so-called "replicator equation" [23]

$$\frac{dx_k}{dt} \;=\; x_k.G_k(\vec{x}) \;; \quad k = 1,\ldots,n$$

with $G_k(\vec{x})$ being of polynomial type. It fulfils the condition

$$x_k = 0 \;\longrightarrow\; \frac{dx_k}{dt} - 0 \;\Longrightarrow\; \frac{d^2 x_k}{dt^2} = 0 \;\Longrightarrow\; \ldots \;; \quad k = 1,\ldots,n\;. \tag{13}$$

As a consequence of (13) the variables $x_k(t)$ do not change sign. Therefore the
trajectories $\vec{x}(t) = (x_1(t),\ldots,x_n(t))$ stay either in the interior of the physically acceptable range of variables, the concentration simplex S_n

$$S_n \;\doteq\; \left\{ x_k \geq 0 \;\forall\; k = 1,\ldots,n \;; \quad \sum_{k=1}^{n} x_k = 1 \right\} \tag{14}$$

or converge to the boundary which, however, is never crossed (Fig. 2).

Let us combine the properties of the gradient system with those of the replicator equation. The potential function $V(\vec{x})$ is linear and hence "hill climbing"
cannot approach a maximum of the potential at finite values of x_k. Instead, the
system converges towards the highest point on the intersection of the physically
accessible part of the concentration space, which is the simplex S_n, and the potential function $V(\vec{x})$. This point is usually mapped onto a corner of S_n and such
an optimal state is characterized as a "corner equilibrium" (Fig. 3). Generally,
we observe selection of the type I_m with the largest net production:

$$I_m : \quad f_m \;=\; \max(f_1,\ldots,f_n)_. \tag{15a}$$

The population becomes homogeneous during the selection process:

$$\lim_{t \to \infty} x_m(t) = 1 \quad \text{and} \quad \lim_{t \to \infty} x_k(t) = 0 \;\forall\; k \neq m\;. \tag{15b}$$

Optimization of the mean net production in the error-free case provides the
scenario for the rare mutation limit of the replication-mutation system. Evolu-

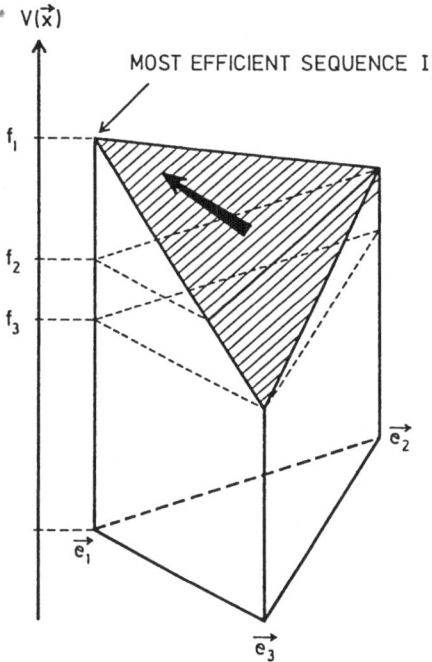

V($\vec{x}$)

MOST EFFICIENT SEQUENCE I₁

f_1

f_2

f_3

$\vec{e}_2$

$\vec{e}_1$

$\vec{e}_3$

Fig. 3: Evolutionary optimization and corner equilibria in error-free replication. The corner equilibrium ($\vec{e}_1$) corresponds to the most efficient sequence, here I_1. Optimization of the potential function $V(\vec{x})$ on S_3 leads to selection of the sequence I_1, because f_1 is the largest of all f-values

tionary adaptation is to be understood then as a succession of selection processes leading to corner equilibria. New, more efficient variants are produced once in a while and replace previously optimal types. The sequence of selected species is characterized by a non-decreasing series of net replication efficiencies f_m.

Let us now turn to the frequent mutation scenario. The dynamics of replication and mutation is described by the differential equation (8). Since the variables x_k are coupled by linear terms and by the nonlinear flux Φ, analytical solutions cannot be obtained straightaway. After removal of the flux by a nonlinear transformation [28,29] the remaining linear equation can be decoupled by standard techniques. The differential equation is transformed and the right eigenvectors $\vec{\ell}_k$ ($k = 1,\ldots,n$) of the value matrix W appear as basis vectors of the new coordinate system. We order the corresponding eigenvalues λ_k according to decreasing absolute values:

$$W.\vec{\ell}_k = \lambda_k.\vec{\ell}_k; \quad k = 1,\ldots,n \quad \text{and} \quad \lambda_1 > |\lambda_2| \le |\lambda_3| \le \ldots \le |\lambda_n| . \qquad (16)$$

Frobenius theorem applies to this case, since all entries of W are positive. Accordingly, the largest eigenvalue λ_1 is non-degenerate and the dominant eigenvector $\vec{\ell}_1$ has exclusively positive components. It represents the stationary distribution of types and lies always in the interior of the concentration simplex S_n. We called it the "quasispecies" [6] in analogy to the notion of species in biology. All other eigenvectors of the value matrix W, $\vec{\ell}_k(k>1)$, lie outside the concentration simplex S_n and thus do not represent physically accessible states of the system.

In order to visualize the evolution of the system described by (8) we consider the transformation of variables in detail:

$$\vec{x}(t) = \sum_{k=1}^{n} x_k(t).\vec{e}_k = \sum_{k=1}^{n} u_k(t).\vec{\ell}_k . \qquad (17)$$

157

The unit vectors in ordinary concentration space are denoted by $\vec{e}_k$. They represent the corners of the concentration simplex. The new variables $u_k(t)$ appear as coefficients of the eigenvectors $\vec{\ell}_k$ which were chosen as basisvectors of the new coordinate system. It is easily verified that the new concentration variables are determined by the differential equation

$$\frac{du_k}{dt} = \left(\lambda_k - \Phi(t)\right) u_k ; \quad k = 1,\ldots,n. \tag{8b}$$

Although (8b) is formally identical with (8a) - it is a replicator equation and therefore (13) holds - there are substantial differences in detail as far as the physical properties of its solutions are concerned.

(1) The variables x_k were relative concentrations and hence non-negative quantities by definition. In contrast, the variables u_k may be positive, zero or negative.

(2) The constants f_k were rate constants of chemical reactions and therefore positive quantities. The λ_k's are the eigenvalues of an unsymmetric matrix and hence may be either real or appear in complex pairs.

The matrix W has exclusively real eigenvalues if the mutation matrix is symmetric [30]: $Q_{ik} = Q_{ki}$. This condition is often fulfilled and therefore we consider only systems with real eigenvalues λ_k here.

Selection is observed in the new variables $u_k(t)$. Starting from some combination of eigenvalues at time $t=0$, $\vec{u}(0)=(u_1(0),\ldots,u_n(0))$, the system converges to the quasispecies, which is the stationary distribution determined by the dominant eigenvector $\vec{\ell}_1$:

$$\lim_{t\to\infty} u_1(t) = 1 \quad \text{and} \quad \lim_{t\to\infty} u_k(t) = 0 \;\forall\; k = 2,3,\ldots,n. \tag{15c}$$

With respect to optimization the behaviour of (8) is more sophisticated. The invariant boundaries of the dynamical system (8b) - $u_k = 0$, $k = 1,\ldots,n$ and $u_j = 0 \;\forall\; j \neq k$ - do not coincide with the boundaries of the physically accessible domain which is the simplex S_n. A three-dimensional example ($n = 3$) is shown in Fig. 4. The intersection of the two triangles shown there splits the simplex S_3 into four regions which are characterized by different signs of the variables u_k ($k=1,2,3$). The trajectories of the dynamical system (8) are confined to the orthant to which they were assigned by the choice of initial conditions, since the variables $u_k(t)$ do not change sign. General results, which are valid independently of dimension n, were derived for two regions [31,32]:

(1) The mean excess production is non-decreasing ($\frac{dV}{dt} \geq 0$) in the positive orthant, $u_k > 0 \;\forall\; k = 1,\ldots,n$.

(2) The mean excess production is non-increasing ($\frac{dV}{dt} \leq 0$) in the orthant defined by $u_1 > 1$ and $u_k < 0 \;\forall\; k = 1,\ldots,n$.

In the remaining regions no general behaviour can be predicted. The mean excess production may increase, decrease or even pass through an extremum.

3.2. The Selection Potential of Second Order Autocatalytic Systems

The most general example of a case(2)-system according to (7) is characterized by a replication rate which consists of a second order polynomial in polynucleotide concentrations and we shall call it therefore the "second order autocatalytic system" here:

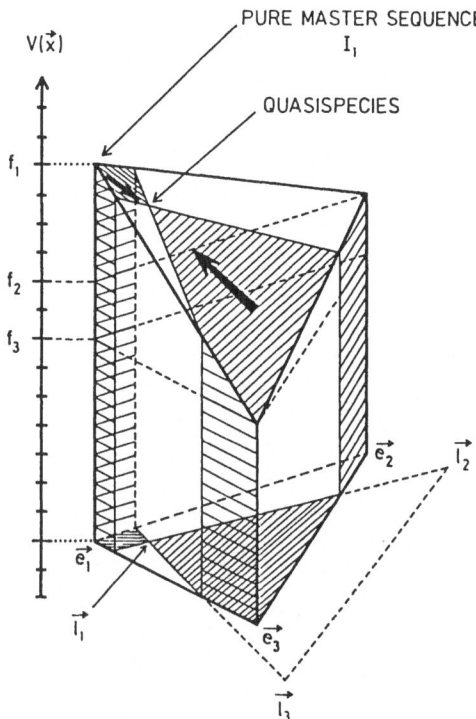

V($\vec{x}$)

PURE MASTER SEQUENCE
I_1

QUASISPECIES

f_1

f_2

f_3

$\vec{e}_2$

$\vec{l}_2$

$\vec{e}_1$

$\vec{l}_1$

$\vec{e}_3$

$\vec{l}_3$

Fig. 4: Evolutionary optimization and corner equilibria in the frequent mutation scenario. The corner equilibrium corresponds to the quasispecies, the dominant eigenvector $\vec{l}_1$ of the value matrix W. Optimization of $V(\vec{x})$ on S_3 is restricted to the horizontally hatched area. Starting from a point in the neighbourhood of the corner $\vec{e}_1$ the potential $V(\vec{x}(t))$ decreases monotonously with time. In the two remaining areas the potential may decrease or increase

$$\frac{dx_k}{dt} = \left(\sum_{j=1}^{n} a_{kj}.x_j - \Phi(t) \right) x_k ; \quad k = 1, \ldots, n. \tag{18}$$

In order to facilitate the analysis equal degradation rate constants, $d_1 = d_2 = \ldots = d_n = d$, are assumed. The constant d is then compensated by the dilution flux. Equation (18) is a replicator equation.

Here we consider only two simple special cases which form generalized gradient systems, when Shahshahani's definition of the gradient, "Grad" is applied [24]:

(1) the multidimensional Schlögl model whose matrix of rate coefficents, A, is diagonal: $a_{kj} = b_k.\delta_{kj}$ and

(2) Fisher's selection equation of population genetics which uses a symmetric matrix A: $a_{kj} = a_{jk}$.

The multidimensional Schlögl model

$$\frac{dx_k}{dt} = \left(b_k.x_k - \Phi(t) \right).x_k ; \quad k = 1, \ldots, n \tag{18a}$$

and its qualititave behaviour were discussed extensively in previous papers

[6,27,31]. We use it here as an illustrative example of a nonlinear replicator equation for which a potential function exists:

$$V(\vec{x}) = \Phi = \sum_{k=1}^{n} b_k \cdot x_k^2 \bigg/ \sum_{k=1}^{n} x_k \ .\tag{19}$$

The properties of the dynamical system can be predicted in full generality from (19). The potential function has an extremum at the equilibrium point

$$\bar{x} = \left(\bar{x}_k = b_k^{-1} \bigg/ \sum_{j=1}^{n} b_j^{-1}; \ k = 1,\ldots,n \right)$$

which lies in the interior of the concentration simplex S_n. This equilibrium point is always unstable because the Hessian matrix has exclusively positive eigenvalues. Consequently, all trajectories converge to the boundary of S_n. There we find n stable equilibrium points which coincide with the corners of the simplex. Apart from special cases with probabilities of measure zero all trajectories starting in the interior of S_n converge asymptotically to one of the corners of the simplex, where one of the species initially present is selected. Every corner of S_n thus represents a corner equilibrium (Fig. 5). We observe a case of multiple stable stationary states in the multidimensional Schlögl model.

The outcome of the selection process is not unique: it depends not only on the rate constants b_k but also on the initial condition, $\vec{x}(0)$. Every asymptotically stable state has its own basin of attraction. The sizes of these basins are

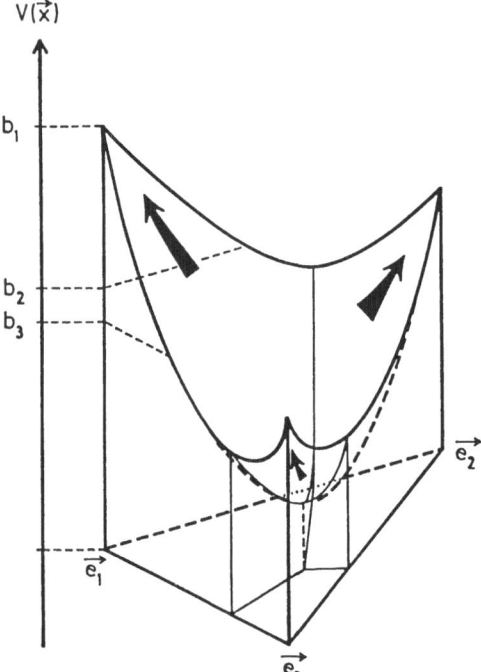

Fig. 5: Evolutionary optimization and corner equilibria in the multidimensional Schlögl model. Every corner of the simplex S_3 represents a corner equilibrium. The basins of attraction are the larger, the higher the value of the rate constants b_k are

determined by the rate constants b_k. From the coordinates of the unstable equilibrium point in S_n follows that the species with the largest rate constant $b_m = max(b_1,...,b_n)$, has the largest basin of attraction.

The dynamics of Fisher's selection equation of population genetics is more complicated. It represents, nevertheless, a generalized Shahshahani gradient system with the potential function $(a_{kj} = a_{jk})$:

$$V(\vec{x}) = \Phi = \sum_{k=1}^{n}\sum_{j=1}^{n} a_{kj}.x_k.x_j \bigg/ \sum_{k=1}^{n} x_k \quad . \tag{20}$$

The extremum of the potential may lie in the interior of the simplex S_n or outside. It may be a maximum or a minimum. This yields a whole collection of different scenarios: the selection equation may sustain one stable equilibrium or multiple stable stationary states which eventually are corner equilibria. We dispense here with details and refer to the literature on theoretical population genetics [33].

It is possible to derive a more general condition for Shahshahani gradients than that of a symmetric matrix of rate constants. The "triangle symmetry relation"

$$a_{ij} + a_{jk} + a_{ki} = a_{ik} + a_{kj} + a_{ji} \tag{21}$$

was shown to be a necessary and sufficient condition for (18) to be a Shahshahani gradient [25,26].

Other special cases of (18) like the hypercycle equation - $a_{kj} = f_k.\delta_{j,k-1}$; $j,k = 1,...,n$; j,k mod n [6] - cannot be transformed into generalized gradient systems. Hypercycles are "permanent": this means that no type which belongs to the cyclic, catalytic reaction network can disappear in the limit $t \to \infty$. All concentrations x_k $(k = 1,...,n)$ remain finite. Hypercycles in small dimensions $(n \leq 4)$ have a unique asymptotically stable fixed point inside the simplex S_n which is the attractor of all trajectories starting in the interior of S_n. Hypercycles in higher dimensions $(n \geq 5)$ sustain stable limit cycles. No chaotic dynamics was observed with hypercycles.

The general equation (18) may lead to undamped oscillations for $n \geq 3$. A chaotic attractor was found by numerical integration for a system with $n = 4$ [34].

In Table 2 we present a summary of the dynamics of second order autocatalytic systems.

4. Value Landscapes

In order to derive a coherent view of the evolutionary process we introduce here one illustrative example of a value landscape which was obtained by computer calculations [35]. The computational procedure is outlined in Table 3. Binary sequences of length $\nu = 70$ are chosen and unknotted two-dimensional structures are computed by means of a standard folding algorithm [36]. Folding, in essence, is based on a thermodynamic minimum free energy criterion. The parameters for base pairing and base stacking were taken from all (G,C) polyribonucleotides [37].

The rate constants d_k and a_k are obtained from the folding patterns according to certain rules which reflect the known properties of small RNA molecules. The rate constants of hydrolytic degradation, d_k, are assumed to increase with increasing numbers of unpaired bases. Thus, d_k is large when the thermodynamic sta-

Table 2: Dynamics of second order autocatalytic systems as described by (18)

System	Rate Coefficients	Long Time Behaviour[*]	Selection Potential
Schloegl Model	$a_{kj} = b_k . \delta_{jk}$	*Selection* *Stable Fixed Points*	$V(\vec{x})$, see (19)
Fisher's Equation	$a_{kj} = a_{jk}$	*Selection or Permanence* *Stable Fixed Point(s)*	$V(\vec{x})$, see (20)
Triangle Symmetry	$a_{ij} + a_{jk} + a_{ki} =$ $= a_{ik} + a_{kj} + a_{ji}$	*Selection or Permanence* *Stable Fixed Point(s)*	$V(\vec{x})$, see (20)
Hypercycle Equation	$a_{kj} = f_k . \delta_{j,k-1}$ $n \leq 4$ $n \geq 5$	*Permanence* *Stable Fixed Point* *Stable Limit Cycle*	*No*
General Case	a_{kj} $n \geq 3$ $n \geq 4$	*Stable Fixed Point(s)* *Stable Limit Cycle(s)* *Chaotic Dynamics*	*No*

(*) Selection and permanence are notions which characterize long-time behaviour of dynamical systems. Selection occurs, if a corner of the concentration simplex is the ω-limit of all trajectories starting from a certain area in the interior of the simplex S_n. All types I_k except the one which is selected go extinct. A dynamical system is permanent if no concentration x_k vanishes. All trajectories starting inside the concentration simplex remain in the interior also in the limit $t \to \infty$. No type which was initially present in the system goes extinct.

bility of a folding pattern is small and vice versa. The "perfect hairpin" - this is the folding pattern with the maximum number of base pairs and optimal stacking - is most resistant to hydrolysis. It is known from RNA replication that only single strands are accepted as templates by the enzyme. Secondary structures of the tem-plate are broken as polymerization proceeds. In our model the replication rate constant a_k, in essence, accounts for the melting of secondary structures. An optimally replicating sequence thus is one which is totally unable to form a secondary structure like the "all 0" or the "all 1" strings, 00...0 and 11...1, respectively. These are just the least stable sequences! Although we do not intend to make the claim that our model is able to predict stabilities and rate constants with quantitatively useful precision, we are convinced, nevertheless, that the most important qualitative features like the complex nature of genotype-phenotype relations are reproduced correctly by the computer simulation.

High net production, $f_k = a_k - d_k$, requires large replication rate constants and small degradation rate constants. In terms of folding patterns these two demands counteract each other. Evolutionary optimization has to find a compromise between high stability and fast replication. The value matrix W is obtained from the rate constants a_k and d_k as well as from a properly chosen mutation matrix Q for binary sequences. Such a matrix which needs only the chain length ν and the "single digit accuracy" q as input parameters was proposed recently [38]:

$$W_{kj} = a_j . q^\nu \left(\frac{1-q}{q} \right)^{d(j,k)} - d_k . \delta_{kj} . \tag{9a}$$

Here, we use $d(j,k)$ for the Hamming distance of the sequences I_j and I_k. Selective values are computed straightaway since $Q_{kk} = q^v$.

Replication and mutation of an ensemble of $N = 3000$ sequences in a flow reactor was simulated as a stochastic process on the computer [35]. The reactor is run under conditions which keep the population size roughly constant within the limits $N \pm \sqrt{N}$. The simulations start from homogeneous populations and are continued until the net production $\bar{F}(t)$ reaches an approximately constant value near the optimum. Different starting sequences and different single digit accuracies q were applied in different runs. Here only a few selected results are presented - for further details see [8] and [35].

Table 3: A computer model of optimization on a value landscape

System	Realization	Process
Digit	$\{0,1\}$	
	$\Downarrow$	*Ligation to String*
Binary Sequence String of $v = 70$ Digits	$\{2^{70} \approx 10^{21} \text{ Sequences}\}$	
Primary Structure	I_k : $011000110100\ldots0$	
	$\Downarrow$	*Folding through Base Pairing*
Secondary Structure		
	$\Downarrow$	*Evaluation*
Molecular Properties	a_k , d_k	
Population $N = 3000$ Sequences	$\left\{\left(10^{21}\right)^{3000} \text{Distributions}\right\}$	
Starting Distribution	$X_k(0);\ k = 1, 2, \ldots, n$	
	$\Downarrow$	*Evolutionary Optimization*
Near Optimal Distribution	$\lim_{t\to\infty} \langle X_k(t)\rangle;\ k = 1, 2, \ldots, n$	

First we summarize some properties of the value landscape:

(1) Nearby sequences - these are sequences with small Hamming distances, $d = 1$ or 2 - sometimes have very different folding patterns. The secondary structure is sensitive to small changes in the genotype.

(2) Selective values in the neighbourhood of an arbitrarily chosen typical sequence show vast scatter. At Hamming distance $d = 1$ we find a few lethal variants as well as some sequences with selective values larger than the average.

In order to study the distribution of rate constants and selective values in sequence space we chose three representative samples of 76000 random binary sequences each, with different ratios of 0|1-digits, 0.286, 0.5 and 0.714, respectively. The qualitative results obtained were practically the same in all three cases.

(3) The distributions of thermodynamic stabilities and degradation rate constants are roughly Gaussian. The expectation values of the distributions depend on the 0|1-ratio. This dependence is interpreted easily: the perfect hairpin can be formed only if the ratio is 0.5 or very close to it. Sequences with smaller or larger ratios of digits are inevitably less stable on the average.

(4) The distribution of replication rates and, of course, also that of selective values is apparently more complicated. We observe vast scatter in the probabilities of nearby lying selective values, which is most probably a consequence of the relatively short chain lengths applied. Secondly, the distributions of selective values show bi- or multimodality. There is a fairly high percentage of lethal variants commonly in the range of a few percent. The remaining distribution is often strongly biased towards higher replication efficiency. We expect and find therefore many sequences which have "near optimal" structures. These structures appear in groups of closely related genotypes which in turn are well separated from each other in sequence space.

The computer simulations allow some conclusions on the dynamics of the evolutionary optimization process. Apart from population sizes, single digit accuracies q are of crucial importance for evolution. High precision of replication implies that only few mutants are steadily produced and therefore the repertoire of variants which are available for optimization is rather poor. The population operates close to the "rare mutation scenario": the optimum is approached in steps which are separated by long "quasistationary" periods of constant mean net production $\bar{F}(t)$. In the computer simulations performed with the highest replication accuracy applied - $q = 0.999$ - the populations never came very close to the optimum. They got stuck in local maxima of the fitness landscape.

Lowering of the single digit accuracy q speeds up evolutionary optimization. Eventually the steps disappear and the optimum is approached smoothly. There is a sharply defined critical q-value, as in the deterministic model [5,6,21,32,38], below which error propagation becomes stronger than the hereditary process. No optimization of the mean net production occurs at error rates than the critical value. The deterministic model strictly applies for infinite populations and yields a critical q-value which is smaller than that obtained by computer simulation for populations of 3000 individuals.

Value landscapes are bizarre objects - nearby genotypes often have highly different selective values - but evolving populations, nevertheless, migrate quite smoothly on them, provided they cover sufficiently large areas in sequence space. The areas populations occupy are the larger, the larger the population sizes and the smaller the single digit accuracies q are. At accuracies lower than the error threshold populations are delocalized in sequence space: they diffuse freely like in random walks. Accordingly, optimization is fastest and the population gets stuck least likely in a local fitness maximum, if it operates with an accuracy just above the critical value.

Acknowledgements

Many stimulating discussions with Prof. Karl Sigmund are gratefully acknowledged. Financial support of this work was provided by the Stiftung Volkswagenwerk (FRG) and the Austrian Fonds zur Förderung der wissenschaftlichen Forschung (Projekt No. 5286).

References

1. R.A. Fisher: <u>The Genetical Theory of Natural Selection</u> (Oxford University Press, Oxford 1930) and 2nd revised ed. (Dover Publications, New York 1958)
2. J.B.S. Haldane: <u>The Causes of Evolution</u> (Harper & Row, New York 1932)
3. S. Wright: <u>Evolution and the Genetics of Populations</u>, Vols. I-IV (The University of Chicago Press, Chicago (1968, 1969, 1977 and 1978)
4. C. Darwin: <u>The Origin of Species</u> (Everyman's Library, Vol. 811, Dent, London 1967). The famous catchphrase of "survival of the fittest" is actually not due to Charles Darwin himself. It has been attributed to H. Spencer and appeared first in the 5th edition of the <u>Origin of Species</u>
5. M. Eigen: Naturwissenschaften <u>58</u>, 465 (1971)
6. M. Eigen, P. Schuster: <u>The Hypercycle - A Principle of Natural Self-Organization</u> (Springer Verlag, Berlin 1979). The booklet is a combined reprint of three papers: Naturwissenschaften <u>64</u>, 541 (1977) and <u>65</u>, 7 and 341 (1978)
7. M. Eigen: Chemica Scripta <u>26B</u>, 13 (1986)
8. P. Schuster: Chemica Scripta <u>26B</u>, 27 (1986)
9. C.K. Biebricher, M. Eigen, W.C. Gardiner: J. Biochem. <u>22</u>, 2544 (1983)
10. C.K. Biebricher, M. Eigen, W.C. Gardiner: J. Biochem. <u>23</u>, 3186 (1984)
11. C.K. Biebricher, M. Eigen, W.C. Gardiner: J. Biochem. <u>24</u>, 6550 (1985)
12. E. Domingo, R.A. Flavell, C. Weissmann: Gene <u>1</u>, 3 (1976)
13. E. Domingo, D. Sabo, T. Taniguchi, C. Weissmann: Cell <u>13</u>, 735 (1978)
14. E. Domingo, M. Davilla, J. Ortin: Gene <u>11</u>, 333 (1980)
15. J. Ortin, R. Najero, C. Lopez, M. Davilla, E. Domingo: Gene <u>11</u>, 319 (1980)
16. S. Fields, G. Winter: Gene <u>15</u>, 207 (1981)
17. D.E. Dykhuizen, D.L. Hartl: Microbiol. Rev. <u>47</u>, 150 (1983)
18. M. Eigen: Ber. Bunsenges. Phys. Chem. <u>89</u>, 658 (1985) This concept of representing genotypes by a point space was first introduced by I. Rechenberg: <u>Evolutionsstrategie</u> (F. Frommann Verlag, Stuttgart 1973)
19. M. Kimura: <u>The Neutral Theory of Molecular Evolution</u> (Cambridge University Press, Cambridge 1983)
20. S. Spiegelmann: Quart. Rev. Biophys. <u>4</u>, 213 (1971)
21. M. Eigen, J. McCaskill, P. Schuster: J. Phys. Chem. (1987), in press
22. T.R. Cech: Science <u>236</u>, 1532 (1987)
23. P. Schuster, K. Sigmund: J. theor. Biol. <u>100</u>, 533 (1983)
24. S. Shahshahani: Memoirs Am. Math. Soc. <u>211</u> (1979)
25. K. Sigmund: In <u>Lotka-Volterra-Approach to Cooperation and Competition in Dynamic Systems</u> , ed. by W. Ebeling, M. Peschel (Akademie Verlag, Berlin 1985) p. 63
26. K. Sigmund: In <u>Dynamical Systems and Environmental Models</u>, ed. by F. Avert (Akademie Verlag, Berlin 1987)
27. P. Schuster, K. Sigmund: Ber. Bunsenges. Phys. Chem. <u>89</u>, 668 (1985)
28. C.J. Thompson, J.L. McBride: Math. Biosc. <u>21</u>, 127 (1974)
29. B.L. Jones, R.H. Enns, S.S. Rangnekar: Bull. Math. Biol. <u>38</u>, 12 (1976)
30. D. Rumschitzky: J. Math. Biol. <u>24</u>, 667 (1987)
31. P. Schuster: Physica <u>22D</u>, 100 (1986)
32. P. Schuster, J. Swetina: Stationary Mutant Distributions and Evolutionary Optimization, preprint (1987)
33. W.J. Ewens: <u>Mathematical Population Genetics</u>, Biomathematics, Vol. 9 (Springer Verlag, Berlin 1979)
34. P. Schuster: Physica Scripta <u>35</u>, 402 (1987)
35. W. Fontana, P. Schuster: Biophys. Chem. <u>26</u>, 123 (1987)
36. M. Zuker, P. Stiegler: Nucleic Acids Res. <u>9</u>, 133 (1981)
37. M. Zuker, D. Sankoff: Bull. Math. Biol. <u>46</u>, 591 (1984)
38. J. Swetina, P. Schuster: Biophys. Chem. <u>16</u>, 329 (1982)

Stochastic and Chaotic Processes
in Biochemical Systems

W. Ebeling, H. Herzel, and L. Schimansky-Geier

Sektion Physik, Humboldt-Universität Berlin,
DDR-1040 Berlin, GDR

1. Introduction

Nonlinear networks of biochemical reactions play a very important role in the regulatory circuits of a living cell [1,2]. Benno Hess and his school played a pioneering role in the exploration of the complicated time patterns which may appear in such networks [3-7]. It was shown that among regular also chaotic and stochastic patterns may be observed [4,5].

In this paper we discuss some peculiarities of complex motions in biochemical systems. Especially we want to study the effect of nonuniformity of the dynamics which is a typical property of biochemical systems on the development of chaos. Furthermore, the effect of noise on the system will be studied.

Special attention will be devoted to Sel'kov-type models which describe regulation mechanisms of biochemical systems [8,9]. The influence of fluctuations will be investigated for the two-dimensional system (1). It is described by the kinetic equations for the densities x, y:

$$dx/dt = \Phi - Bx - xy^2 + g(x,y)\ \theta(t)$$
$$dy/dt = A\ (xy^2 - y) + h(x,y)\ \Omega(t)$$

$$; \qquad \Phi \rightleftharpoons X \xrightarrow{\oplus} Y \rightarrow \qquad (1)$$

Here the terms $g\ \theta(t)$ and $h\ \Omega(t)$ model the influence of noise. $\theta(t)$ and $\Omega(t)$ are random variables with zero mean. We have to note that in biochemical systems with dissipation, noise is necessarily present due to the validity of fluctuation-dissipation relations and due to the fact that the real particle numbers are always integers (shot noise). Following Onsager, all processes which produce entropy are necessarily connected with fluctuations. Besides thermal noise and shot noise, a fluctuating in- and output and "hidden" reactions are possible sources of random perturbations. In other words, strictly speaking, models which do not take into account the influence of noise are of mathematical interest only. Any physical model of biochemical processes should obligatorily include fluctuations.

We shall study in sections 3 and 4 the influence of noise in some detail. Simple approximations for the natural noise will also be derived there. Further it is shown by calculating the stationary probability distribution, that even additive noise modifies the Hopf bifurcation significantly [10]. Due to the coexistence of a limit cycle and a stable node in system (1) a drastic amplification of fluctuations may appear. An extended model containing in addition to (1) a reversible deposition of x into an inactive form z exhibits deterministic chaos [11]:

$$dx/dt = \Phi - Bx - xy^2 - Exy + z + g\ \theta(t)$$
$$dy/dt = A\ (xy^2 - y + D) + h\ \Omega(t)$$
$$dz/dt = F\ (Exy - z)$$

$$; \qquad (2)$$

Chaotic dynamics can be characterized by an exponential instability of trajectories in the long-term average. However, studying the actual growth of perturbations large deviations from the average behavior are observed which we refer to as "nonuniform dynamics" [12-16]. In section 2 we introduce and review some quantities describing nonuniformity. Moreover, we suggest that strong nonuniformity might be a typical property of chemical systems.

Section 4 is devoted to effects of fluctuations on chaotic dynamics. It turns out that typically essential properties of chaos persist in the presence of noise. However, it is shown that the separation of nearby trajectories is significantly modified due to noise.

2. Peculiarities of Biochemical Chaos

The phase space coordinates of chemical systems are restricted to the positive conus since negative concentration values make no sense (in contrast to mechanical or electronic devices). This restriction often leads to a triangular shape of limit cycles [17].

Another peculiarity of temporal self-organization in chemical systems is a permanent change of the dominant reaction mechanism during the time. Near the axes linear reactions dominate, whereas at higher concentration values nonlinear reactions are "switched on" [17]. As a result quite different "velocities" in phase space may occur since reaction rates vary widely. Furthermore, large fluctuations of the stability properties of trajectories can be observed, termed "nonuniformity" [16]. This characteristic property of chemical systems will be discussed now in more detail.

For convenience our systems of consideration are written as follows:

$$d\underline{x}/dt = \underline{f}(\underline{x}). \tag{3}$$

Infinitesimal deviations $\underline{q}(t)$ from the solution $\underline{x}(t)$ are governed by the linearized equations

$$d\underline{q}/dt = J(\underline{x})\underline{q} \qquad \text{with} \quad J_{ij}(\underline{x}) = df_i/dx_j. \tag{4}$$

For almost all initial deviations $\underline{q}(0)$, the mean growth of the Euclidean norm $\|\underline{q}(t)\|$ gives the maximum Lyapunov exponent λ_1 [18]:

$$\lambda_1 = \lim_{t \to \infty} 1/t \, \ln\{\|\underline{q}(t)\| / \|\underline{q}(0)\| \}. \tag{5}$$

Deterministic chaos is characterized by $\lambda_1 > 0$, i.e. perturbations grow exponentially in the long-term average. Such an exponential growth is indicated in Fig. 1 by dashed lines (note the logarithmic scale). The deviations of the actual growth (full lines) from the average behavior shall be discussed. A comparison of the celebrated LORENZ model [19] (see Fig. 1a) with our biochemical system (Fig. 1b) reveals that the latter exhibits rather strong deviations from a purely exponential growth. A first implication of such nonuniformity can be understood from a comparison of Fig. 1b and Fig. 1c which correspond to chaos and a "periodic window", respectively. During some few oscillations (the characteristic period is about 10) the behavior is quite similar and, thus, the exponential instability in Fig. 1b is not essential on these time scales.

Now we briefly discuss a quantitative description of nonuniform dynamics. The quantity of central interest is the error growth rate over a finite time τ:

$$l_i(\tau) = \ln \{\|\underline{q}(t_i+\tau)\| / \|\underline{q}(t_i)\| \}. \tag{6}$$

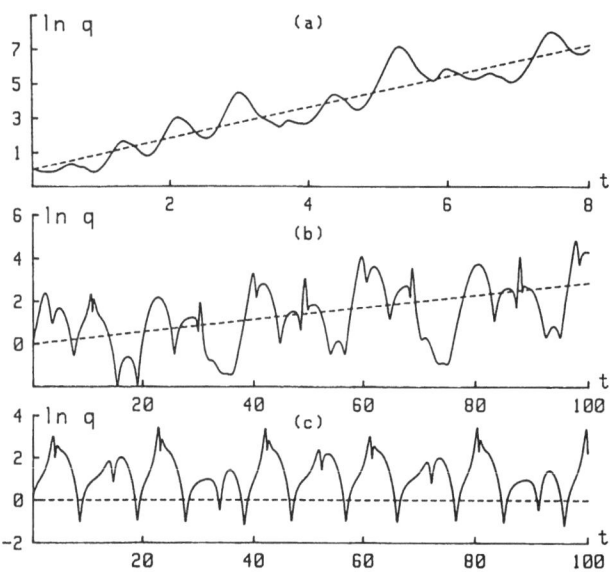

Fig. 1: Typical evolution of infinitesimal perturbations $\|q\|$ (see (4)). The dashed lines indicate the mean growth. (a) Lorenz model (parameters as in [19]); (b) Equations(2): chaotic regime (A = 4, B = 0.35, D = 0.1, E = 1.5, Φ = 1); (c) Equations (2): 6-period limit cycle (E = 1.5095)

Equation (5) implies that the mean growth is related to the maximum Lyapunov exponent λ_1:

$$\langle l_i(\tau)\rangle = \lambda_1\tau. \tag{7}$$

Here averages $\langle\cdots\rangle$ are performed as time averages over sufficiently long trajectories. The standard deviation

$$\Delta l(\tau) = \{\langle l_i^2(\tau)\rangle - \langle l_i(\tau)\rangle^2\}^{1/2} \tag{8}$$

may serve as measure of nonuniformity over an arbitrary time scale τ. In the limiting cases $\Delta l(\tau)$ is governed by previously defined quantities:

(i) $\tau \to 0$: $\Delta l(\tau) \approx NUF\cdot\tau$ (9)

               (with NUF = Non-Uniformity-Factor [12, 15])

(ii) $\tau \to \infty$: $\Delta l(\tau) \approx \sqrt{2D\cdot\tau}$ (10)

               (with D = "diffusion constant" [13, 14]).

In nonuniform chaotic systems (we assume NUF > λ_1) two time-scales can be distinguished with respect to the stability properties: on short time-scales nonuniformity dominates ($\Delta l(\tau) > \langle l_i(\tau)\rangle$) whereas on large scales the behavior is mostly governed by the Lyapunov exponent λ_1. In order to separate these scales we introduce a time τ_c such that

$$\Delta l(\tau) < \langle l_i(\tau)\rangle = \lambda_1\tau \qquad \text{for all} \quad \tau > \tau_c. \tag{11}$$

The behavior of $\Delta l(\tau)$ and the meaning of the quantities defined above are illustrated in Fig. 2. For a computation of $\Delta l(\tau)$ it is sufficient to evaluate the

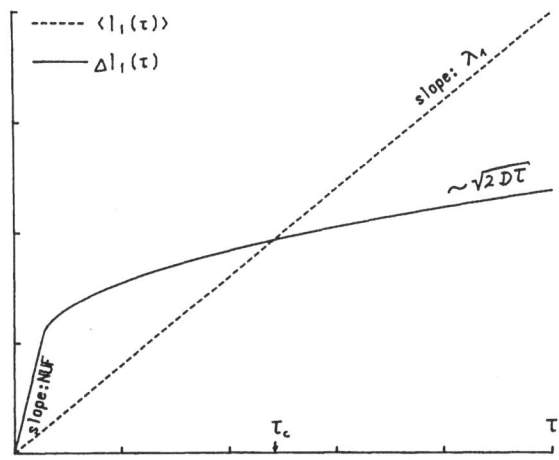

Fig. 2: Schematic illustration of measures of nonuniformity (see (6)-(11))

correlation function of a "Local Divergence Rate" [15] and to use Equation (2.9) of Ref. 14.

The reviewed quantities measuring nonuniformity were obtained for several models [12-16, 20]. It turns out that standard models (Lorenz model, Henon map, logistic map) are weakly nonuniform, whereas all considered chemical systems (Equations (2), models of OLSEN [21] and GOLDBETER [22], BZ map [20]) are extremely nonuniform. This statement is illustrated in the Figs. 3 and 4. As in Fig. 1 the growth of a small perturbation q(t) is presented during about 10 "periods" of the underlying biochemical system. For the peroxidase-oxidase reaction [21] as well as for the coupled autocatalytic enzyme reactions [22] large deviations from a purely exponential error growth occur.

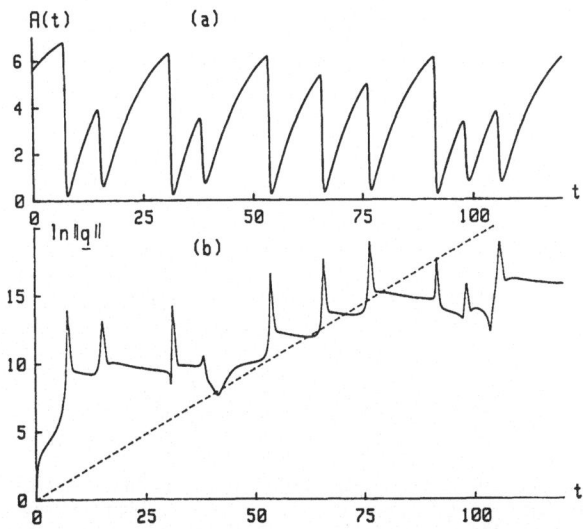

Fig. 3: Chaotic dynamics in a model of peroxidase-oxidase reaction (parameters as in Fig. 3b of Ref. [21]). (a) Chaotic oscillations of A corresponding to O. (b) Nonuniform growth of small perturbation $\|q\|$ (full line) and mean growth: $\|q(t)\| = \|q(0)\| \exp(\lambda_1 t)$ (dashed line)

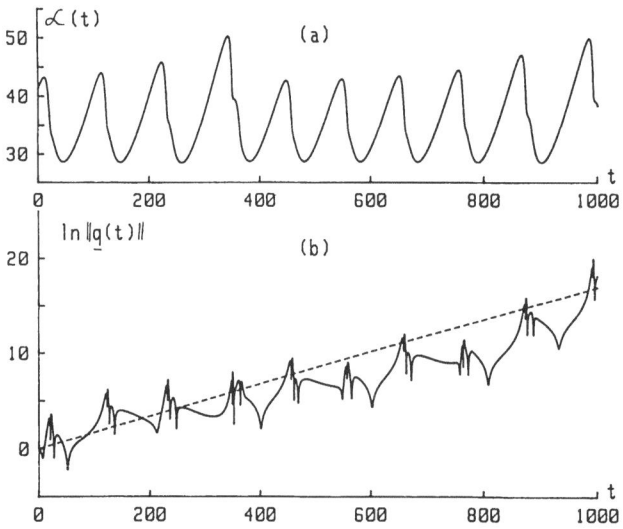

Fig. 4: Chaotic dynamics of an enzyme reaction model (parameter as in Fig. 7 of Ref. [22]). (a) Chaotic oscillations of the substrate. (b) Nonuniform growth of a perturbation $\|q\|$ governed by (4) (full line). The dashed line indicates the mean exponential growth

Now we briefly discuss peculiarities of chemical strange attractors. It was emphasized in Ref. 13 that nonuniform dynamics is intimately related to a complicated structure of the attractors ("inhomogeneous fractals"). A measure of inhomogeneity is the difference between the information dimension D_1 and the correlation exponent D_2. Whereas these dimensions nearly coincide for the Lorenz model [13, 23]

$$D_1 = 2.0584 \pm 0.0007 \quad , \quad D_2 = 2.05 \pm 0.01, \tag{12}$$

they differ significantly for our biochemical system [24] (parameters as in Fig. 1b):

$$D_1 = 2.115 \pm 0.003 \quad , \quad D_2 = 2.02 \pm 0.06. \tag{13}$$

These results confirm our assumption that biochemical attractors are typically inhomogeneous.

So far we have not studied the influence of external noise on chaos. There seems to be a relation between nonuniformity and the sensitivity to perturbations, since in nonuniform biochemical systems nontrivial effects of fluctuations were found [16, 20].

3. The Influence of Fluctuations on the Hopf-Bifurcation and Amplification of Noise

In this section effects of noise in the two-dimensional model (1) are considered. Deterministically in the two-dimensional system for B = 0, Φ = 1 a limit cycle arises at A = 1 which loses its stability around A = 1.234 [9,26]. First we discuss the stationary probability distribution near this Hopf-bifurcation. It is well known that fluctuations play an important role near bifurcation thresholds [17, 27].

The effect of shot noise on Sel'kov-type systems was studied in earlier work [28]. Here we shall study only continuous models. For such models exists a general condition on the noisy parts in (1)

$$g(x,y) = 0 \quad \text{if} \quad x = 0 \; ; \quad h(x,y) = 0 \quad \text{if} \quad y = 0. \tag{14}$$

In other words, the normal component of the noise should vanish on the boundary of the positive cone, this is to avoid negative concentrations.

The concrete form of the noise functions may be very different depending on the physical conditions and especially on the influence of noise on the parameters in the model. However, let us underline that there exists always a minimal noise strength which is called in the following natural noise. The natural noise is due to the discreteness of particle numbers and the dissipation in the system.

Let us consider the simplest case that all these random perturbations can be modelled by additive short correlated fluctuations. Thus the noisy part of (1) is assumed as Gaussian white noise. Let us give a lower estimate of the value of the noise intensity. Therefore, we consider shot noise in equilibrium neglecting the estimated contributions caused by thermal noise, by nonequilibrium effects and by perturbations from outside.

Equilibrium is reached if the in- and output of X are in detailed balance. Fluctuations are described by the equilibrium distribution

$$P(X,Y) = \delta(Y) \; X_0^X \exp (-X_0) \; / \; X!$$

with the mean equilibrium particle number

$$<X> \; = X_0 = \; \Phi/ \; B.$$

Since $<\delta X \; \delta X> = X_0$ it results for the density

$$<\delta x \; \delta x> = D_0 = \Phi \; / \; (\; V \; B \;).$$

Here V is the effective reaction volume.

Otherwise, modelling the fluctuations as the result of Gaussian white noise in (1), D_0 is just the minimal noise intensity of the noisy part in the x-direction. Also in equilibrium we see, $<Y> = Y = 0$. The y-component vanishes exactly without any fluctuating contribution. Following this consideration the simplest Langevin-equation to model fluctuations has to invoke stochastic source terms in the kinetic equation for the x-component. Therefore we introduce

$$dx/dt = 1 - xy^2 + g \; \theta(t) \; ; \quad <\theta(t)> = 0$$
$$\tag{15}$$
$$dy/dt = A \; (xy^2 - y) \quad ; \quad <\theta(t) \; \theta(t')> = 2 \; D \; \delta(t-t')$$

where g = 1 if x > 0 and g = 0 on the axes x = 0. Here D is an effective noise intensity and it holds necessarily $D > D_0$. These Langevin-equations correspond uniquely to a Fokker-Planck-equation for the probability distribution P(x,y;t). We will be interested in the stationary solution $P(x,y;t \to \infty) = P^0(x,y)$ near to the bifurcation threshold. An approximative formula was obtained locally around the singular point x = y = 1 using a solution approach for the stationary Fokker-Planck-equations with nonvanishing probability fluxes [29]. Introducing appropriate abbreviations

$$u = 1 - x + (1-y)/A \quad \text{(deviations from the}$$
$$v = y - 1 \qquad\qquad \text{focus } x = y = 1) \tag{16}$$

$$a = A - 1 \qquad\qquad \text{(bifurcation parameter)}$$

and supposing sufficiently strong noise ($a^2 < D$) the following stationary probability distribution was found:

$$P^0(x,y) \approx \exp \{2(u-v) + (a - 4D)(u^2 + v^2)/2D + 3uv/4 - (u^2 + v^2)^2/16D$$

$$+ O(D, a^2, aD ; u^5, u^4v,..., v^5; u^6,...)\}. \tag{17}$$

We discuss the topological changes of $P^0(x,y)$ for increasing a as a proper characteristic of "stochastic bifurcations" [30]. Roughly seen, a stable focus corresponds to an elliptic one-peak distribution, whereas a limit cycle corresponds
to a probability crater. However, as was pointed out [31], the situation is more
complicated in a close vicinity of the bifurcation threshold. The effect of the
fluctuations creates modified bifurcation scenarios. For mechanical oscillators
between the one-peak and crater distribution for $a \approx 0$ two-peak distributions
without deterministic counterparts were found [32]. For Sel'kov-type oscillators
we observe the following:

i) Slightly below the threshold the maximum of the probability distribution is
   shifted proportionally to $D^{1/3}$.
ii) At $a = a_c = 3 (D / 4)^{2/3}$ minimum and a saddle appear. The saddle and the maximum are located on the crest of the crater.

Thus, at a_c a skew crater is created. The bifurcation threshold is shifted. Figure 5 illustrates this scenario which is the typical picture of a stochastic
Hopf-bifurcation for chemical systems. We point out that despite the simple
additive noise the deterministic predictions are not only smeared out but significant modifications of the transition to the limit-cycle-regime occur.

Now let us investigate the effect of fluctuations in the regime of developed
oscillations (a >> 0). As can be seen from Fig. 1c, also in nonchaotic systems
contraction and expansion of small perturbation can change rapidly. Especially
near separatrices an anomalous enhancement of fluctuations is observable. Since
coexistence of attractors is of general importance in biochemical systems
[5,11,22], noise-induced jumps between coexisting attractors might be of comparable importance to chaos itself.

In Fig. 6 we present an example of such noise amplifier. Deterministically a
limit cycle and a stable node (x = 1/B; y = 0) coexist (see Fig. 6a). Due to relatively small fluctuations, which are visualized at the top of Fig. 6b, the separatrix is crossed and nonperiodic oscillations result as shown in Fig. 6b. They
can easily be confused with truly chaotic behavior. Our example contains some
warning to identify any large irregular oscillations as deterministic chaos, not
thinking about effects of fluctuations always present. Only a careful time series
analysis allows to distinguish chaos from amplified fluctuations [18,33].

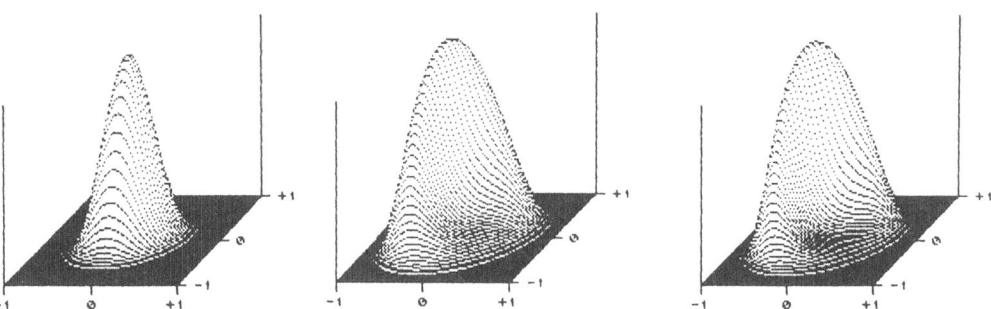

Fig. 5: Evolution of the stationary probability distribution near the Hopf-bifurcation (see (17))

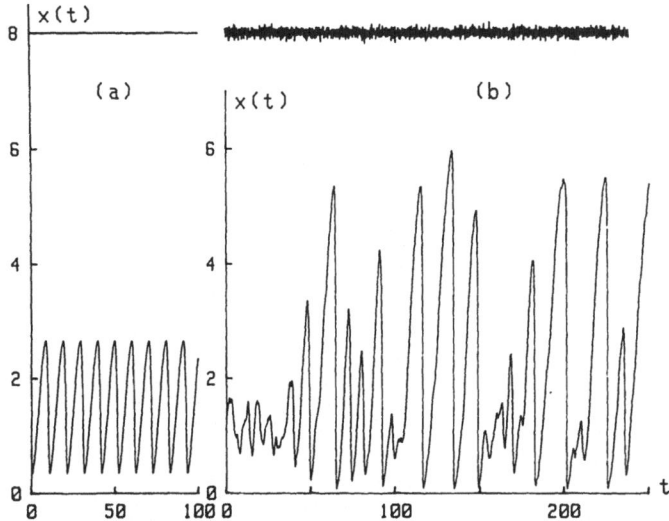

Fig. 6: Noise-induced oscillations in model (1) (A = 0.9, B = 0.125) with fluctuating source terms in both equations. (a) Coexistence of periodic and stationary behavior in the absence of fluctuations. (b) Irregular oscillations due
to noise

4. The Influence of Noise on Chaotic Dynamics

Effects of noise on chaos were subjected to intense investigations [16,18,
20,24,34], and it turned out that despite the instability of trajectories statistical measures like, e.g., the Lyapunov exponent are relatively robust. Figure 7
exemplifies this statement: the dependence of the maximum Lyapunov exponent λ_1
on the parameter E is smoothed by noise. This result implies the following effects:

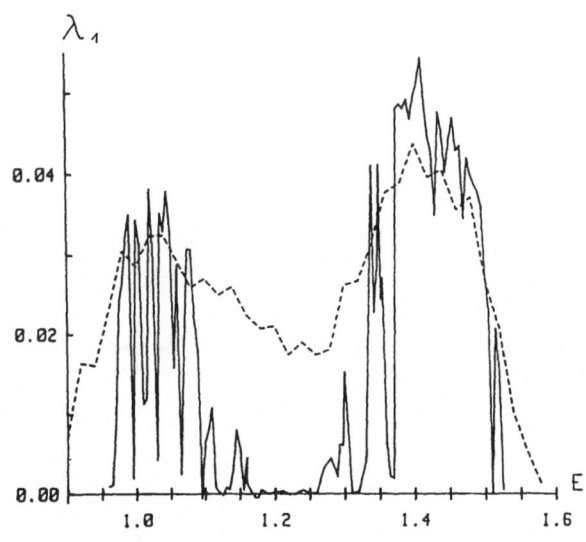

Fig. 7: Maximum Lyapunov exponent λ_1 versus parameter E.
Neighbouring points have been
joined by lines to guide the eye.
Integration times: 6000. Full
line: without noise; $\Delta E = 0.005$.
Dashed line: x(t) and y(t) disturbed by noise (D = 1.25×10^{-6});
$\Delta E = 0.02$

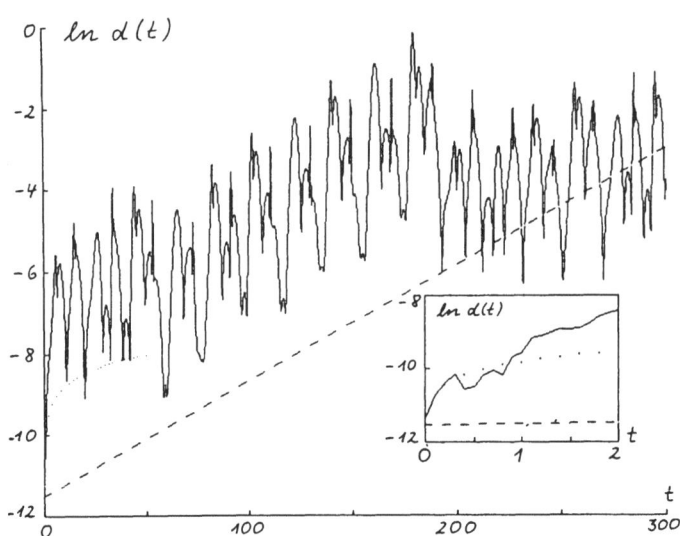

Fig. 8: Typical evolution of the separation distance d(t) between two chaotic trajectories in the presence of noise ($d(0) = 10^{-5}$; $D_x = D_y = 5 \times 10^{-10}$). Dashed line: exponential law $d(t) = d(0)\exp(\lambda_1 t)$. Dotted line: diffusive separation $d(t) = d(0) + \sqrt{2(D_x + D_y)t}$. A magnification of first time units is inserted to demonstrate the applicability of the diffusion law

  i) the thresholds are shifted,
 ii) periodic windows disappear,
iii) the orbital instability expressed by $\lambda_1 > 0$ is typically "robust" against
     small fluctuations.

    More details (especially concerning the effects of noise on next-amplitude maps) can be found elsewhere [16,24]. Here we want to discuss the meaning of a positive Lyapunov exponent in the presence of noise. As in the deterministic case the maximum Lyapunov-exponent can be computed via linearization (see (3)-(5)) [35]. However, in the presence of fluctuations the separation of nearby trajectories cannot be described by linear stability analysis only. In order to analyze the separation of real trajectories influenced by different realizations of random perturbations, it is useful to distinguish three scale regions:

  i) diffusion-like behavior up to scales comparable with the amplitude of
     the fluctuations,
 ii) exponential separation on intermediate scales,
iii) saturation on scales comparable with the attractor size.

    These considerations are illustrated in Fig. 8 where a typical separation of initially adjacent trajectories is given. Obviously, the Lyapunov-exponent λ_1 is an appropriate measure of the error growth only if the second scale region exists, i.e. for sufficiently small noise intensities.

    We note, that due to nonuniformity originally small fluctuations can be amplified drastically, which complicates the distinguishment of the above-mentioned scale regions.

5. Conclusions

In the paper we used specific models to discuss peculiarities of biochemical chaos and effects of noise. Special attention was devoted to Sel'kov-type models exhibiting sustained oscillations, coexistence of attractors, period doubling and chaos. Analyzing the chaotic properties of chemical dynamics it was found that chemical chaos seems to be characterized by large deviations from average behavior. On the one hand, a very nonuniform separation of near-by orbits takes place referred to as "nonuniform dynamics" and, on the other hand, attractors are "inhomogeneous". In order to quantify nonuniform error growth it turned out to be useful to study the statistics of growth rates (see (6)-(11)). Nonuniformity may have a lot of interesting implications, of which only a few have been considered in this paper. Then we discussed effects of short-correlated fluctuations. These are of particular importance near bifurcations and if transitions between coexisting attractors are possible. Thus we studied the evolution of the stationary probability distribution at the Hopf-bifurcation and found significant changes of the bifurcation scenario. We observed large nonperiodic oscillations due to noise-induced crossing of the separatrix.

In section 4 the influence of fluctuations on chaotic dynamics was mainly analyzed with the aid of Lyapunov exponents. The meaning of the maximum Lyapunov exponent for the separation of adjacent trajectories was studied. For this purpose it was useful to distinguish three time scales: dominance of noise for short times, exponential separation on an intermediate time scale, and saturation on large scales. For this consideration as well as for the discussion of nonuniformity a graphical illustration of the actual error growth was illuminating (see Figs. 1, 3, 4, 8).

Summarizing we emphasize that a deeper insight into effects of noise and nonuniform dynamics would help to build a bridge between theoretical models and experimental evidence of complex motions in biochemical systems.

References

1. J.G. Reich, E.E. Sel'kov: Energy Metabolism of the Cell (Academic Press, London 1981)
2. L. Rensing, N.I. Jaeger (eds.): Temporal Order, Springer Ser. Syn., Vol. 29 (Springer Verlag, Berlin, Heidelberg 1984)
3. B. Hess, B. Chance: Naturwissenschaften 46, 248 (1959)
4. B. Hess, A. Boiteux: Ber. Bunsenges. Phys. Chem. 84, 346 (1980)
5. B. Hess, M. Markus: Ber. Bunsenges. Phys. Chem. 89, 642 (1985)
6. S.C. Müller, Th. Plesser, B. Hess: In Temporal Order, ed. by L. Rensing, N.I. Jaeger (Springer Verlag, Berlin, Heidelberg 1984) p. 194
7. M. Markus, S.C. Müller, B. Hess: Ber. Bunsenges. Phys. Chem 89, 651 (1985)
8. E.E Sel'kov: Eur. J. Biochem. 4, 79 (1968)
9. Th. Schulmeister, E.E. Sel'kov: stud. biophys. 65, 121 (1977)
10. A.V. Tolstopyatenko, L. Schimansky-Geier: In Selforganization by Nonlinear Irreversible Processes, ed. by W. Ebeling and H. Ulbricht (Springer, Berlin, Heidelberg 1986) p. 76
11. Th. Schulmeister: stud. biophys. 72, 205 (1978)
12. J.S. Nicolis, G. Mayer-Kress, G. Haubs: Z. Naturforsch. 38a, 1157 (1983)
13. P. Grassberger, I. Proccacia: Physica 13D, 34 (1984)
14. H. Fujisaka: Prog. Theor. Phys. 71, 513 (1984)
15. G. Haubs, H. Haken: Z. Phys. B 59, 459 (1985)
16. H. Herzel, W. Ebeling, Th. Schulmeister: Z. Naturforsch. 42a, 136 (1987)
17. G. Nicolis, I. Prigogine: Self-Organization in Nonequilibrium Systems (Wiley, New York 1977)
18. J.P. Eckmann, D. Ruelle: Rev. Mod. Phys. 57, 617 (1985)
19. E.N. Lorenz: J. Atmos. Sci. 20, 130 (1963)
20. H. Herzel, B. Pompe: Phys. Lett. 122A, 121 (1987)
21. L.F. Olsen: Phys. Lett. 94A, 454 (1983)
22. A. Goldbeter, O. Decroly: Am. J. Physiol. 245, R478 (1983)
23. P. Grassberger, I. Proccacia: Physica 9D, 189 (1983)
24. H. Herzel, Th. Schulmeister: Syst.Anal.-Mod.-Sim. 4, 105 (1987)

25. B. Pompe, J. Kruscha, R.W. Leven: Z. Naturforsch. <u>41a</u>, 801 (1986)
26. W. Ebeling, H. Herzel: stud. biophys. <u>98</u>, 147 (1983)
27. W. Ebeling, R. Feistel: <u>Physik der Selbstorganisation und Evolution</u> (Akademie-Verlag, Berlin 1982)
28. R. Feistel, W. Ebeling: Physica <u>93A</u>, 114 (1978)
29. L. Schimansky-Geier, A.V. Tolstopyatenko, W. Ebeling: Phys. Lett. <u>108A</u>, 329 (1985)
30. H. Engel-Herbert, W. Ebeling, H. Herzel: In <u>Temporal Order</u>, ed. by L. Rensing, N.I. Jaeger (Springer, Berlin, Heidelberg 1984) p. 144
31. H. Malchow, L. Schimansky-Geier: <u>Noise and Diffusion in Bistable Nonequilibrium Systems</u> (Teubner-Verlag, Leipzig 1986)
32. W. Ebeling, H. Herzel, W. Richert, L. Schimansky-Geier: ZAMM <u>66</u>, 141 (1986)
33. J. Kurths, H. Herzel: Physica <u>25D</u>, 165 (1987)
34. H. Herzel, W. Ebeling, L. Schimansky-Geier, E.E. Sel'kov: In <u>Lotka-Volterra-Approach to Cooperation and Competition in Dynamical Systems</u>, ed. by W. Ebeling and M. Peschel (Akademie-Verlag, Berlin 1985)
35. L. Arnold: In <u>Fluctuations and Sensitivity in Nonequilibrium Systems</u>, ed. by W. Horsthemke, D.K. Kondepudi (Springer, Berlin, Heidelberg 1984)

Protein Complexity

H. Frauenfelder

Department of Physics, University of Illinois at Urbana-Champaign,
1110 West Green Street, Urbana, IL 61801, USA

1. Introduction

Complexity and order are easy to understand, but difficult to define rigorously. Up to a short time ago, nearly all physicists were mainly concerned with systems of small complexity. The two prototypes of simple systems are the perfect gas (complete disorder) and the ideal solid (perfect order). We can assign complexity 0 to both of these systems, because no information can be stored either in an ideal gas or a perfect solid. Studies of such simple systems have contributed greatly to the foundations of physics. All truly interesting systems, from proteins to brains and from languages to computers, are neither fully ordered nor completely disordered and they possess complexity. Within the past few decades, physicists have turned their attention to complex systems, much to the amusement of life scientists who have known for many years that living systems are extremely complex and that the exploration of such complex systems is exciting. In any case, here we are, trying to understand systems like proteins. In addition to the intrinsic reward, the work on biological molecules has brought me the pleasure of getting to know the life scientists who investigate similar problems but with a much longer history and much deeper insight. It is a particular pleasure to dedicate this paper to Benno Hess, who has contributed so much to the field of complexity and who always sees through the fog of unimportant details to recognize the important aspects.

Proteins are typically built from a few hundred amino acids and consist of the order of a few thousand atoms. We are interested here in globular proteins in which the primary polypeptide chain folds into a tertiary, roughly spherical structure with a diameter of a few nm. The protein is usually surrounded by a hydration shell, consisting of the order of 1000 water molecules. The protein consequently is a complex many-body system, intimately coupled to the solvent. A protein is often called disordered, but the disorder is of the same character as encountered in Beethoven's Opus 133. 3.5 Gy of R&D have made proteins into superbly constructed miniature machines that perform most functions of life.

In our studies, we attempt to find the physical concepts that govern protein function. To do so, we have selected a relatively simple family, the heme proteins. We study a simple process, namely the association and dissociation of small molecules such as dioxygen (O_2) and carbon monoxide (CO) and mainly use three techniques, flash photolysis [1], X-ray diffraction [2], and conformational labelling [3]. To evaluate the data, we introduce a simple model, describe the observed phenomena quantitatively, and establish connections to other parts of physics.

Among the heme proteins that we use, myoglobin (Mb) occupies a unique position. It consists of 153 amino acids, has a molecular weight of about 18 kdalton, and contains about 1200 nonhydrogen atoms. Its globular structure encloses a small organic molecule, heme. At the center of the heme group, an iron ion acts as the "catalytic center". The main function of Mb is simple - it reversibly binds O_2 and CO at the iron ion. The structure of Mb is well known and its various properties have been widely studied; it can be considered as a test case, the "hydrogen atom of biology" [4].

Here we will show that the experiments on Mb yield remarkably rich information; even though Mb is a rather simple protein it displays a surprising range of phenomena. Mb has been investigated for many years, but we are still only at a beginning and it will require many more experiments before we can classify and understand the states and motions of Mb and connect this information to function.

2. Experimental Techniques

To explore protein complexity, we use the binding of O_2 and CO to myoglobin. From the many experiments that yield insight into the properties of Mb, we select three, conformational labelling, flash photolysis, and X-ray diffraction. Conformational labelling is based on the fact that CO can bind to Mb in at least three different orientations. The orientations differ in the CO stretching frequency [5,6,7] and they have been seen in X-ray diffraction [8]. The intensities of the bands are very sensitive to the protein conformation and to external influences such as pressure and pH. The ratio of the bands can consequently be used as indicators for conformational changes. In flash photolysis, a well-known technique, the liganded heme protein (MbCO) is placed in a cryostat. The bond between the CO and the heme iron is broken with a laser flash and the subsequent rebinding of the CO is observed. X-ray diffraction, of course, yields the protein structure. In addition, however, the Debye-Waller factor provides dynamic information [9,10].

3. Endless Processes

A clear and unambiguous sign for protein complexity appears when the rebinding of CO to Mb after flash photolysis is observed at temperatures below about 180 K [1]. Denote with N(t) the survival probability, i.e. the probability that a Mb molecule has not rebound a ligand at the time t after photodissociation. In a simple system, N(t) is expected to be exponential in time, $N(t) = \exp(-kt)$. The low-temperature binding of CO and O_2, however, is not exponential in time, but approximately follows a power law, $N(t) = (1+t/t_0)^{-n}$, where t_0 and n are temperature-dependent parameters. Because this result is at first so unexpected, we have checked it in a wide variety of different proteins, for instance leghemoglobin [11] and horse radish peroxidase [12], and we have monitored rebinding at various wavelengths in the uv, the visible, and the ir [3]. All results are consistent and prove that the nonexponentiality is not an artifact.

Nonexponential phenomena, sometimes called "endless processes", have a long history [13]. They were first observed and described by W. Weber in Göttingen who followed a suggestion by Gauss [14]. They turn up in a wide range of problems such as mechanical creep, discharge of capacitors, dielectric relaxation, phosphorescence, luminescence, radiation damage, NMR, dynamic light scattering, remnant magnetization in spin glasses, and photosynthesis, and are always characteristic of complexity.

Where is the complexity in heme proteins that produces the nonexponential time dependence of binding at low temperatures? To discuss one possible explanation, we note that the binding of CO to the heme iron is already a complicated process. Before binding, the heme group is domed and the iron has spin 2 and lies out of the mean heme plane. In the bound state, MbCO, the heme is planar, the iron has spin 0 and is in the heme plane: Binding involves a motion of the heme, the iron, and the CO. It is plausible that these motions give rise to an energy barrier which must be overcome by the system before the Fe-CO bond is formed. In the simplest case, the rate k for binding is then given by an Arrhenius relation, $k = A \exp(-H/RT)$, where H is the barrier height. If H is the same in all Mb molecules, k has a unique value and rebinding is exponential in time. If, however, different Mb molecules possess different barrier heights H, the nonexponential rebinding is explained. Mb molecules with small barriers rebind quickly, those with large barriers rebind slowly. If we denote with g(H)dH the probability of finding a Mb molecule with barrier height between H and H+dH, the survival probability N(t) becomes

$$N(t) = \int dH \, g(H) \, \exp(-k(H)t), \tag{1}$$

where k(H) is related to H by an Arrhenius relation [1]. From the experimentally observed function N(t), g(H) and the preexponential A can be determined by inversion of (1).

4. Conformational Substates

While a nonexponential time dependence implies complexity, the cause of the complexity is not always clear. The structure of proteins suggests a molecular origin for the case of ligand binding. Protein folding is unlikely to lead to a unique tertiary structure in which each atom occupies exactly the same spatial position. Small changes in the structure, of the arrangement of the weak bonds, and of the water molecules on the outside of a protein are unlikely to change the total binding energy by much. The small changes in the structure may, however, affect a local property like the activation energy H. We therefore assume that a given protein can exist in a large number of conformational substates (CS) [1,9]. All CS have the same overall structure, but differ in local arrangements. Proteins perform the same function in all CS, but possibly with different rates.

The temperature dependence of the nonexponential time dependence of ligand binding provides information about the structure of the energy space. If CS indeed exist, they should be observable also in other properties. Spatial information, in particular, should be obtainable through X-ray diffraction. As indicated above, the local structure of the same protein should be slightly different in different CS. In an ensemble we therefore can expect that the same atom does not always occupy the same equilibrium position, but that small deviations from the average position occur. An atom in a protein therefore should have a larger mean-square-deviation (msd) than an atom in a periodic crystal. Indeed, systematic studies of the Debye-Waller factor in proteins indicate that the msd is much larger than in a crystalline solid and, moreover, varies from atom to atom [2,9,10].

5. Equilibrium Fluctuations and Functional Motions

The existence of CS leads to new features. In order to perform a function, a protein must be able to exist in more than one state. Myoglobin, for instance, can be in the liganded (MbCO) or deoxy (Mb) state. Since each of these states can assume a very large number of CS, two different types of motions can be distinguished. Equilibrium fluctuations (EF) lead from one CS to another; nonequilibrium motions lead from one state to another. We denote the nonequilibrium motions as fims, for functionally important motions. EF and fims are not totally independent, but are related by fluctuation-dissipation theorems [15,16,17].

6. Proteinquakes

To study EF and fims we take recourse to an analogy with earthquakes. In an earthquake, a stress is relieved; the sudden release of strain energy leads to permanent deformations and to the emission of shear and pressure waves. Properties of the earth are investigated through the observation of the deformations and waves. Similar phenomena occur in proteins: Stress is relieved when the bond is broken by the laser flash and the protein then changes from the liganded to the unliganded structure. We call the rearrangement after the bond breakage a proteinquake [18]. By observing suitable spectroscopic markers, the progress of the quake can be followed.

The proteinquake following photodissociation of MbCO reveals a number of phases; the quake does not occur in one smooth motion, but in a sequence of at least three steps. This observation implies a complexity exceeding the one originally postulated to explain the nonexponential time dependence of ligand binding. A possible interpretation of the result is to assume that conformational substates are hierarchically arranged. Substates of the first tier, denoted by CS^1, are mainly responsible for the nonexponential time dependence. Each substate of the first tier is again subdivided into substates of the second tier (CS^2). These are energy valleys separated by smaller barriers than the one separating the CS^1. Each CS^2 is again subdivided by smaller barriers into substates of the third tier (CS^2). The details of the arrangement of the CS or, in other words, the details of the energy hypersurface are most likely very complicated. An enormous amount of experimental work will be needed to explore the full structure.

7. Conformational Labelling

More information about the complexity and sensitivity of proteins comes from conformational labelling [3]. As pointed out in Section 2, CO binds to the heme iron in Mb in at least three different orientations with different CO stretching frequencies [3,5,6,7]. Here we consider only two bands, A_0 at 1966 cm^{-1} and A_1 at 1946 cm^{-1}. With decreasing temperature, the intensity ratio $r_0 = A_0/A_1$ increases down to a temperature T_{sg}; below, r_0 is constant. T_{sg} depends on the solvent; for a 75% glycerol-water solvent it is about 180 K. We interpret this temperature dependence by postulating that the different orientations correspond to substates with different energies and entropies; A_0 is lower in energy than A_1 by a few kJ/mol. Above T_{sg}, the substates interconvert freely; below, they are frozen. Near T_{sg}, freezing occurs slowly and is nonexponential in time. We call the transition near 180 K a <u>slaved glass transition</u>; glass transition because of the similarities with an ordinary glass transition [19] and slaved because the temperature where it occurs depends on the solvent. The protein appears to be slaved to the solvent and solvent and protein must be considered to be one system. Again, the complexity of the protein is expressed through the nonexponential time dependence. In ligand binding, the progress of binding is nonexponential; in the equilibration of the A substates, protein relaxation is nonexponential.

8. Outlook

In the present short review, only a few of the most important concepts have been discussed. Possibly the most important result of the dynamics experiments is the concept of conformational substates or, in other words, the appearance of multiple ground states - many energy valleys in the energy hypersurface of proteins. This result implies on the one hand that proteins are intrinsically complex and that no simple model can represent all major features of a protein well. On the other hand, the appearance of multiple ground states relates biomolecules to other complex systems such as glasses and spin glasses. Theories and models for these simpler systems may therefore help in the construction of adequate treatments for protein dynamics and function. A second important result, well known also from many other experiments, is the fact that proteins cannot be considered alone; protein and surroundings together form the important system. Both experimentally and theoretically we are only at a beginning in the understanding of protein dynamics. Even for a simple protein like myoglobin, we have not yet classified all motions and we know little about the connections between dynamics and structure, and dynamics and function. For more complex proteins such as bacteriorhodopsin, we know even less. This situation is reassuring; it will give Benno Hess many apportunities to apply his deep understanding and knowledge of complex and even chaotic phenomena and to contribute even more to this rapidly growing field.

Acknowledgements

This work was supported in part by U.S. National Institute of Health grants GM 18051 and GM 32455, National Science Foundation grant DMB 82-09616, and Office of Naval Research grant N00014-86-K-00270.

References

1. R.H. Austin, K.W. Beeson, L. Eisenstein, H. Frauenfelder: Biochemistry 14, 5355 (1975)
2. H. Frauenfelder, G.A. Petsko, D. Tsernoglou: Nature 280, 558 (1979)
3. A. Ansari, J. Berendzen, D. Braunstein, B.R. Cowen, H. Frauenfelder, M.K. Hong, I.E.T. Iben, J.B. Johnson, P. Ormos, T.B. Sauke, R. Scholl, A. Schulte, P.J. Steinbach, J. Vittitow, R.D. Young: Biophys. Chem. 26, 337 (1987)
4. R.E. Dickerson, I. Geis: In Hemoglobin: Structure, Function, Evolution and Pathology (Benjamin/Cummings 1983)
5. M.W. Makinen, R.A. Houtchens, W.S. Caughey: Proc. Natl. Acad. Sci. USA 76, 6042 (1979)
6. W.S. Caughey, H. Shimada, M.G. Choc, M.P. Tucker: Proc. Natl. Acad. Sci. USA 78, 2903 (1981)
7. H. Shimada, W.S. Caughey: J. Biol. Chem. 257, 11893 (1982)
8. J. Kuriyan, S. Wilz, M. Karplus, G.A. Petsko: J. Mol. Biol. 192, 133 (1986)
9. H. Hartmann, F. Parak, W. Steigemann, G.A. Petsko, D. Ringe Ponzi, H. Frauenfelder: Proc. Natl. Acad. Sci. USA 79, 4967 (1982)
10. G.A. Petsko, D. Ringe: Ann. Rev. Biophys. Bioeng. 13, 331 (1984)
11. F. Stetzkowski, R. Banerjee, M.C. Marden, D.K. Beece, S.F. Bowne, W. Doster, L. Eisenstein, H. Frauenfelder, L. Reinisch, E. Shyamsunder, C. Jung: J. Biol. Chem. 260, 8803 (1985)
12. W.Doster, S.F. Bowne, H. Frauenfelder, L. Reinisch, E. Shyamsunder: J. Mol. Biol. 194, 299 (1987)
13. J.T. Bendler: J. Stat. Phys. 36, 625 (1984)
14. W. Weber: Götting. Gel. Anz. p. 8 (1835), Annalen der Physik und Chemie (Poggendorf) 34, 247 (1835)
15. L. Onsager: Phys. Rev. 37, 405 (1931)
16. H.B. Callen, T.A. Welton: Phys. Rev. 83, 34 (1951)
17. R. Kubo: Rep. Prog. Phys. 29, 255 (1966)
18. A. Ansari, J. Berendzen, S.F. Bowne, H. Frauenfelder, I.E.T. Iben, T.B. Sauke, E. Shyamsunder, R.D. Young: Proc. Natl. Acad. Sci. USA 82, 5000 (1985)
19. J. Jäckle: Rep. Prog. Phys. 49, 171 (1986)

Chemical Turnover and the Rate
of Heat Production
in Complex Reaction Systems

Th. Plesser[1] *and I. Lamprecht*[2]

[1]Max-Planck-Institut für Ernährungsphysiologie, Rheinlanddamm 201,
 D-4600 Dortmund 1, Fed. Rep. of Germany
[2]Institut für Biophysik, Freie Universität Berlin,
 Thielallee 63, D-1000 Berlin 33, Germany

1. Introduction

Chemical turnover, the formation and modification of chemical compounds, constitutes the basis for many processes going on in a broad spectrum of organized objects covering the cellular level in living matter as well as the production facilities in the chemical industry. The complexity of the processes depends on the available input and the required output; it may just consist of one simple first order reaction or a long sequence of intermediate steps with branching points and reaction loops.

The analysis of the equilibrium properties of an unknown reaction pathway has to clarify at least three major aspects. In a first approach the various distinct reaction steps, their reaction orders, and the participating chemical species have to be identified, together with the succession of the chemical conversions taking place. The result of that work is a scheme of coupled reactions which naturally demands, for the second aspect, the determination of the rate constants or, in enzyme catalyzed reactions, the rate laws. Thirdly, the mechanism as well as the full set of kinetic parameters have to be consistent with the first and second law of thermodynamics.

New concepts and extended sets of system variables have to be formulated for conditions under which a reaction pathway performs real work or transduces chemical energy in a finite interval of time. When time comes into play the equilibrium variables have to be complemented by the corresponding flows, e.g. : concentrations by matter flows, energy by power, and entropy by the rate of entropy production [1,2]. Energy dissipation and the rate of entropy production are coupled with the efficiency of the whole process or parts of it [3-5]. Efficiency again points to the intrinsic regulatory properties of the system [6-8]. Beyond the merely technical interpretation of the just mentioned keywords they may be of central importance for an unifying description of self-organizing systems. Theories of their fascinating dynamics, the balance between the formation and decomposition of structures and patterns in time and space are formulated and reviewed, for example, in [9,10] and in this book.

This article, however, is restricted to the subject of chemical turnover detectable by microcalorimetric enthalpy and heat flow experiments. Their relevance for the analysis of chemical and biochemical systems exhibiting periodic reaction dynamics is discussed.

The enthalpy change ΔH in a system reacting under constant pressure in a closed system is given by the amount of heat exchanged with the surroundings $\Delta H = \delta Q$. The appropriate devices for the determination of heat are calorimeters [11]. Modern heat flow calorimeters have a particular advantage for the investigation of dynamic systems, in that they measure the rate of heat production from which the total heat exchanged with the surroundings can be obtained by integration.

The usefulness of heat flow calorimetry is demonstrated by two examples: the degradation of sugar by the glycolytic pathway [12] and the oxidation of aniline by bromate without a catalyst [13,14].

Glycolysis, the oldest and most common pathway for the energy supply in living matter, is a reaction sequence of eleven enzyme-catalyzed steps performing the transformation of glucose into pyruvate and ATP or, as a special but important extension in yeasts, the fermentation of glucose into CO_2 and ethanol accompanied with the phosphorylation of ADP. Here we report on calorimetric experiments with a cytoplasmic medium extracted from yeast cells containing the full set of glycolytic enzymes and metabolites [15].

The second example is an uncatalyzed chemical reaction. We chose the oxidative bromination of aniline (Aromatic compound) by potassium Bromate in an Acid environment ("ABA" reaction [13,14]). In such systems less oscillations are observed than in catalyzed reactions. Common to many ABA systems is a long preoscillatory period [13,14] which may last from minutes to hours and during which bromoderivates accumulate in the system. Aniline is an exception in the sense that the preperiod is energetically less pronounced than for other aromates and that the oscillations start with low amplitudes and increase to a maximum amplitude shortly before the cessation of the oscillations.

2. Methods and Materials

The cytoplasmic medium used for the experiments was extracted from commercial baker's yeast according to a procedure published in [16]. The protein content was adjusted to about 47 mg/ml which corresponds to approximately one third of the concentration in the cytoplasm of intact yeast cells. For the calorimetric experiments the following mixture was used: 430 μl extract, 45 μl 1M potassium phosphate, 15 μl 20 mM NAD and 15 μl 15 mM AMP. In a series of batch experiments increasing amounts of trehalose were applied by addition of appropriate volumes of 0.7 M trehalose. NAD, AMP and trehalose were dissolved in 0.1 M potassium phosphate buffer.

The optimum composition of the ABA system was 2 mM aniline and 50 mM $KBrO_3$ in 1 M sulfuric acid. The aniline-to-bromate ratio was varied around the optimum value rendering changes in the length of the preperiod, the number of oscillations and the amplitudes. In both the glycolytic and the ABA system, an intensive stirring of the vessel content by means of a pneumatic pump was essential to ensure homogeneity throughout the sample. All chemicals used were of analytical grade.

The experiments were performed with a modified isothermal isoperibolic batch calorimeter, type "Triflux" from Thermoanalyse/Grenoble with a sensitivity of 81.9 mV/W. The vessels of this differential (twin) instrument had an active volume of 1.2 ml.

The apparatus was modified for the simultaneous measurement of the rate of heat production together with one out of three other signals: optical density, electrode potential, or gas pressure [17,18].

Most of the data, recorded with a double-channel recorder and a microprocessor, were processed by the program package PERSDEC (PERiodic Signal DEConvolution) for the deconvolution of periodic signals by Fast Fourier Transform [19], since the inertia of the calorimeter and the electrode causes a relaxation time comparable to the period of the chemical reactions under investigation. The optical signal is assumed to be instantaneous so that no deconvolution procedure was applied to it.

3. Results

3.1. Glycolysis in a Cytoplasmic Medium Extracted from Yeast

In a series of experiments with a cytoplasmic medium extracted from yeast cells the amount of the applied substrate, the disaccharide trehalose, was increased from 0 to 42 μmol. The rate of heat production as a function of time was recorded for up to 18 hours when it was ascertained that the signal had merged with the baseline. Fig. 1 shows the rate versus time for two examples out of eleven experiments. The heat rate is normalized to a sample volume of 1 ml, since due to the various amounts of trehalose added the experiments differed in their final sample volumes.

The heat in the first experiment (Fig. 1A) is produced exclusively by the fermentation of endogenous sugar reservoirs. The solid curve clearly exhibits two characteristic parts of the heat production rate: an oscillating segment with about 8 high amplitude oscillations on a decreasing baseline and a non-oscillating segment in which the rate of heat production drops to about zero. The dashed line is the integral of the solid curve and represents the total heat pro-

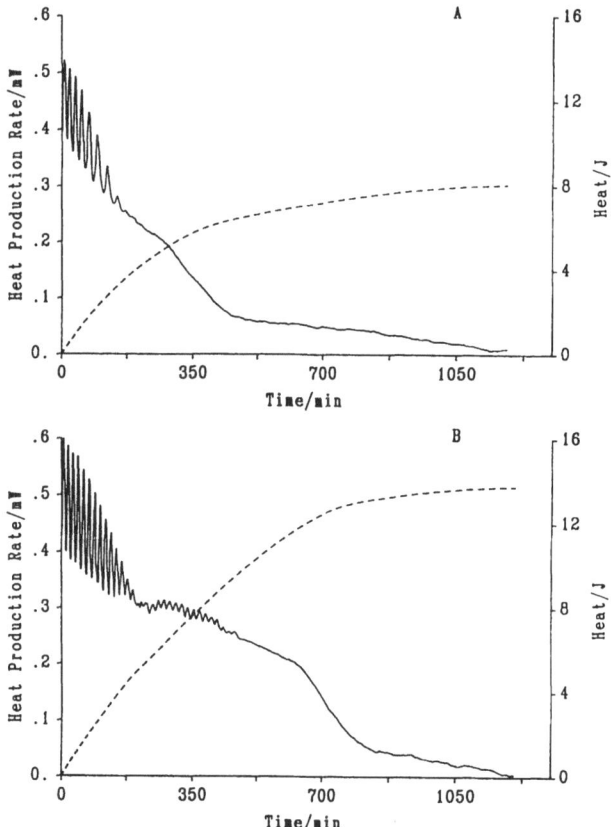

Fig. 1: Rate of heat production during sugar metabolism in a cytoplasmic medium extracted from yeast cells. (A) Metabolism of endogenous sugar; no addition of trehalose. (B) Metabolism of endogenous sugars and of 21 μmol trehalose added to the sample. The solid lines exhibit the rate of heat production (left graduation) and the dashed lines describe the integrated rate, i.e. the total heat produced (right graduation)

duced as a function of time. Figure 1B displays the data of an experiment where 21 μmol of trehalose were added to the extract sample. The curves are similar to those in Fig. 1A except for the train of small amplitude oscillations between the high amplitude oscillations at the beginning and the non-oscillatory segment at the end.

From ten such experiments we calculated by integration the total heat produced. As shown in Fig. 2, it is within a certain range a linear function of the added amount of trehalose. For presentation of the data the amount of trehalose is converted into a more general graduation, the concentration of glucose in a 1 ml sample. This has the advantage that reservoirs of other carbohydrates may easily be included. A linear least squares fit of the data up to 102 mM glucose (dashed line in Fig. 2) results in an enthalpy change of ΔH=-139.7 kJ per mole of glucose fermented into two moles of ethanol and CO_2. The expected enthalpy difference for the reaction

$$\text{Glucose} \rightarrow 2 \text{ Ethanol} + 2 \text{ } CO_2 \tag{1}$$

can be calculated from the heat of formation of the involved compounds [20, 21] resulting in ΔH=-138.3 kJ/mol, if an aqueous solution is considered as the reaction medium. A value of ΔH=-126 kJ/mol is found in the literature, if a buffered medium is assumed for CO_2 [21]. The dotted line in Fig. 2 corresponds to this enthalpy value.

Due to the agreement of the higher enthalpy values all further calculations in this paper were done with ΔH=-138.3 kJ/mol as the literature enthalpy value for (1). The enthalpy for the splitting of one mole of trehalose into two moles of glucose is negligible as has been estimated from the data for the splitting of maltose [21]. The amount of heat contributed to the total heat by the phosphorylation of ADP and dephosphorylation of ATP is approximately zero since the adenine nucleotide pool is kept at equilibrium by the action of the non-glycolytic enzymes adenylate kinase and ATPase.

The intersection of the fitted line with the ordinate at about 14 kJ/ml reflects the heat produced by the the fermentation of endogenous sugar reservoirs equivalent to about 102 μmol glucose per ml extract. The exceptionally low heat value at 150 mM glucose (Fig. 2) has to be confirmed by new experiments.

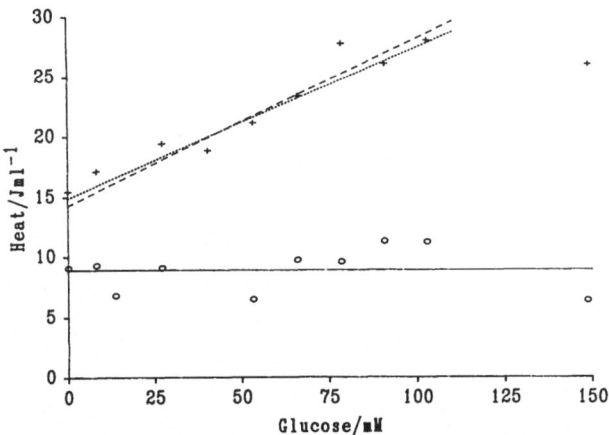

Fig. 2: Total heat produced in a cytoplasmic medium extracted from yeast cells as a function of the amount of trehalose added. The trehalose content is given as concentration of glucose in 1 ml. +: total heat produced; the dashed line corresponds to an enthalpy change of ΔH=-139.7 kJ/mol and the dotted line to ΔH=-126 kJ/mol of glucose. o: heat produced during the time period of steady chemical turnover as shown in Fig. 1

It is known from the theory of dynamic systems and from experiments with periodically metabolizing yeast extracts [22,23] and model calculations thereof [24-26] that the domain of the oscillatory state of the system has sharp boundaries at which the system bifurcates and branches into another state, from an oscillatory to a steady state or vice versa. In our experimental system a pool of sugar substrates is metabolized through the glycolytic pathway. The sugar input is regulated mainly by the trehalose splitting enzyme trehalase. During the consumption of substrate the pool reaches a sugar concentration at which the glucose influx necessary for the oscillatory state can not be maintained. This critical concentration is an intrinsic parameter of the system and therefore independent of the amount of sugar at the beginning of the experiment. Thus we expect that the amount of heat produced after cessation of the oscillations is the same in all experiments. This is clearly shown in Fig. 2 by the data with the symbol (o). The solid line indicates the average value of 8.9 kJ/ml or a concentration of 64.4 mM glucose.

As a consequence of the existence of a critical concentration the glycolytic flux should be the same at the end of the oscillations in all experiments. In a first approximation the flux can be calculated from the rate of heat production P per ml by $P/\Delta H$. The data presented in Fig. 3 support our expectation and give an average glycolytic flux at the lower bound of the oscillatory domain of 0.21 mM/min. HESS et al. [27] found by direct infusion of glucose a boundary at 0.33 mM/min in an extract with a protein content of 60 mg/ml. Correction for the protein content reduces this number to 0.26 mM/min, slightly higher than the critical value for the trehalase regulated glucose influx.

The results presented above show that the sugar in an experimental sample is totally metabolized and converted into ethanol and CO_2 confirming the biochemical analysis of the metabolites and reaction products [27,28]. Due to this conservation and the knowledge of the ΔH value for reaction (1) we are able to calculate the sugar concentration in units of glucose at any point of time of the experiment. In addition, the glycolytic flux can be calculated at any point of time from the rate of heat production by $P/\Delta H$ as already given above. This allows for the elimination of time as plotted in Fig. 4 for the data shown in Fig. 1. The graphs can be considered, at least in the non-oscillating part, as reaction kinetics from which a K_m value, the concentration of half maximum velocity, can be determined. From the experiments an average K_m value of 32 mM glucose or 16 mM trehalose can be estimated. This value has to be compared with a K_m of about 5 mM

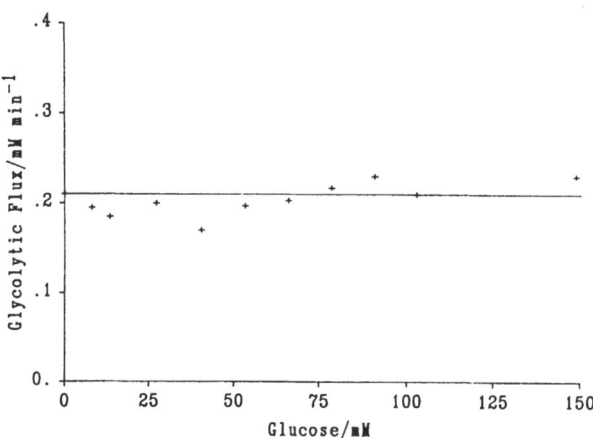

Fig. 3: Glycolytic flux at the transition point between the oscillating and non-oscillating state as a function of added trehalose. The trehalose content is given as concentration of glucose in 1 ml. The solid line exhibits the average flux of 0.21 mM/min calculated from the heat power with $\Delta H = -138.3$ kJ/mol glucose

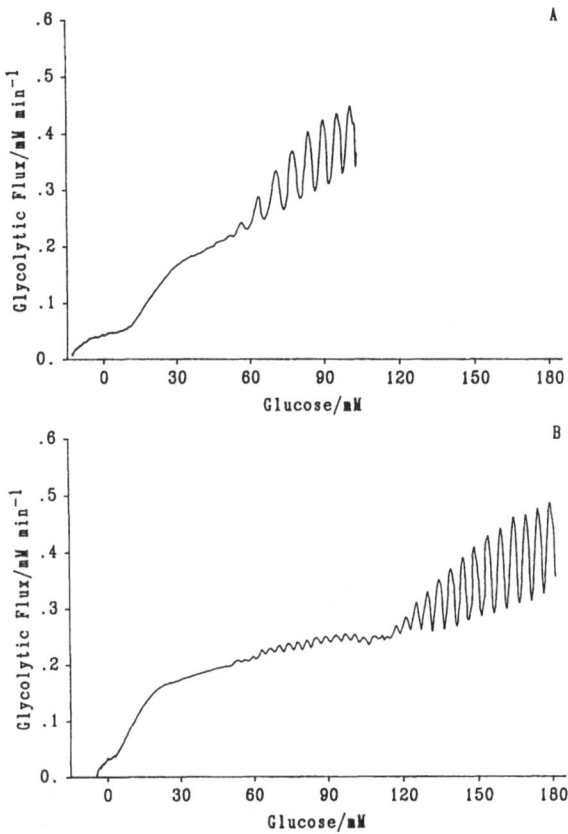

Fig. 4: Glycolytic flux plotted over the total sugar concentration given in glucose units. The actual sugar concentration is calculated from the integrated rate
of heat production with an enthalpy change ΔH=-138.3 kJ per mole glucose. (A) No
addition of trehalose; only endogenous sugar reservoirs are fermented. (B) Addition of 21 μmol trehalose

trehalose as found for the enzyme trehalase by KRÜGER and HESS [29] and also
reported in [30].

In Fig. 4B the linear decrease of the baseline of the 14 high amplitude
oscillations is followed by an almost horizontal train of low amplitude oscillations. The length of this part depends on the amount of trehalose added to the
sample whereas the number of high amplitude oscillations is independent of the
added trehalose for amounts larger than 10 μmol. An explanation of this behaviour
may be found in the careful analysis of the activation mechanisms of trehalose by
ORTIZ et al. [30]. These authors confirm and corroborate earlier experiments that
two forms of the trehalose splitting enzyme trehalase exist, an active and a
"cryptic" inactive form. It is reported that extracts from S.cerevisiae contain
only 19% of active trehalase. This state of the trehalase has a fivefold higher
activity compared to the cryptic one. If we assume this number to be valid in
our extract preparations at the beginning of an experiment and if we correlate
the linear decrease of the baseline of the high amplitude oscillations with the
inactivation of the trehalase by the action of a phosphatase, then we have to interpret the nearly constant average flux level during the low amplitude oscillations as the maximum rate of the cryptic trehalase flux. Under these assumptions
the ratio between the glycolytic flux at the beginning of an experiment and the

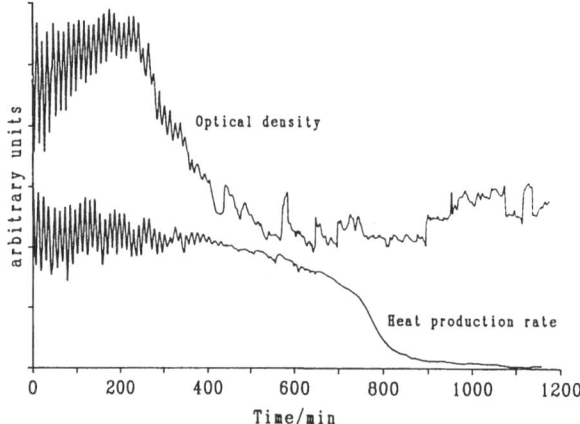

Fig. 5: Simultaneous measurements of the rate of heat production and the optical density at a NADH specific wavelength in a cytoplasmic medium extracted from yeast cells

maximum "cryptic" flux should be 0.19 x 5 + 0.81 = 1.76. In our experiments a ratio of 1.69 is found, a remarkable agreement. This somehow speculative interpretation of the data has to be substantiated by experiments dedicated to a supposed stimulation of the glycolytic flux by a cAMP mediated trehalase activation [30].

More information about the regulatory coupling between different parts of the glycolytic pathway can be extracted from experiments in which the rate of heat production is measured simultaneously with the optical density of the sample at a NADH specific wavelength. An experiment of this type with an added trehalose content of 42 μmol is shown in Fig. 5. It is not clear yet whether the trends in the NADH trace are a general behaviour or specific for this experiment. Figures 6A and B show an interval of 200 minutes of the oscillatory range. In A the maxima of the two curves seem to indicate a phase shift between the two signals. This difference is merely the result of the inertia of the calorimeter having a relaxation time of about two and a half minutes compared to a glycolytic oscillation period of about 12 minutes. After deconvolution of the calorimetric data with the measured response function of the instrument a coincidence of the extrema is observed (Fig. 6B).

3.2. ABA Reactions

The heat signal of most ABA reactions is clearly separated into a pre-oscillatory period in which the aromatic compound is consumed and first bromoderivates appear, an oscillatory period governed by two kinetic states of the system, and a post-period when the reaction has left the oscillatory window. In that range chemical turnover - and thus heat production - continues for a considerable time interval. The chemical processes predominant in these periods are very complex and by far not understood [14]. The calorimetric signals of the various ABA reactions are quite different. While with phenol or sulfanilic acid a strong exothermic pre-period occurs [17], aniline exhibits a pre-period of nearly constant rate of heat production (Fig.7). The length of this period depends on the experimental conditions and increases from a few seconds to several hours with an increasing aniline-to-bromate ratio. The number of oscillations is much smaller than in catalyzed periodic reactions. As with phenol or sulfanilic acid [17] it increases with the concentration of the aromatic compound. It exhibits up to 12 oscillations in most cases and may go up to more than 20 at a properly chosen aniline-to-bromate ratio. In contrast to other ABA reactions studied by us so far , the

188

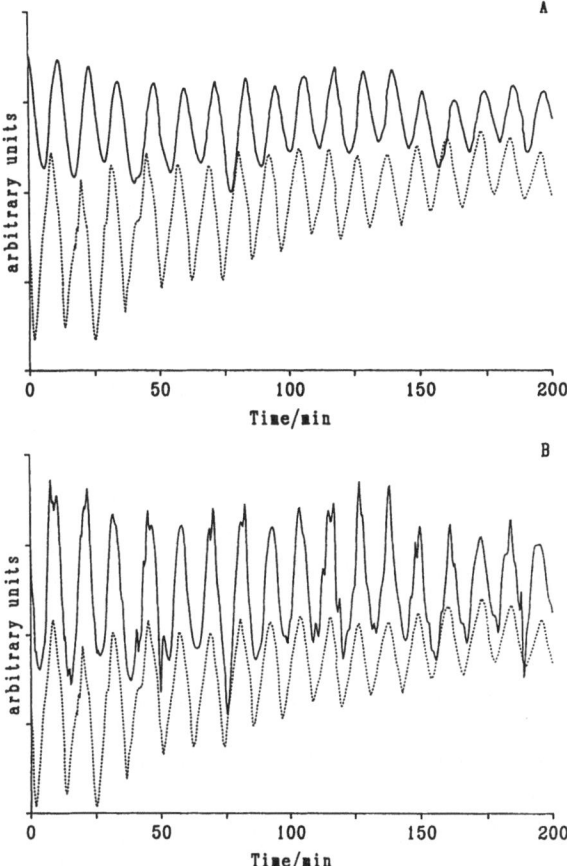

Fig. 6: Simultaneous measurements of the rate of heat production and the optical density at a NADH specific wavelength in a cytoplasmic medium extracted from yeast cells. (A) Measured signals in the oscillatory range of Fig. 5; solid line: rate of heat production; dotted line: NADH signal. (B) Solid line: rate of heat production after deconvolution with the resolution function of the instrument; dotted line: NADH signal as in (A)

oscillations start with small, slowly growing amplitudes up to a maximum amplitude and decrease in a much faster manner. This effect is less pronounced after applying the deconvolution procedure, as can be seen by comparing Figs. 7A and 7B. The number of periods up to the maximum amplitude is always larger than that after the maximum.

In the same way the period length increases approximately by a factor of 2 from the first oscillation to the period with maximum amplitude and returns lateron to the initial value. The shortest periods are in the order of 2 minutes. In some experiments there is a gap without oscillations after the maximum, leading to an increase in period length by a factor of more than 3, such as if the oscillations broke step.

The heat production is strongly modulated in the oscillatory range with rates returning almost to zero. This is clearly demonstrated by the deconvoluted calorimeter signal in Fig. 7B. Detailed analysis of the simultaneously measured electrode potential (Fig. 7B) reveals that the maxima of the heat rate coincide pre-

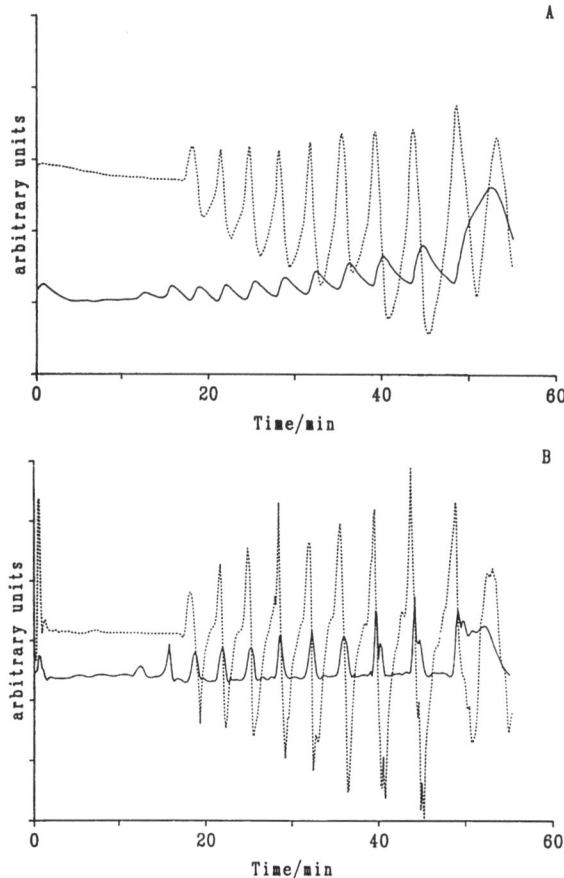

Fig. 7: Oxidation of aniline by potassium bromate in an acid environment. (A) Measured signals; solid line: rate of heat production; dotted line: electrode potential. (B) Solid line: deconvolution rate of heat production; dotted line: deconvoluted electrode potential

cisely with the fastest change during the downward phase of the potential. Further calorimetric and potentiometric experiments relevant to this question are in progress.

4. Discussion

The measurement of heat rates with modern microcalorimeters is one of the few techniques for the continuous non-invasive recording of rates. The data presented in this paper from calorimetric experiments with a cytoplasmic medium extracted from yeast cells show that the amount of trehalose offered to the glycolytic pathway is in total converted into ethanol and CO_2. The transition from the oscillatory to the steady state domain is coupled with critical values of sugar concentration and metabolic flux. This observation supports the view of glycolysis to be a dynamic system able to undergo bifurcations. The fact that the critical flux is found to be the same in these experiments in which the inflow of glucose is regulated by an enzyme, namely trehalase, and in experiments from the late sixties when the critical lower bound flux was determined by the infusion of glucose [27], demonstrates the high reliability of this experimental model system for the transduction of free energy by a biochemical reaction pathway with complex regulatory properties.

Calorimeters are widely used in the life sciences for the determination of the relation between energy and growth of whole organisms [31] on the one hand and for the precise measurement of the very elementary processes as for example the enthalpy of binding [32] on the other hand. The glycolytic system in a cytoplasmic medium extracted from yeast cells seems to be of intermediate complexity for a thorough calorimetric study. This biochemical reaction pathway is advantageous in that it is composed of distinct enzyme catalyzed reaction steps which can be studied by calorimetry and other means with high precision independently of the whole pathway. Detailed calorimetric studies are already published for the enzymes hexokinase [33], pyruvate kinase [34], and alcohol dehydrogenase [35]. Calorimetric data for other glycolytic reactions are vague [36].

In addition to the measurement of biophysical parameters and the analysis of the data in the light of the dynamics of complex systems these investigations support also biochemical work related to the regulation of carbohydrate reservoirs by the enzyme trehalase.

The situation is not so advantageous for oscillatory chemical rections, e.g. the aniline oxidation, the second example reported here. The reaction network of the chemical oscillators cannot be split so easily into segments which may be investigated separately.

The disadvantage of calorimetry not to be selective for the various heat sources in an object under investigation can be reduced by simultaneous measurement of the rate of heat production and at least one selective system signal. Moreover, quantitative computer simulations of non-selective signals as the rate of heat production, electrode potential or even better pH, based on the contributions of individual reaction steps, may significantly improve our experience with the regulatory mechanisms in complex chemical and biochemical reaction networks.

Dedication and Acknowledgements

This article is dedicated to Professor Benno Hess on the occasion of his 65th birthday.

We thank Mrs. I. Beckmann and Mr. A. Jordan for valuable assistance in the laboratory and Mrs. R. Hübner for the careful preparation of the figures.

References

1. P. Glansdorff, I. Prigogine: Thermodynamic Theory of Structure, Stability and Fluctuations (Wiley-Interscience, New York 1971)
2. A. Katchalsky, P. Curran: Nonequilibrium Thermodynamics in Biophysics, 3rd ed. (Harvard University Press, Cambridge 1974)
3. P.H. Richter, P. Rehmus, J. Ross: Progr. Theor. Phys. 66, 385 (1981)
4. S.R. Caplan, A. Essig: Bioenergetics and Linear Nonequilibrium Thermodynamics in the Steady State (Harvard University Press, Cambridge 1983)
5. J. Stucki: In Metabolic Compartmentation, ed. by H. Sies (Academic Press, London, New York 1982) p. 40
6. Th. Bücher, W. Rüssmann: Angew. Chemie 75, 881 (1963)
7. B. Hess: In Energy Transformation in Biological Systems, Elsevier Excerpta Medica (North Holland, Amsterdam, Oxford, New York 1975) p.369
8. H. Haken: Advanced Synergetics, Springer Ser. Syn.,Vol. 20 (Springer,Berlin, Heidelberg 1983)

9. G. Nicolis, I. Prigogine: <u>Self-Organization in Nonequilibrium Systems</u> (Wiley-Interscience, New York 1977)
10. <u>Self-Organization - Autowaves and Structures far from Equilibrium</u>, ed. by V.I. Krinsky, Springer Ser. Syn.Vol. 28 (Springer,Berlin, Heidelberg 1984)
11. W. Hemminger, G. Höhne: <u>Calorimetry-Fundamentals and Practice</u> (Verlag Chemie Weinheim, Weinheim 1984)
12. A. Lehninger: <u>Biochemistry</u> (Worth Publishers, New York 1970)
13. M. Orban, E. Körös: J. Phys. Chem. <u>82</u>, 1672 (1978)
14. M. Orban, E. Körös, R.M.Noyes: J. Phys. Chem. <u>83</u>, 3057 (1979)
15. Th. Plesser, S.C. Müller, B. Hess, I. Lamprecht, B. Schaarschmidt: FEBS Lett. <u>189</u>, 42 (1985)
16. B. Hess, A. Boiteux: Hoppe-Seyler's Z. Physiol. Chem. <u>349</u>, 1567 (1968)
17. I. Lamprecht, B. Schaarschmidt, Th. Plesser: Thermochim. Acta <u>112</u>, 95 (1987)
18. I. Lamprecht, B. Schaarschmidt, Th. Plesser: Thermochim. Acta <u>119</u>, 175 (1987)
19. K.H. Müller, Th. Plesser: Thermochim. Acta <u>119</u>, 189 (1987)
20. E.A. Battley: Physiologia Pl. <u>13</u>, 192 (1960)
21. R.C. Wilhoit: In <u>Biochemical Microcalorimetry</u>, ed. by H.D. Brown (Academic Press, New York 1969) p.305
22. N. Minorsky: <u>Nonlinear Oscillations</u> (R.E. Krieger Publishing Company, Huntington, New York 1974)
23. B. Hess, A. Boiteux: Annu. Rev. Biochemistry, <u>40</u>, 237 (1971)
24. A. Goldbeter, R. Lefever: Biophys. J. <u>12</u>, 1302 (1972)
25. J.G. Reich, E.E. Selkov: <u>Energy Metabolism of the Cell</u> (Academic Press,London 1981)
26. D. Erle, K.H. Mayer, Th. Plesser: Math. Biosci. <u>44</u>, 191 (1979)
27. B. Hess, A. Boiteux, J. Krüger: In <u>Advances in Enzyme Regulation</u>, Vol. 7 (Pergamon Press, Oxford, New York 1969) p.149
28. S.B. Jonnalagadda, J.U. Becker, E.E. Selkov, A. Betz: Biosystems <u>15</u>, 49 (1982)
29. J. Krüger, B. Hess: Arch. Mikrobiol. <u>61</u>, 154 (1968)
30. C.H. Ortiz, J.C.C. Maia, M.N. Teman, G.R. Braz-Padrao, J.R. Mattoon, A.D. Panek: J. Bacteriol. <u>153</u>, 644 (1983)
31. <u>Biological Microcalorimetry</u>, ed. by A.E. Beezer (Academic Press, New York 1980)
32. <u>Thermodynamic Data for Biochemistry and Biotechnology</u>, ed. by H.-J. Hinz (Springer, Berlin, Heidelberg 1986)
33. R.N. Goldberg: Biophys. Chem. <u>3</u>, 192 (1975)
34. R.L. Cheer, G.R. Hedwig, I.D. Watson: Biophys. Chem. <u>12</u>, 73 (1980)
35. K. Burton: Biochem. J. <u>143</u>, 365 (1974)
36. S. Minakami, C.-H. De Verdier: Eur. J. Biochem. <u>65</u>, 451 (1976)

Pattern Formation by Coupled Oscillations: The Pigmentation Patterns on the Shells of Molluscs

H. Meinhardt and M. Klingler

Max-Planck-Institut für Entwicklungsbiologie,
Spemannstraße 35, D-7400 Tübingen, Fed. Rep. of Germany

The diversity and beauty of the pigmentation patterns on the shells of snails and bivalved molluscs invites to construct models to understand their formation. The similarity of patterns in unrelated species on the one hand and the diversity in closely related species on the other encourages the assumption that most of them are generated by a common mechanism.

Pattern formation on the shells of molluscs proceeds in most species in a strictly linear manner since new pattern elements are added only along the growing edge of the shell. The second dimension is a protocol of what has happened at the growing edge as function of time. The shell pattern represents, so to say, a space-time plot. Frequent are pigmentation lines or ridge-like structures which are oriented perpendicular, parallel or oblique to the growing edge.

Keeping in mind the space-time character of the shell pattern, lines perpendicular to the growing edge indicate the formation of a spatial periodic pattern of pigment production along the edge which is stable in time. Other patterns indicate that pigment deposition oscillates. A particular cell produces pigment only during a certain time interval and enters then into an inactive (refractory) period until the next pigment production takes place. A synchronous oscillation in pigment production leads to lines which are parallel to the shell margin. Oblique lines originate from travelling waves of pigment production. Such waves arise if pigment-producing cells trigger their neighboring cells so that - after a certain delay - these cells also start to produce pigment.

In an earlier paper, the basic idea of using reaction-diffusion systems to model patterns of mollusc shells has been published [1]. A model, based on similar principles but focused on the involvement of the nervous system, has been proposed by ERMENTROUT et al. [2]. A more detailed modeling of shell patterning will appear elsewhere [3].

Oscillations and travelling waves can be generated with reaction-diffusion systems containing an autocatalytic and an antagonistic component. Similar ingredients are required to produce spatial patterns which are stable in time [4,5]. Thus, the pattern of mollusc shells seems to be a special application of general pattern forming mechanisms. The diversity of patterns found in nature provides an inroad for the investigation of the range of possible patterns which can be generated by modifications of a basic mechanism.

The nonlinear partial differential equations (1a) and (1b) describe an interaction, which is able to generate the basic shell patterns:

$$\frac{\partial a}{\partial t} = \rho s a^{2}{}^{*} - \mu a + D_{a} \frac{\partial^{2} a}{\partial x^{2}} \tag{1a}$$

This is a modified version of a paper presented at the International Symposium on Biomathematics, Kyoto 1985, which will appear in Lecture Notes in Biomathematics (Springer)

$$\frac{\partial s}{\partial t} = \sigma - \rho s a^{2*} - \nu s + D_s \frac{\partial^2 s}{\partial x^2} \qquad \text{where} \tag{1b}$$

$$a^{2*} = \frac{a^2}{1 + \kappa a^2} + \rho_o .$$

According to these equations, the autocatalytic activator is a molecule which controls pigment production; a has a non-linear feedback on its own production which saturates at high a concentrations. The autocatalysis is limited due to the depletion of a substrate s, which is consumed in the course of autocatalysis; σ describes the constant production rate of s; μ and ν are the decay rates; D_a and D_s are the diffusion constants; and ρ_o is a small basic activator production.

Alternatively, the reaction antagonizing the autocatalysis may be accomplished by an inhibitor. According to (2), an inhibitor h is produced under the control of the activator which inhibits the activator autocatalysis.

$$\frac{\partial a}{\partial t} = \frac{\rho(a^{2*} + \rho_o)}{h} - \mu a + D_a \frac{\partial^2 a}{\partial x^2} \tag{2a}$$

$$\frac{\partial h}{\partial t} = \rho a^{2*} - \nu h + D_h \frac{\partial^2 h}{\partial x^2} + \rho_1 \quad , \qquad \text{where} \tag{2b}$$

$$a^{2*} = \frac{a^2}{1 + \kappa a^2} \quad .$$

Depending on the parameters, the basic types of shell patterns mentioned above can be simulated with both equations. Simulations of more complicated patterns, for instance, those containing intersections or bifurcations of pigment lines require systems in which at least three substances interact. By computer simulations we have shown that the overall patterns as well as many fine details of shell patterning can be understood on the basis of these mechanisms. The details of the simulations can be derived from the computer program which has been used for these simulations [6].

Lines perpendicular to the growing edge

The formation of stripes perpendicular to the growing edge (Fig. 1a) requires, as mentioned, that groups of cells permanently produce pigment, and that the pigment-producing regions are separated from each other by regions in which pigment production never occurs. Such patterns can be generated by local autocatalysis and long ranging inhibition. Required is that the diffusion of the antagonistic substance is much higher than that of the activator, i.e. $D_s \gg D_a$ or $D_h \gg D_a$. The resulting patterns will be stable in time if the antagonistic reaction has a shorter time constant, i.e. $\sigma > \nu$ or $\nu > \mu$, otherwise oscillations are likely to occur (see below). Small random fluctuations are sufficient to initiate the pattern formation from a homogeneous initial state. Fig. 1 shows a corresponding simulation with (2).

Lines parallel to the growing edge

This pattern results if the cells underneath the growing edge deposit pigment more or less simultaneously for certain time intervals and if these pigment-producing phases are separated by phases in which no pigment production takes place. In a reaction scheme containing an (autocatalytic) activator and an antagonistic substance, oscillations can occur if the antagonistic reaction follows a change in the concentration of the activator too slowly, i.e. if $\sigma < \mu$ (1) or $\nu < \mu$ (2)

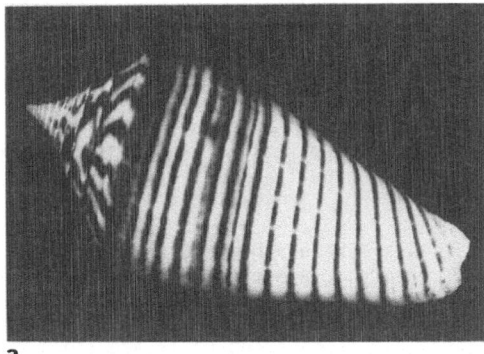
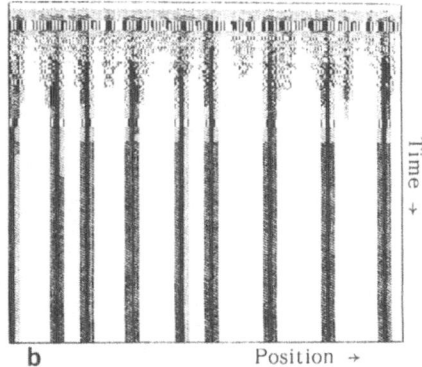

a   b   Position →

Fig. 1: Stripes parallel to the direction of growth. This pattern indicates a stable periodic pattern of pigment-producing and non-producing regions. (a) Natural pattern: shell of *Conus macronatus*. (b) Pattern formation by the interaction of an autocatalytic activator and its rapidly diffusing antagonist, in this example a diffusible inhibitor (2). In this and the following simulations, the density of dots indicates the activator concentration as function of time. High activator concentration is assumed to cause pigment production. Calculated with (2), $\rho = 0.03 \pm 5\%$ random fluctuations, $\mu = 0.03$, $\rho_0 = 0.02$, $D_a = 0.01$, $\nu = 0.05$, $D_s = 0.4$

[7]. Stripes parallel to the growing edge appear if the oscillation of the cells occurs in synchrony. Such a synchronization results if the activator shows a substantial diffusion (i.e. if $D_a \gg D_s$ or $D_a \gg D_s$), since a faster oscillating cell can advance the phase of a delayed neighbor due to the activator exchange. Usually, the pigment lines are not precisely parallel to the growing edge (and the growth lines), indicating that - despite the fact that neighboring cells are in a similar phase - a substantial phase difference can accumulate over the total extension of the growing edge. Figure 2 shows a natural pattern together with a simulation.

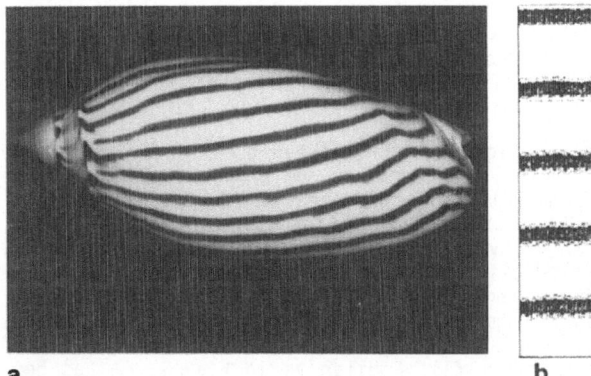
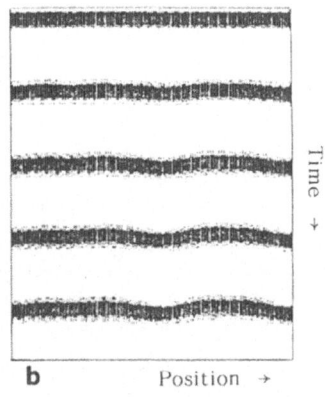

a   b   Position →

Fig. 2: Stripes parallel to the growing edge. This pattern indicates an oscillatory pigment production. (a) Pattern on the shell of *Amoria ellioti*. (b) Model: assumed is an activator - inhibitor system (2). The inhibitor has a longer time constant than the activator; thus, oscillations can occur. High diffusion of the activator enforces a near-synchronization between neighboring cells. Calculated with (2), $\rho = 0.03$, $\mu = 0.03$, $\rho_0 = 0.02$, $D_a = 0.1$, $\nu = 0.05$, $D_h = 0$ and $\rho_1 = 0.0075$

Oblique lines resulting from travelling waves

If the activator diffusion is smaller than that required for the formation of
parallel pigment lines, travelling waves can occur. A just activated cell can
"infect" its neighbor which becomes, with some delay, also activated and so on.
The resulting pigment pattern on the shells consists of oblique lines. The larger
the speed of the wave (in relation to the speed of shell growth) the smaller the
angle between the line and the growing edge.

Characteristic for many shell patterns is a W-like arrangement of the oblique
lines. The Λ-element of this pattern results from a spontaneous trigger of a cell
followed by an infection of both neighbors, and so on. In other words, a spontaneous trigger of a cell gives rise to two diverging oblique lines. If two such
lines merge, both lines become extinct because all neighboring cells are still in
the refractory period, leading to the V element of the pattern (see Fig. 3).
According to this view, the formation of the oblique lines on the shells of many
molluscs has much in common with the formation of travelling waves in aggregating
slime molds [8], except that in molluscs the pattern formation occurs along a
line and the second dimension is a record of the temporary pattern.

Formation of branches originating from oblique lines

The formation of a branch indicates the sudden formation of a backward wave. This
requires two features. (i) The refractory period must be short in relation to the
time interval between one activation and the next spontaneously occurring activation. (ii) Occasionally, the activation of a small group of cells lasts longer
than the refractory period of a neighboring cell such that a backward infection
can take place. In the simulation of the tent-like pattern of the shell of *Olivia
porphyria L.* (Fig. 3), the interaction has been designed in such a way that
branching occurs if the number of travelling waves becomes too few. (The number
of travelling waves becomes reduced due to the mutual annihilation of two travelling waves by collision.) To generate a branch, it has been assumed that a hormone-like, homogeneously distributed substance is produced by the activated cells
which has an inhibiting influence on the inhibitor decay. Thus, whenever the number of activated cells becomes too small, the inhibitor life time becomes short-
ened such that the cells are shifted from the oscillatory mode into a steady
state. After passing the refractory period of a neighboring, still oscillating
cell, a backward wave is initiated. This causes an increase of the number of
travelling waves and thus an increase in the inhibitor lifetime. All cells return
to the oscillating mode and the formation of the branch is completed. This global
control of branching leads to the formation of several branches at a particular
time, in agreement with the natural pattern.

In a pattern forming process in which a switch between oscillatory and steady
state activator production is involved, the appearance of branches depends criti-
cally on the ratio of the maximum activator concentration during the autocatalytic burst and the steady state activator production. If the latter is low,
during branch formation many cells have to switch from the oscillatory to the
steady state mode of activator production in order to increase the hormone concentration in such a way that the system returns to the oscillatory mode. The effect can be seen by a comparison of simulation of the pattern on *Conus marmoreus*
(Fig. 4) and that of *Olivia porphyria* (Fig. 3). The simulation in Fig. 4 is based
on an activator-depleted substrate model (1). Similar as in Fig. 3, it has been
assumed that all activated cells produce a hormone. In this case, the hormone inhibits substrate production. Whenever the number of activated (pigment-producing) cells becomes too low, the substrate production is increased with the consequence that the activated cells remain activated such that they will infect their
neighbors. Many cells must become activated before a switching back to the oscillatory mode is possible. This transition is marked by the beginning of the white
triangles. The resulting pattern consists of a pigmented, triangle-like branching
pattern which enframes white triangles in between.

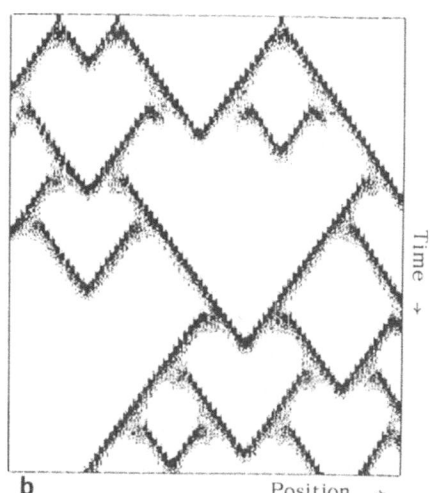

a

b

Time →

Position →

Fig. 3: Formation of branches. (a) A shell of *Olivia porphyria L.* Oblique lines indicate the formation of travelling waves of pigment production. If two waves meet each other, both become extinct. A branch indicates the sudden formation of a backward wave. (b) Model: branching occurs whenever the number of travelling waves drops below a certain threshold value by a shift of the cells from the oscillatory into a steady state activator production until, a backward wave is initiated. (Calculated with (2) and $\rho = \mu = 0.1$, $\kappa = 0.25$, $D_a = 0.015$, $D_h = 0$, $\nu = 0.014$, $\partial R/\partial t = 0.1$ (Σ a/n - R), R is uniformly distributed, n: total number of cells). In (2), ν is substituted by ν/R and $1/h$ by $1/(0.1+h)$. A higher ν value would lead to a higher density of lines

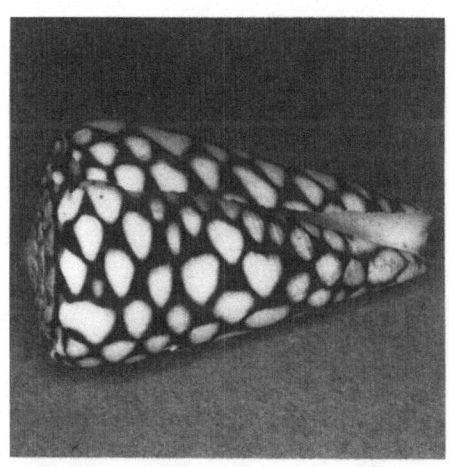
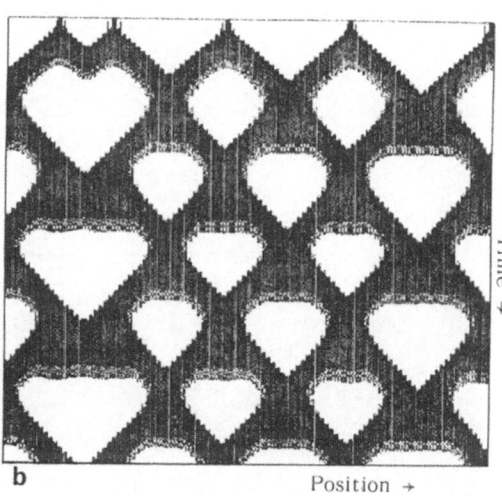

a

b

Time →

Position →

Fig. 4: White triangles on a pigmented background. (a) *Conus marmoreus.*
(b) Model: assumed is an activator-substrate model (1) with the addition that the substrate production is inhibited by a hormone-like substance R. Whenever the number of travelling waves becomes too small (R-concentration too low), cells begin a steady state activator production for a prolonged period which allows the infection of neighbors. If the number of activated cells is sufficient the steady state activator production becomes instable and the cells again begin to oscillate (top edge of the white triangles). Calculated with (1), σ is replaced by σ/R; $\rho = 0.1 \pm 5$ % random fluctuations, $\mu = 0.1$, $\kappa = 3$, $\nu = 0.005$, $\sigma = 0.038$, $D_s = 0$. In addition, $\partial R/\partial t = 0.05$ (Σ a/n - R); n is the total number of cells

Crossings of oblique lines

Pigmentation lines which cross each other can be found on shells of *Tapes litter-atus L*. (Fig. 5). Crossings of oblique lines can be regarded as the formation of two backward waves at the position where two waves collide. Similar as in the case of branch formation, the backward waves can result from cells which remain activated for a prolonged period due to a shift into the steady state. In the simulation Fig. 5b, an activator-substrate mechanism has been assumed with para-meters such that a cell, once activated, would remain in a steady state. An addi-tional diffusible inhibitor is produced by the activated cells (3a-c):

$$\frac{\partial a}{\partial t} = \frac{\rho s a^{2*}}{1 + \gamma h} - \mu a + D_a \frac{\partial^2 a}{\partial x^2} \tag{3a}$$

$$\frac{\partial h}{\partial t} = \nu(a-h) + D_h \frac{\partial^2 h}{\partial x^2} \tag{3b}$$

$$\frac{\partial s}{\partial t} = \sigma - \frac{\rho s a^{2*}}{1 + \gamma h} - \epsilon s \qquad \text{where} \tag{3c}$$

$$a^{2*} = \frac{a^2 + \rho_o}{1 + \kappa a^2} \quad .$$

A travelling wave results since each newly activated cell extinguishes the activation of the preceding cell. If two waves collide, no newly activated cell is available to extinguish the activation of the cells at the point of collision. These cells would remain in the steady state until one or two backward waves are triggered. The newly activated cells of the backward waves extinguish the pro-longed activation of the cells at the point of collision.

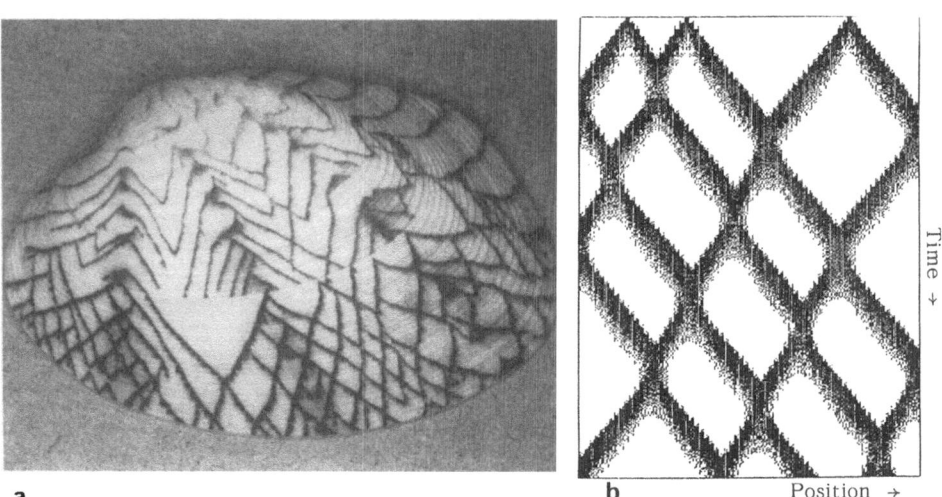

a   b   Position →

Fig. 5: Crossings of pigmentation lines. (a) Pattern on *Tapes spec*. Crossings indicate that two waves do not annihilate each other at a point of collision but either penetrate each other, or the collision initiates two backward waves.
(b) Model: assumed is a three-component system; an activator-substrate mechanism tuned to produce a steady state activation, plus a diffusible inhibitor. Calcu-lated with (3), $\rho = 0.08 \pm 5\%$ random fluctuations, $\mu = 0.08$, $D_a = 0.01$, $\kappa = 0.5$, $\gamma = 3$, $\nu = 0.02$, $D_h = 0.4$, $\rho_o = 0$, $\sigma = 0.1$, $\epsilon = 0.005$

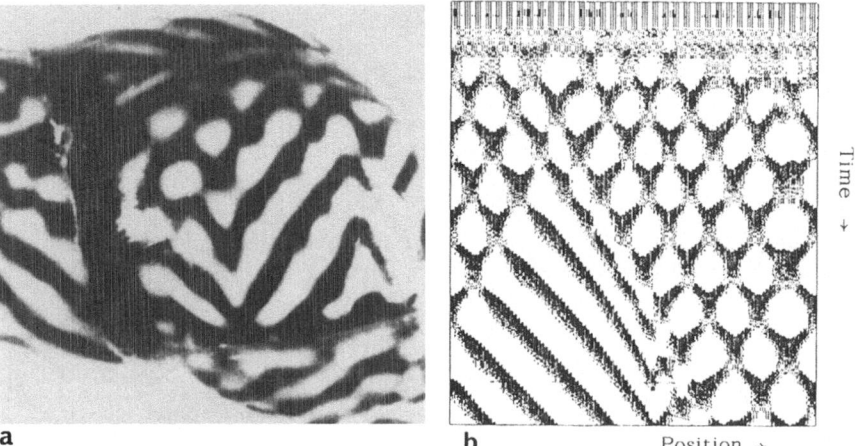

Fig. 6: Meshwork and checkerboard-like pattern. (a) Patterns on shells of *Bankivia fasciata*. (Photograph kindly supplied by J. Campbell, see [2]). The patterns can consist of oblique lines of either orientation or in superimposition, forming a meshwork-like pattern. (b) Model: assumed is one autocatalytic reaction, antagonized by a non-diffusible substrate which becomes depleted during autocatalysis, plus a diffusible inhibitor. Calculated with (3) and ρ = 0.05 ± 7.5 % random fluctuations, μ = 0.05, ρ_o = 0.005, κ = 1, γ = 2, D_a = 0.01, D_h = 0.4, σ = 0.03, ϵ = 0, ν = 0.028

Formation of meshwork-like patterns

The formation of a meshwork- or checkerboard-like pattern requires a spatial periodic and simultaneously a time-wise periodic pattern. In addition, the pigmented and non-pigmented regions must have the same extension in both dimensions, space and time.

The mechanism outlined for the generation of crossings, based on the superposition of two inhibitory reactions, is also suited to generate a meshwork-like pattern. One inhibitory substance diffuses rapidly, generating in this way a periodic spatial pattern (see Fig. 1). The second inhibitory substance does not diffuse but has a long time constant. It causes therefore a periodic pattern in time (see Fig. 2). In the example of (3), this inhibition results from the depletion of a substrate. The superposition of both inhibitory effects leads to patterns which are periodic along the space as well as along the time coordinate. However, a series of oblique lines also has this feature. Therefore, depending on the parameter, the one or the other type of pattern is formed and even an instability can lead to a transition between the two types of pattern (Fig. 6). This ambiguity in the pattern is typical for shells of *Bankivia fasciata* (see [2]).

Superposition of a spatially stable and temporally periodic pattern

Many shell patterns can be explained under the assumption that they are generated by the superposition of two patterns, one which is stable in time and a second one which is based on oscillations. The stable pattern controls the parameters of the oscillations. Two examples will be given. In the two simulations, Fig. 7 and 8, it has been assumed that the production rate of the substrate (σ in (1)) is a function of the position: $\sigma = \sigma(x)$. In the simulation of the shell pattern of *Amorina undulata L.* (Fig. 7) the cells oscillate faster in the region of high σ compared with those in a region of low σ. Instead of lines parallel to the grow-

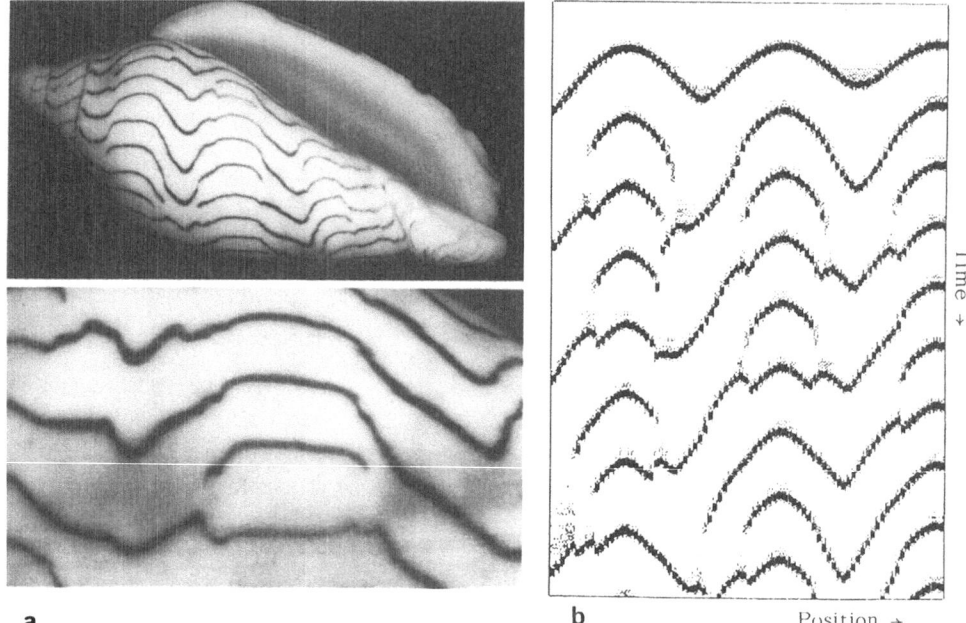

a   b   Position →

Fig. 7: Wavy lines. (a) Shell of *Amorina undulata*. (b) Simulations: Assumed is a stable pattern of $\sigma(x)$ (substrate production) which leads to a higher oscillation frequency in regions of high σ. A moderate activator diffusion enforces a reduction of the phase difference among neighboring cells. If the phase difference is not too large, a just activated cell can activate its somewhat delayed neighbors. The steepness of the lines is an indication of the readiness of a cell to become activated. If the phase difference becomes too big, a trigger of the neighboring cells is not possible; the pigmentation line terminates abruptly. These cells skip one oscillation. Calculated with (1), $\rho = 0.1 \pm 2\%$ random fluctuations, $\mu = 0.01$, $\rho_o = 0.01$, $\kappa = 0.5$, $D_a = 0.004$, $\nu = 0$, $\rho_1 = 0.01 \text{-} 0.039$

ing edge (Fig. 2), the lines obtain a wavy shape. Since the cells oscillate with different frequencies, an increasing phase difference between the oscillating cells would accumulate with time. However, diffusion of the activator leads to a partial synchronization of the oscillating cells as long as the phase difference does not become too large. Otherwise, the synchronization fails with the consequence that a line of pigmentation terminates.

The model reproduces correctly fine details of the pattern of *Amorina undulata*. If a pigmentation line terminates, the next pigmentation line has a wriggle which points towards the terminated line. According to the model the cells which fail to become activated (since the phase difference has been too large) remain unactivated for a prolonged period. This can lead to a spontaneous trigger and thus to a Λ-like pattern element.

In the simulation of the shell pattern of *Cyprea ziczac L.* (Fig. 8) the cells in the regions of high substrate production (high $\rho(x)$) remain in a steady state. Thus, in that region pigment is deposited permanently. Stripes parallel to the direction of growth appear. These permanently activated cells trigger the cells between these stripes which are in the oscillating mode. Thus, a series of travelling waves (oblique lines) are initiated at regular intervals at both sides of the steady state line. Since two oppositely moving waves extinguish each other, the oblique lines have the shape of a V.

200

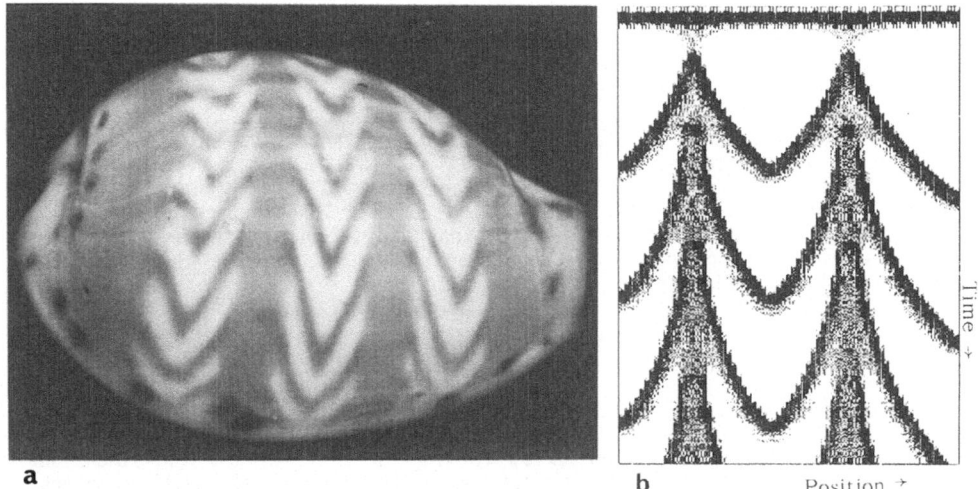

a   b   Position →

Fig. 8: Fish-bone-like pattern resulting from the combination of permanent and oscillatory pigment deposition. (a) Shell of *Palmadusta ziczac*. (b) Model: assumed is an activator-substrate model. In regions where substrate production $\rho(x)$ is high, cells enter into a steady state activation, causing the pigmented bands parallel to the direction of growth. These bands of activated cells trigger periodically travelling waves (V-elements). Calculated with (1), $\rho = 0.1$, $\kappa = 0.5$, $\mu = 0.1$, $D_a = 0.01$, $\sigma_{max} = 0.09$, $D_s = 0.05$, $\nu = 0.002$

A very frequent pattern on shells consists of columns of dots or patches. Attempts to simulate these patterns by a two-component system (Equations (1) or (2)) have shown that this is possible only in a very narrow range of parameters

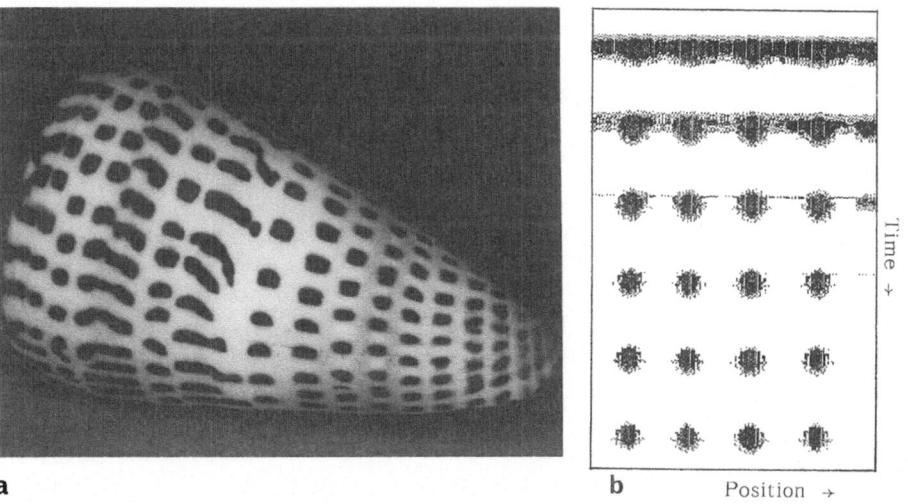

a   b   Position →

Fig. 9: Rows of patches. (a) Shell of *Conus litteratus*. (b) Model: assumed is an oscillating activator-inhibitor system (1) and a positive feedback of $\underline{a}$ on ρ. With each oscillation, the originally homogeneous ρ distribution becomes more and more patterned which allows the separation into individual spots. Calculated with (2), $\rho_o = 0.03 \pm 2.5\%$ random fluctuations, $\mu = 0.03$, $\rho_o = 0.3$, $D_h = 0.4$, $\nu = 0.01$, $\rho_1 = 0.001$ and $\partial\rho/\partial t = 0.001$ (a-ρ)

since the autocatalytic activation proceeds too rapid to allow that some cells suppress other cells in the neighborhood. However, a simulation is possible under the assumption that a spatially stable pattern similar to those shown in Fig. 1 exists which controls the oscillation frequency. This stable pattern can be generated by the oscillating pattern itself. For the simulation of the pattern of *Conus litteratus* (Fig. 9), it has been assumed that the activation has a positive feedback on the source density ($\rho(x)$ in (1) and (2)). Since the change of ρ is assumed to be much slower than that of the activator, a somewhat increased $\underline{a}$-concentration during one activator burst will lead to a long-lasting ρ-increase and thus to an advantage in the next oscillation, and so on. Therefore, cells which produce a bit more activator have a better chance to suppress their neighbors during the next oscillation.

Conclusion

The overwhelming richness of patterns on shells of molluscs can be explained by relatively simple reaction-diffusion mechanisms acting along one axis and their permanent record along the other axis, forming in this way the two-dimensional shell pattern. Most of the patterns are explicable by nearest-neighbor interactions, in some cases they have to be supplemented by an evenly distributed hormone-like substance. By computer simulations, one can only uncover the logic behind the pattern formation. It is not yet clear how these principles are biologically realized.

References

1. H. Meinhardt: J. Embryol. exp. Morph. <u>83</u> (Supplement), 289 (1984)
2. B. Ermentrout, J. Campbell, G. Oster: The Veliger <u>28</u>, 369 (1985)
3. H. Meinhardt, M. Klingler: J. theor. Biol. (1987), in press
4. A. Gierer, H. Meinhardt: Kybernetik <u>12</u>, 30 (1972)
5. H. Meinhardt: <u>Models of Biological Pattern Formation</u> (Academic Press, London 1982)
6. H. Meinhardt, M. Klingler: In <u>Lecture Notes in Biomathematics</u> (1987), in press
7. H. Meinhardt, A. Gierer: J. Cell Sci. 15, 321 (1974)
8. G. Gerisch: Curr. Top. Dev. Biol. <u>3</u>, 157 (1968)

Nonlinear Polymerization of Fibrinogen

R. Rigler[1], *U. Larsson*[1,2], *and B. Blombäck*[2]

[1]Department of Medical Biophysics, Karolinska Institute,
 Box 60400, S-10401 Stockholm, Sweden
[2]Coagulation Research, Karolinska Institute,
 Box 60400, S-10401 Stockholm, Sweden

Conversion of fibrinogen into fibrin by a polymerization process is coupled to periodic conformational transitions which are induced by the catalytic cleavage of fibrinogen. During the polymerization a regular 3-dimensional network is formed. Its pore size, as well as the oscillatory transitions, depends on the activity of thrombin, which catalyses the cleavage of fibrinogen. The formation of the fibrin network appears as an ideal model system for studying the formation of ordered 3-dimensional biological structures from a random distribution of structural elements.

1. Introduction

Fibrinogen is an elongated protein consisting of 3 peptide chains in a symmetric arrangement (Fig. 1). It has been an object of continuing interest since its conversion into fibrin is of central importance for the mechanism of blood clotting. Different structures for the fibrinogen molecules have been suggested mainly from studies by electron microscopy /1,2/ of small angle X-ray scattering /3/ and small angle neutron scattering /4,5/ as well as of the hydrodynamic properties /6-8/:

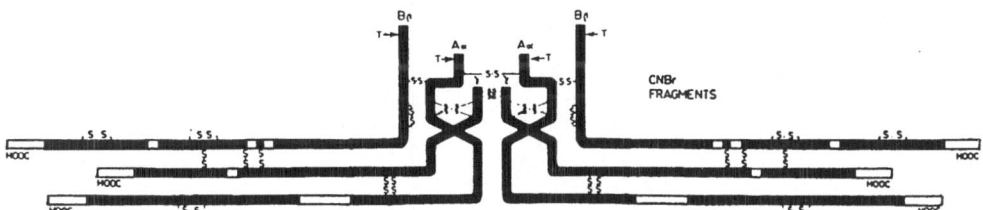

Fig. 1: Structure of the fibrinogen molecule deduced from chemical studies. Cleavage sites indicated by arrows at the top of Aα and Bβ chains /9/.

Fibrinogen is an elongated molecule of 450 nm length and 9 nm cross section with varying degrees of flexibility. The conversion of fibrinogen to fibrin is initiated by limited proteolysis of the molecule. Thrombin releases fibrinopeptide A (FPA) and fibrinopeptide B (FPB) from fibrinogen while a small venom enzyme, Batroxobin, releases FPA only, leading to differences in the polymerization and gelation procedure /10,11/.

2. Early Stages of Polymerization

A sensitive way to study fibrinogen polymerization is an analysis of the time-dependent hydrodynamic properties as evaluated by dynamic laser light scattering. From the fluctuations of coherently scattered laser light information on translational (D_T) and rotational diffusion constants (D_R) can be obtained by the auto-correlation function $G(\tau)$ /13/

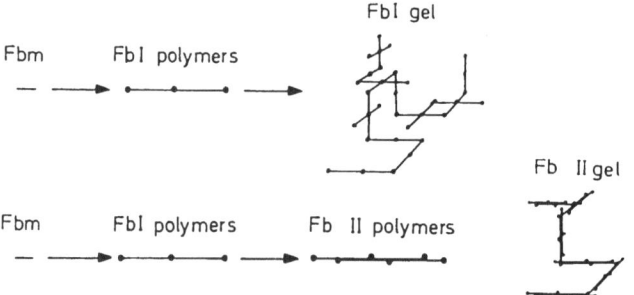

Fig. 2:Scheme of the fibrin polymerization as induced
by Batroxobin (Fibrin I gel) or thrombin (Fibrin II gel)/11/

$$G^2(\tau) = (A_1\exp(-D_Tq^2t)+A_2\exp(-(D_Tq^2+6D_R)t))^2+c \ .$$

From the dependence of $G(\tau)$ on the wave vector q, D_T and D_R can be obtained /8/.
For fibrinogen alone the contributions of D_R are small but measurable and
increase with increasing length of the fibrin oligomer during polymerization /8/.
Both D_T and D_R depend on size and shape of fibrinogen as well as on the frictional
force in the diffusion process.

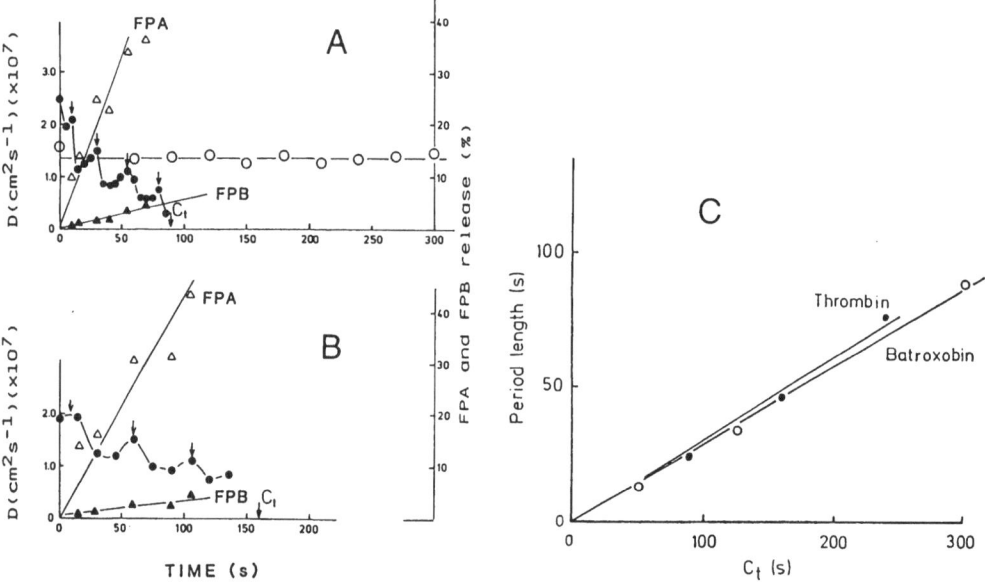

TIME (s)

Fig. 3:The time dependence of the apparent diffusion constant (D) (●)
from DLS experiments performed on human fibrinogen, 2 mg ml^{-1}, in Tris-
imidazole buffer, pH 7.4, 20 mM CaCl$_2$ activated by different concentrations of thrombin: (A) 0.4 NIH-units ml^{-1}; (B) 0.2 NIH-units ml^{-1}
(final concentrations). The release of FPA (△) and FPB (▲) is also
shown for each sample as a percentage of the total fibrinopeptide
released upon 100% activation. The clotting time is shown by an arrow
marked C_t. The time intervals between two adjacent arrows above the
peak values of D are the period length. A buffer blank (0.15 M NaCl)
is shown in panel A (○); (C) Period length vs clotting time, C_t,
for thrombin (●) and Batroxobin (○) induced polymerizations /8/

204

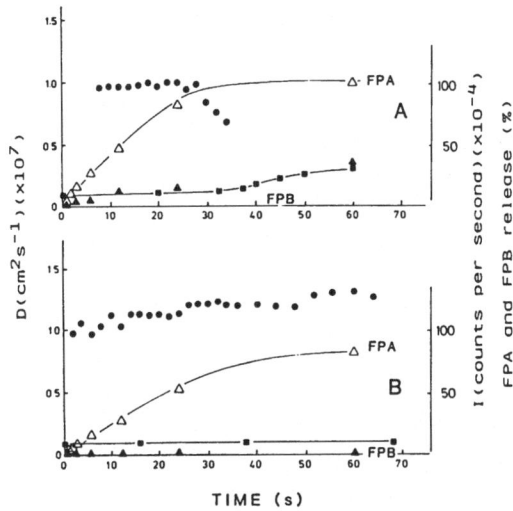

Fig. 4:The time dependence of the apparent diffusion constant /D/ (●), scattered intensity /I/ (■), FPA- (△) and FPB- release (▲) from DLS experiments performed on a dysfunctional fibrinogen, fibrinogen Aarhus, 2 mg ml^{-1}, in Tris-imidazole buffer, pH 7.4, 20 mM CaCl$_2$ activated by: (A) 0.17 NIH-units m.$^{-1}$ thrombin,(B) 0.10 B-units ml^{-1} Batroxobin /8/

In order to follow the polymerization process we have measured the time depen-dence of the apparent diffusion constant D_{app} which can be regarded as a weighted mean of contributions from translational and rotational processes. It was discovered that the value of D_{app} of fibrinogen in the course of polymerization showed periodic oscillations superimposed on a general decrease due to increasing molecular size. These periodic fluctuations clearly exceed stochastic perturbations, moreover their frequency is related to the concentration of Thrombin as well as of Batroxobin (Fig. 3).

With enzyme concentrations catalyzing a slow polymerization of fibrinogen slow frequency oscillations are found, while high enzyme concentrations leading to rapid polymerization result in high frequencies. It is interesting no note that in dysfunctional fibrinogen such as fibrinogen Aarhus /12/ with abnormal fibrin poly-merization no such oscillations can be observed (Fig. 4).

We have suggested, as explanation for the periodic fluctuations in the observed diffusion properties, conformational transitions due to bending and flexing of the fibrinogen molecules in concert with release of fibrinopeptides A and B during the polymerization /8/. The oscillations in the diffusion properties are observed during the linear phase of increased static light scattering. In normal fibrinogen this period is followed by a non-linear increase in the scattered intensity during the gelation phase when the fibrin network is formed. In fibrinogen Aarhus this n network formation apparently is disturbed (Fig. 4).

3. Formation of a 3-Dimensional Fibrin Lattice

Usually fibrin appears as a rather irregular structure. However, if the polymeriza-tion process can occur undisturbed a regular 3-dimensional network is formed as seen by scanning electron microscopy /14/ (Fig5c,d). Special precautions were taken to stabilize the gel structure under the freeze drying process required for imaging. Though⁻ the network is somewhat disturbed due to the shrinkage process in freeze drying,a highly ordered lattice of cubic and/or hexagonal arrangement can be seen. The size of the lattice unit varies between ca 5-10 µ (diameter), which means a size even suited for light microscopic observation. A regular 3-dimensional network similar to that observed by SEM is found in the phase contrast microscope (Fig. 5 a,b). Furthermore, the pore size depends on the enzyme concentration /14/. Low enzyme concentrations were found to cause the formation of large pore sizes and vice versa.

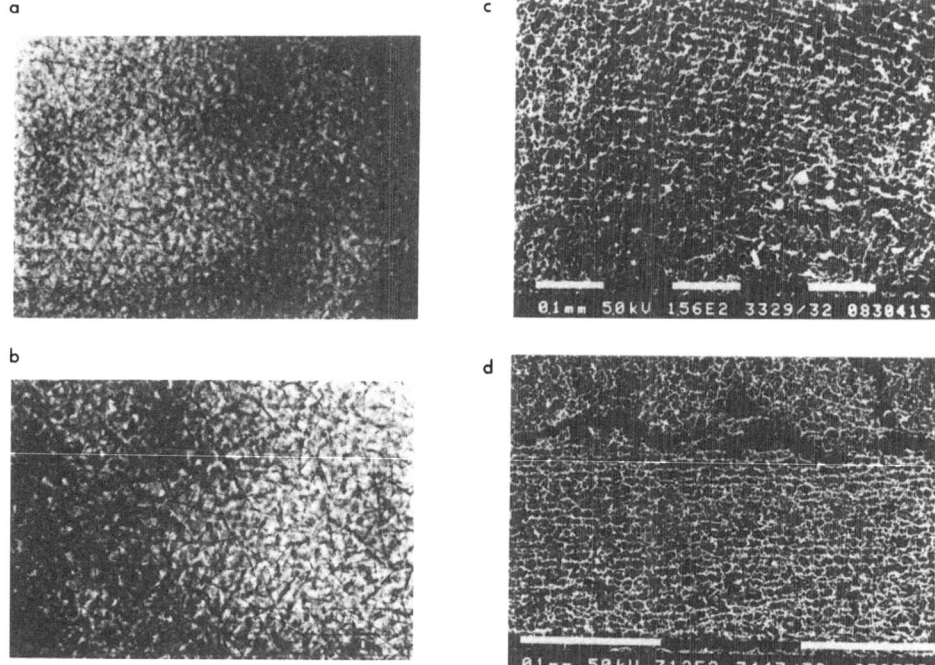

Fig. 5: Micrographs from light microscopy on fibrin gels (2.24 g/l fibrinogen) in 0.04 M Tris-imidazole buffer, pH 7.4, containing 20 mM CaCl$_2$ and (a) 0.15 NIH-units/ml; (b) 0.01 HIH-units/ml and micrographs from scanning electron microscopy on freeze dried fibrin gels (∿2g/l fibrinogen) in Tris-imidazole buffer, pH 7.4, containing 20 mM CaCl$_2$ and dextran at two different enzyme concentrations, (c) and (d) /14/. White bars = 100 μm

Light microscopy has moreover the advantage that the formation of the fibrin structure can be visualized continuously during the polymerization process. With the resolution of the light microscope a pentamer fibrinogen polymer would be expected to be visualized.

4. Oscillations and Formation of Regular Structures

Polymerization of fibrinogen constitutes an irreversible process involving various non-equilibrium steady states, a prerequisite for the formation of dissipative structures /15/. Oscillations coupled with the space order have been found for chemical as well as biochemical systems involving autocatalytic steps. In one of the most wellknown (Belusov-Zhabotinski) reaction time oscillations of the concentrations of Ce^{3+} and Ce^{4+} are observed which also lead to a transient or dissipative space order. With regard to biochemical systems, oscillations during glycolysis coupled to the formation of regular patterns have been described in detail /16,17/.

Although the observed oscillations in fibrinogen polymerization have been attributed to periodic conformational transitions in the course of an irreversible process driven by a gradient in chemical activity, we cannot exclude that part of the signal fluctuations we observe are due to an oscillating spatial arrangement of fibrinogen molecules. Fibrinogen is of a size that oligomers will diffract laser light, and variations in size as well as concentrations will cause fluctuations in laser light.

Obviously there exists a link between the frequency of the observed oscillation and the formation of the final pore size in the fibrin gel, which is also
supported by the inability of the abnormal fibrinogen Aarhus to undergo periodic
oscillations and most likely also form a fibrin network. There is no reason for
the formation of a regular periodic fibrin structure from stereochemical considerations, rather one would expect a random network of longitudinally polymerized
fibrinogen with a random array of cross bridges. It is thus very likely that the
formation of the periodic fibrinogen network is a consequence of chemical oscillations in space and time.

5. Outlook

Fortunately early stages of fibrinogen polymerization can be observed with the
light microscope as well as by the diffraction of laser light. Thus the formation
of ordered 3-dimensional structures like the fibrin lattice from a random arrangement of structural elements can be followed up under various conditions.

The fibrinogen-fibrin transition appears to be a very interesting model
for the study of formation of ordered biological structures and it may also apply
to other polymerizing proteins involved in e.g. the formation of the cytoskeleton,
such as tubulin.

6. Acknowledgement

This work was supported by grants from the Swedish Medical Research Council, the
Swedish Natural Science Research Council, the National Institute of Health,
Bethesda, Maryland and the K. and A. Wallenberg foundation.

7. References

1. C.E. Hall and H.S. Slayter: J. Biophys. Biochem, Cytol. 5, 11-15 (1959)
2. L. Bachmann, W.W. Schmitt-Fumian, R. Hammel and K. Lederer: Makromol. Chimie
   176, 2603-2618 (1975)
3. K. Lederer and R. Hammel: Makromol. Chem. 176, 2619-2639 (1975)
4. G. Marguerie and H.B. Stuhrmann: J. Mol. Biol. 102, 143.156 (1976)
5. U. Larsson, R. Rigler, B. Blombäck, K. Mortensen and R. Bauer: In Structure
   Dynamics and Function of Biomolecules, ed. by A.Ehrenberg et al., Springer
   Series in Biophysics, Vol. 1 (Springer Series in Biophysics), p. 152-158
6. P. Wiltzius, W. Känzig. V. Hofmann and P.W. Straut: Biopolymers 20, 2035-2049
   (1981)
7. A.U. Acuna, J. Gonzalez-Rodrigues, M.P. Lillo and K. Razi Naqvi:
   Biophysical Chemistry 26, 63-70 (1987)
8. U. Larsson, B. Blombäck and R. Rigler: Biochim. Biophys. Acta in press (1987)
9. B. Blombäck, P.H. Hogy, B. Gårdlund, B. Hessel and B. Kudryk: Throm. Res. 8,
   Suppl. II, 329-346 (1976)
10. T.C. Laurent and B. Blombäck: Acta Chem. Scand. 12, 1875-1877 (1958)
11. B. Blombäck and M. Okada: Thromb. Res. 25, 251-257 (1982)
12. B. Hessel, S. Stenbjerg, B. Judryk, L. Therkildsen and B. Blombäck: Throm. Res.
    42, 21-37 (1986)
13. B.J. Berne and R. Pecora: Dynamic Light Scattering, J.Wiley & Sons, New York
    1976.
14. B. Blombäck, M. Okada, B. Forslind and U. Larsson: Biorheology 21, 93-104
    (1984)
15. P. Glansdorff and I. Prigogine: Thermodynamic Theory of Structure, Stability
    and Fluctuations. Wiley-Interscience, London 1971
16. B. Hess et al.: In Membranes, Dissipative Structures and Evolution, ed. by
    G.Nicolis and R.Lefever, Advances in Chemical Physcis, Vol. 29 (Wiley,
    New York), p. 137
17. S.C. Müller, Th. Plesser and B. Hess: Naturwissenschaften 73, 165-179 (1986)

Part IV

Cellular and
Intercellular Organization

Dynamic Phenomena in Molecular and Cellular Biology

L.A. Segel

Department of Applied Mathematics, The Weizmann Institute of Science, Rehovot 76100, Israel

Over many decades, and with accelerating pace recently, many fascinating dynamic phenomena have been revealed by mathematicians and other theoreticians. The mathematical side is well reviewed in the book by GUCKENHEIMER and HOLMES [1]. The series of books on Synergetics (e.g. HAKEN [2]) stresses qualitative nonlinear phenomena common to many sciences.

A central question for theoretical and experimental biologists is whether the various nonlinear phenomena are of major importance in the biological sciences. I shall concentrate on molecular and cellular biology, in which I am particularly interested. (These areas have certainly interested Benno Hess, who has made salient contributions to their elucidation.) But it should not be forgotten that biology contains fascinating intramolecular dynamics, as well as dynamics on supracellular levels of organisms and populations.

I will proceed by discussing a key special case, presenting a brief survey of dynamic phenomena that affect the cellular slime molds. As a framework, it is first necessary to review briefly the life cycle of these organisms (see for example RAPER [3]). We shall concentrate on the most studied species *Dictyostelium discoideum*. One can start with the spores, a dormant stress-resistant stage. If conditions are appropriate, the spores will germinate, releasing slime mold amoebae. These creatures are about 20 microns in diameter, and move about by extending pseudopods. They eat bacteria, and as long as food is present will continue multiplying by cell division. Under ideal circumstances, such divisions can occur every three hours.

When the food supply is exhausted, a wide variety of biochemical and phenomenological changes are triggered. In particular, after about 8 hours, the cells shift from an undirected random movement to a "purposeful" aggregation toward more or less evenly spaced foci. An aggregate of 10,000 to 100,000 cells moves about for some time in a worm-like form, heading towards light, moisture and other favorable conditions. After a time, the worm-like slug stops and goes through a sequence of controlled cell movements that eventually result in the production of a lollipop shaped structure. An assemblage of dead cellulose-filled stalk cells support a spheroid composed of a large number of spores. Typically about 20% of the cells become stalk, no matter what the size of the initial cell assemblage. If the spores are somehow carried away to a more favorable location, the cycle begins again.

It has been established for some time that the aggregation of the cells is mediated by a chemical called cyclic-AMP (cAMP). The cells are *chemotactic* toward this chemical, that is they move towards relatively high concentrations of it. Laboratory experiments have revealed the following sequence of events. From zero to 6 hours after starvation, the cells typically secrete a very small amount of cAMP, if any. After 6 hours, they become excitable. That is, if cAMP is added to the extracellular medium in superthreshold amounts, the cells respond by synthesizing cAMP internally and then secreting it. After 8 hours, a collection of cells will rhythmically secrete cAMP with a period of 5 to 10 minutes. This secretion pattern is believed to play a dominant role in the aggregation of *Dic-*

tyostelium discoideum. Certain cells become autonomously pulsing centers. These secrete cAMP, to which a concentric ring of cells responds (because they perceive a positive cAMP gradient) by moving towards the center. At the same time they themselves are triggered to secrete cAMP, which passes to the next outermost ring of cells, and so the process continues.

In considering theoretical analyses of cellular slime mold behavior, let us begin with the area which was historically examined first, collective patterned movement. KELLER and SEGEL [4] incorporated both diffusive-like random movement and chemotaxis in an equation for the amoebae density a (where for simplicity we consider only a single spatial variable x):

$$\frac{\partial a}{\partial t} = - \frac{\partial}{\partial x} \left[- \mu \frac{\partial a}{\partial x} + \chi a \frac{\partial \rho}{\partial x} \right] . \tag{1a}$$

A second equation for the attractant density ρ embodied secretion, destruction, and diffusion:

$$\frac{\partial \rho}{\partial t} = fa - k\rho + D \frac{\partial^2 \rho}{\partial x^2} . \tag{1b}$$

It was shown that the onset of aggregation could be explained from these equations as an instability of the uniform state brought about by a slow increase in the cAMP secretion rate f and the chemotactic sensitivity χ. Many further analyses along these general lines were carried out for example by HAGAN and COHEN [5] who showed that suitably generalized equations could account for the spiral aggregative patterns that are sometimes seen. Simulations (PARNAS and SEGEL [6]) seem to become necessary when details of the complex behavior are taken into account. Notable is the work of MacKAY [7] who beautifully demonstrated the stream formation which is frequently observed. It is perhaps time to incorporate much new knowledge into such simulations, particularly concerning adaptation (see below).

It seems fairly clear that an instability mechanism does indeed operate for simple slime mold species without the excitability feature, while much is understood concerning how spontaneous oscillation and excitability lead to aggregation in *Dictyostelium discoideum.* Nonetheless, fundamental questions remain open. In *Dictyostelium,* for example, is aggregation essentially an instability in a near-uniform cell population or is it primarily induced by certain special cells who then somehow inhibit nearby cells from becoming independent oscillators (see WADDELL [8])? How is the observed "territory size" determined?

Research on collective cell movement in the slime mold has been stimulated recently by the experimental and theoretical investigations of slug movement by ODELL and BONNER [9] (also see BONNER et al. [10]). This work includes an elegant theory for how differential chemotaxis can overcome what may be termed the "beggar thy neighbor" problem. Each cell must get a purchase on its surroundings in order to move forward, but in doing so it pushes its neighbors backward. If all cells do this, how can there be overall progress? Other ideas for solving this problem have been put forward, albeit only in descriptive form, by WILLIAMS et al. [11].

The fact that stalk-spore proportions are already evident in the slug has inspired a large number of theoretical papers. A fundamental outstanding problem here is whether the proportions are first determined in a mixture of cells, followed by sorting out of the two different cell types. Or, alternatively, are the two cell types established at once in their eventual order, mediated perhaps by some sort of chemical gradient, with the majority of prestalk cells in the anterior 20% of the slug? The latter approach is a continuation of a line beginning with TURING [12], GMITRO and SCRIVEN [13], NICOLIS and PRIGOGINE [14], and GIERER and MEINHARDT [15] with more recent contributions exemplified by the papers of OTHMER and PATE [16] and KHAIT and SEGEL [17]. Examples of sorting models are those of MEINHARDT [18], SEKIMURA and KOBUCHI [19], and GRINFELD and SEGEL [20].

Let us turn now to the level of molecular physiology. We have already mentioned that slime mold cAMP oscillations play an important part in cell aggregation. In addition, they play many roles in further cell development. To give just one example, even though aggregation in the primitive slime mold *D. minutum* is mediated by the continuously secreted chemoattractant folic acid, yet pulses of cAMP govern fruiting body formation (SCHAAP et al. [21]). These oscillations were studied originally by GOLDBETER and SEGEL [22], who proposed a model for *D. discoideum* cAMP secretion that is closely related to the one for glycolysis earlier studied (for example) by GOLDBETER and LEFEVER [23] and BOITEUX et al. [24]. The model concerns dimensionless concentrations of the substrate of the chemical reaction α (intracellular ATP), intracellular cyclic-AMP concentration β, and extracellular cyclic-AMP concentration γ. The equations studied are as follows:

$$d\alpha/d\tau = 1 - \sigma\phi(\alpha,\gamma), \tag{2a}$$

$$d\beta/d\tau = q\sigma\phi(\alpha,\gamma) - k_t\beta, \tag{2b}$$

$$d\gamma/d\tau = k_t h^{-1}\beta - k\gamma, \quad \text{where} \tag{2c}$$

$$\phi(\alpha,\gamma) \equiv \frac{\alpha(1 + \alpha)(1 + \gamma)^2}{L + (1 + \alpha)^2(1 + \gamma)^2} . \tag{2d}$$

The equation for ATP describes a constant input together with an extracellular cAMP-influenced synthesis of intracellular cyclic-AMP. There are terms that represent secretion of intracellular cyclic-AMP into the extracellular medium and also the observed destruction of extracellular cAMP by the enzyme phosphodiesterase.

The model (2) is capable of explaining many observed dynamical phenomena. In particular it predicts relay or excitability, with thresholds for stimuli as well as absolute and relative refractory periods. The major point of the analysis comes in the postulation of a *developmental path* in parameter space, wherein slow changes of parameters result in automatic shifts in behavior - from passivity, to excitability, to autonomous oscillation.

There have been strong interactions between theory and experiment. For example experiments did not reveal the ATP oscillations that are a concomitant of model (2). To meet this objection Goldbeter and Segel were able to change parameter values so that the amount of ATP oscillation fell within the 10% error of the experiments.

Such a quantitative change in the model, however, was not sufficient to overcome another experimental finding. In the formulation above, the Goldbeter-Segel theory predicts that transitions to excitability and autonomous oscillation will occur upon the increase in activity of adenylate cyclase (the enzyme that catalyzes cAMP synthesis) and of phosphodiesterase (the enzyme that catalyzes external cAMP destruction), followed by a decrease in extracellular cyclase concentration. Such a decrease would follow from the known secretion of an adenylate cyclase inhibitor. However, secretion of the inhibitor turns out not to be timed properly to bring about a shift from relay to oscillation. Nonetheless, GOLDBETER and SEGEL [25] showed that their model could be accommodated to these facts by a qualitative change, namely the introduction of a term - $k\alpha$ to equation (2a).

It is natural to ask, why be concerned with such details when it is almost certain that the model (2) is incorrect in its oversimplification of the linkage between the extracellular cAMP receptor and the adenylate cyclase? The answer is that phase plane analyses show that only certain broad qualitative features of the model are required to yield the major qualitative behaviors that have been mentioned. It is a reasonable hope that more accurate models will retain these same general qualitative features.

Nonetheless, the model as it stands is falsifiable by experiments. DEVREOTES and STECK [26] demonstrated adaptation processes in cAMP secretion that were not explicable by model (2). This discovery initiated several theoretical works on adaptation. These works stand by themselves as providing possible explanations for adaptation on the molecular level, but the adaptation models can also be incorporated into the previous models, in essence by replacing the function Φ in (2) by a more complex mechanism. At present three such models seem in principle capable of explaining all of the qualitative features observed [27-31]. Decisive progress here awaits further experiments.

Let us summarize the dynamic phenomena which are exemplified in the cellular slime mold and on which significant theoretical and experimental research has been carried out: (i) various patterns in collectives of spatially isolated cells organized by chemical signals; (ii) regulated collective cell motion in tightly packed assemblages; (iii) patterned differentiation into two cell types; (iv) major qualitative shifts in cell signaling behavior.

In addition to the work already done, there are other categories of research that must be pursued in order to understand the basic dynamic behavior of the cellular slime molds. These include the following: (v) the complex organized cell movement that leads from the simple prestalk-prespore pattern in the slug to the final structure of the fruiting body with its mass of spores atop a delicate tapering stalk; (vi) on the individual cell level - the organization of cytoskeletal elements in such a way to give the observed gross cell motion, including chemotaxis.

Let us now turn to some general remarks concerning tasks for the future. Most important perhaps is to provide many indubitable examples of *important* dynamic behavior in biology. When I started writing this survey, I was not positive that we theorists could point decisively to any such examples. But it seems clear that the slime mold amoebae exemplifies more than half a dozen important dynamic phenomena. To mention just one more example, in yeast glycolysis, unequivocal demonstrations of several fascinating dynamical behaviors have been provided by Benno Hess and his associates. The functions of these behaviors have not yet been unequivocally demonstrated, but strong arguments have been advanced that the oscillations incr se efficiency (SCHELL et al. [32]).

The cellular slime mold exemplifies classical dynamic behaviors of steady state solutions with their domains of attraction, autonomous oscillations, excitability, and pattern-forming spatial instabilities. What about chaos? Here, in addition to hoping for a strong verification that such a phenomenon exists in various biological preparations, one could ask two further major questions. How much "biological variability" is in fact due to identifiable chaos rather than a mere concatenation of many uncontrollable factors? Also, is biological chaos "planned"? (For example, is there a chaotic motion generator underlying the random walk of bacteria and other creatures as they search for food?)

A subject in its infancy is the investigation of dynamic diseases, disorders that come about from failure of normal dynamics. WINFREE's [33] discussion of rotating waves in intestinal wall, heart muscle and brain provides several examples of such matters. (ELKAIM et al. [34] have even suggested that certain pathologies in family interactions can have dynamic bases.) Yet another topic is the evolution of dynamism: can one find in simple creatures a progression over the ages from simpler to more complex dynamic behavior? These and other topics suggest that we are only at the beginning of our understanding of the role of dynamics in biology.

References

1. J. Guckenheimer, P. Holmes: Nonlinear Oscillations, Dynamical Systems, and Bifurcations of Vector Fields (Springer, Berlin 1983)

2. H. Haken: Synergetics: An Introduction, 2nd ed. (Springer, Berlin 1978)
3. K. Raper: The Dyctyostelids (Princeton University Press, Princeton 1984)
4. E.F. Keller, L.A. Segel: Nature 227, 1365 (1970)
5. P.S. Hagan, M.S. Cohen: J. theor. Biol. 93, 881 (1981)
6. H. Parnas, L.A. Segel: J. theor. Biol. 71, 185 (1978)
7. S. MacKay: J. Cell Sci. 33, 1 (1978)
8. D.R. Waddell: J. Embryol. exp. Morph. 70, 75 (1982)
9. G.M. Odell, J.T. Bonner: Phil. Trans. Roy. Soc. London B312, 487 (1986)
10. J.T. Bonner, H.B. Suthers, G.M. Odell: Nature 323, 630 (1986)
11. K.L. Williams, P.H. Vardy, L.A. Segel: BioEssays 5, 148 (1986)
12. A.M. Turing: Phil. Trans. Roy. Soc. London B237, 37 (1952)
13. J. Gmitro, L. Scriven: In Intracellular Transport, ed. by K.B. Warren
    (Academic Press, New York 1966), p. 221
14. G. Nicolis, I. Prigogine: Self-Organization in Nonequilibrium Systems (Wiley,
    New York 1977)
15. A. Gierer, H. Meinhardt: Kybernetik 12, 30 (1972)
16. H.G. Othmer, E. Pate: Proc. Natl. Acad. Sci. USA 77, 4180 (1980)
17. A. Khait, L.A. Segel: J. theor. Biol. 110, 135 (1984)
18. H. Meinhardt: Differentiation 24, 191 (1983)
19. T. Sekimura, Y. Kobuchi: J. theor. Biol. 122, 325 (1986)
20. M. Grinfeld, L.A. Segel: J. theor. Biol. 121, 23 (1986)
21. P. Schaap, T.M. Konijn, J.M. van Haastert: Proc. Natl. Acad. Sci. USA 81,
    2122 (1983)
22. A. Goldbeter, L.A. Segel: Proc. Natl. Acad. Sci. USA 74, 1543 (1977)
23. A. Goldbeter, R. Lefever: Biophys. J. 12, 1302 (1972)
24. A. Boiteux, A. Goldbeter, B. Hess: Proc. Natl. Acad. Sci. USA 72, 3829 (1975)
25. A. Goldbeter, L.A. Segel: Differentiation 17, 127 (1980)
26. P.N. Devreotes, T.L. Steck: J. Cell Biol. 80, 300 (1979)
27. A. Goldbeter, J.L. Martiel: In Rhythms in Biology and Other Fields of Appli-
    cation, ed. by M. Cosnard, J. Demongeot, A. LeBreton (Springer, Berlin 1982)
    p. 173
28. H.G. Othmer, P.B. Monk, P.E. Rapp: Math. Biosci. 77, 35 (1985)
29. H.G. Othmer, P.B. Monk, P.E. Rapp: Math. Biosci. 77, 79 (1985)
30. L.A. Segel, A. Goldbeter, P. Devreotes, B.E. Knox: J. theor. Biol. 120, 151
    (1986)
31. M. Barchelon: M.Sc. Thesis, Dept. of Applied Math., Weizmann Institute of
    Science (1987)
32. M. Schell, K. Kundu, J. Ross: Proc. Natl. Acad. Sci. USA 84, 424 (1987)
33. A. Winfree: The Geometry of Biological Time (Springer, Berlin 1980) Ch. 14D
34. M. Elkaim, A. Goldbeter, E. Goldbeter-Merinfeld: J. Social Biol. Struct. 10,
    21 (1987)

Thermodynamics of Energy Conversion in the Cell

J.W. Stucki

Pharmakologisches Institut,
Friedbühlstraße 49, CH-3010 Bern, Switzerland

1. Exergetic Efficiency of Oxidative Phosphorylation

The aim of every evolved and adapted biological system should consist in extracting the quality of energy to do work, i.e. the exergy [1], from the available foodstuff in possibly the most efficient manner. This work may consist in muscular contraction, ion transport across a membrane, synthesis and assembly of the building blocks of the organism, etc. Common to all these processes is that they are driven by the energy-rich ATP. The most important source of ATP is oxidative phosphorylation which is localized within the mitochondria. In this process the exergy liberated from the combustion of reducing equivalents is converted into the formation of ATP. Therefore, oxidative phosphorylation can be considered as an exergetic energy converter. The aim of this paper is to give insight into some mechanisms by which the cell can optimize the exergetic efficiency of oxidative phosphorylation.

The quality of energy to do work or, in other words, the directed energy, has been called exergy in contrast to randomized thermal motion which is called entropy [1,2]. These two quantities are connected in the Maxwell-Gouy-Stodola equation

$$- \dot{\Lambda} = TS > 0 , \tag{1}$$

which shows that in every irreversible process exergy is converted into entropy. Note that for isothermal systems the exergy Λ is identical to the Gibbs free energy ΔG. The exergetic efficiency η of oxidative phosphorylation is defined as the ratio of output and input power [1,3,4]

$$\eta = - \frac{J_p X_p}{J_o X_o} , \tag{2}$$

where J_p and J_o are the net rates of ATP production and oxygen consumption and X_p and X_o are the phosphate potential and the redox potential of the oxidizable substrates, respectively. This exergetic definition of the efficiency is formally identical to the entropic definition. Conceptually, however, it seems to be more natural to keep the exergetic definition in mind, since an analysis based on the entropic definition would be tantamount to attempting to measure the efficiency of a modern society by the amount of waste it produces.

2. Optimal Efficiency and Conductance Matching

Many biological energy converters, in particular membrane bound processes such as oxidative phosphorylation, have been found experimentally to obey relations between flows and forces which are reminiscent of the classical Onsager relations as far as linearity and symmetry are concerned [4,5].

$$J_p = L_p X_p + L_{po} X_o \tag{3}$$

$$J_o = L_{po}X_p + L_oX_o \ .$$ (4)

The L_{ij} are the phenomenological coefficients of oxidative phosphorylation which have been determined by suitable measurements. In principle, these coefficients can be interpreted on the basis of kinetic models, as was done for example in mosaic nonequilibrium thermodynamics [6]. For the present purpose it is, however, sufficient to take them as purely phenomenological quantities. It is important to stress that the experimentally determined linearity and symmetry of oxidative phosphorylation is only valid within a narrow range of the phosphate potential from 10 to 16 kcal and may be due to a pseudolinearity and pseudosymmetry of the system far from equilibrium [7]. Therefore this linearity and symmetry does not imply thermodynamic coupling in the sense of Onsager as may apply to some systems close to equilibrium. Nevertheless, the validity of the relations (3,4) in our system entitles us to analyze the efficiency of oxidative phosphorylation by using the formalism of nonequilibrium thermodynamics and the theory of linear energy converters.

A detailed analysis has shown that linear and symmetric energy converters are characterized by an optimal efficiency which is a pure function of the degree of coupling $q = L_{po}/\sqrt{L_pL_o} \in (0,1)$ [3]:

$$\eta_{opt} = [q/(1 + \sqrt{1-q^2})]^2 \ .$$ (5)

This feature is illustrated in Fig. 1 where the efficiency is plotted as a function of the force ratio $x = ZX_p/X_o$ where $Z = \sqrt{L_p/L_o}$ is the phenomenological stoichiometry. In order to clamp oxidative phosphorylation at the steady state of optimal efficiency it is necessary to attach a load to this energy converter in the form of irreversible ATP utilizing reactions such as they occur in the cell. The necessary and sufficient condition for the load conductance L_l allowing an operation of oxidative phosphorylation at the state of optimal efficiency, irrespective of the degree of coupling, turned out to be [3,4]

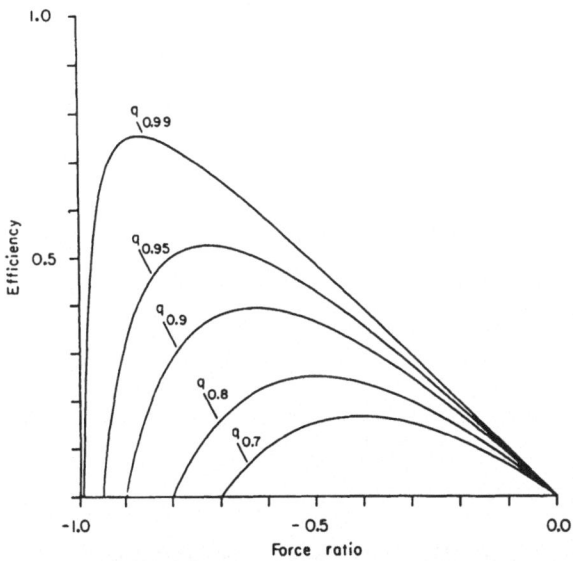

Fig. 1: *Efficiency as a function of the force ratio.* The efficiency η as defined in (2-4) is plotted as a function of the force ratio x for different values of the degree of coupling q as indicated in the figure

$$\frac{L_1}{L_p} = \sqrt{1-q^2} \; . \tag{6}$$

Several experiments in vivo have suggested that in a living cell oxidative phosphorylation operates at optimal efficiency and that hence the condition of conductance matching of the load was fulfilled [8,9], although it was completely unknown thus far, how the cell could sense this particular state.

3. Superstability of Optimal Efficiency

The state of optimal efficiency of oxidative phosphorylation is not characterized by an extremum of exergy decrease or entropy production [4]. Therefore I investigated whether conductance matching implied some distinguished dynamical properties of the system, in particular stability with respect to external fluctuations of the load conductance.

A sufficient proof of the globally asymptotic stability of oxidative phosphorylation with an attached load was already given in earlier work in the form of a Lyapunov function [10]. Hence the system is stable for all values of q and L_1 which implies that its dynamic equations are characterized by eigenvalues with negative real parts only [11]. However, this qualitative statement tells nothing about the relative stability of one steady state as compared to another. The basic idea behind relative stability is as follows: the more negative an eigenvalue of a particular state is, the more rapid perturbations are damped out and the more stable this state behaves towards external fluctuations. Consequently, the preferred steady state of the system is the one characterized by the most negative eigenvalues.

A detailed analysis of the eigenvalues of the efficiency $\lambda_\eta = \partial\dot{\eta}/\partial\eta$ revealed that at the state of optimal efficiency, i.e. when the condition of conductance matching (6) was fulfilled, these eigenvalues tended to diverge to $-\infty$. The detailed stability analysis of this system is too involved to be reproduced here and will be published elsewhere [12]. But Fig. 2, where λ_η is plotted as a func-

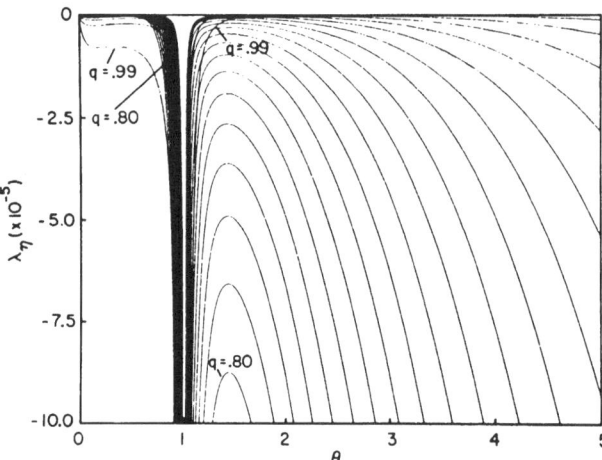

Fig. 2: *Superstability of the state of optimal efficiency.* The eigenvalue λ_η as described in the text was plotted as a function of the parameter θ which is a measure of the distance from conductance matching. Note that at $\theta=1$ the conductance is matched, else not. The matched state shows the most negative eigenvalues, implying biggest stability

tion of the parameter θ which measures deviation from conductance matching, summarizes the final result of this analysis. Starting with (6) we can define $L_1 = L_p \theta \sqrt{1-q^2}$ with $\theta \epsilon [0,\infty]$ and obviously for $\theta = 1$ we have conductance matching, else not. The eigenvalues are strictly negative over the whole range of q and θ, but clearly near $\theta = 1$ they tend to $-\infty$ which means that the state characterized by conductance matching, which by definition is the state of optimal efficiency of oxidative phosphorylation, is superstable. This implies that in a living cell within a fluctuating environment the energy converter is locked into this mode of optimal performance in a quite natural way. In other words, in a living cell oxidative phosphorylation will try to operate at the state of optimal efficiency, whenever the energy demands of the cell will permit to do so. If the load conductance is chronically mismatched, then the only alternative is that the mitochondria change the degree of coupling such that the condition of conductance matching in (6) is again fulfilled. Distinguished values of q and possible mechanisms for the regulation of this parameter are discussed in the next section.

4. Regulation of the Degree of Coupling

Assuming that in a living cell conductance matching is fulfilled due to superstability of the efficiency we are left with the question about appropriate values of q. At first sight q = 1 might appear to be the best solution because then energy conversion could be accomplished without any losses and wastes. However, this would imply that $L_1 = 0$ in order to fulfill (6). This alternative is certainly of no interest for a living organism, since the absence of a load naturally corresponds to the dead state. Actually, a fully coupled energy converter is similar to a Carnot machine which exhibits maximal efficiency at the state of thermodynamic equilibrium where all flows and forces vanish by definition. For these reasons we have to require that q < 1 which is also amply supported by experimental observations in different living tissues and organisms.

What particular value q has to assume depends entirely on the nature of the output required from the energy converter. Figure 3 shows several output characteristics plotted as a function of q. These can be constructed by using the transformation $\alpha = \arcsin(q)$ and then evaluating the output superfunction [4,13]

$$\Omega = \tan^n\left(\frac{\alpha}{2}\right)\cos(\alpha), \tag{7}$$

where varying the exponent n = 1,2,3,4 will produce the output functions in the order mentioned below. For example, the cell might be interested in having a maximal net rate of ATP production at optimal efficiency $(J_p)_{opt}$ which implies the value $q_f = 0.786$. Alternatively, instead of maximizing output flow one might maximize output power at optimal efficiency which becomes possible at $q_p = 0.91$. If in addition to these brute force maximizations the additional constraint of minimal energy costs is imposed on these output functions, then we get the economic net flow $(J_p \eta)_{opt}$ at $q_f^{ec} = 0.953$ and the economic output power $(J_p X_p \eta)_{opt}$ at $q_p^{ec} = 0.972$. These theoretically predicted values have indeed been measured in vivo. The value q_f for example, was measured in Na^+ transport in epithelial cells [14] or in growing bacteria [9] where maximization of net flows are obviously of major interest. On the other hand, the values q_f^{ec} and q_p^{ec} were determined in livers in vivo in starved and fed rats, respectively [8]. A liver in a fed rat is in a metabolic resting state and maximization of economic output power seems to be the best choice under these circumstances. In contrast, in a starved rat the liver has to produce glucose from low carbon precursors by way of gluconeogenesis and here the emphasis of energy conversion is shifted to the maximization of an economic flow of ATP production. These experiments not only confirm the theoretically predicted values of q but, in addition, show that the degree of coupling is not a constant but a parameter subject to metabolic regulation.

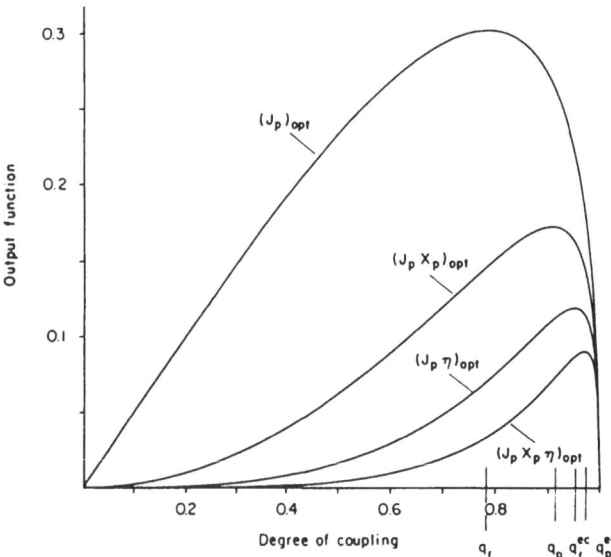

Fig. 3: *Maximization of different output functions as a function of the degree of coupling.* The output functions $(J_p)_{opt}$ to $(J_p X_p \eta)_{opt}$ as described in the text are plotted as a function of q showing maxima at the distinguished values of this parameter

An indication of what might regulate q stems also from these experiments. It was found that there is a correlation between the free fatty acid concentration and the degree of coupling insofar as fatty acids lowered q [8]. This fact was also known from experiments with isolated rat liver mitochondria. Here it was observed that the most potent decouplers are fatty acids with chain lengths of 14 (myristic acid) and 16 (palmitic acid) carbon atoms [15].

Recently the acylation of cellular proteins by fatty acids has become a subject of great scientific interest and activity, especially in the context of cell transformation [16]. Interestingly this acylation exhibits the same chain-length specificity as the depression of the degree of coupling in mitochondria mentioned above. It was therefore tempting to investigate whether 1) specific acylation by myristic and palmitic acid also occurs in mitochondria and 2) whether such an acylation could be responsible for the regulation of q.

In Fig. 4 the results of an experiment with highly purified mitochondria are depicted. The mitochondria were first incubated in the presence of unlabelled fatty acids with chain length from C_8 (octanoate) to C_{22} (laurate). The unlabelled acids were then chased with myristic-1-^{14}C acid. The autoradiography of an SDS gel shows acylation in the range of chain-lengths from C_{10} to C_{14}. It is important to note that prior to the acylation the free fatty acids have to be activated to acyl-CoA esters. Therefore in this experiment the observed specificity may be blurred by the kinetics of this activation.

Long term incubations with rat liver mitochondria so far have led to the following results:

1) The acylation of mitochondrial proteins via an ester bond by myristic and palmitic acid is reversible.

2) After hydrolysis of the labelled proteins with hydroxylamine the unmetabolized acids were recovered.

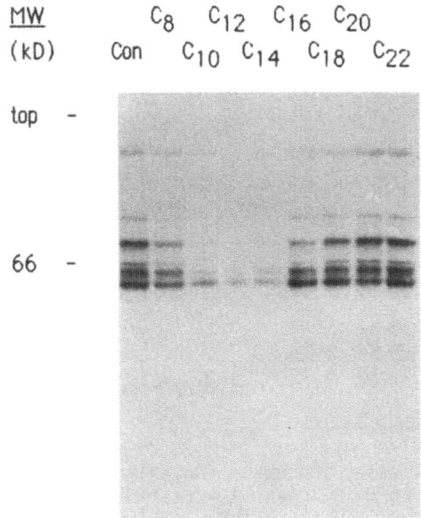

MW
(kD)

top -

66 -

bottom -

Con C$_8$ C$_{10}$ C$_{12}$ C$_{14}$ C$_{16}$ C$_{18}$ C$_{20}$ C$_{22}$

Fig. 4: *Autoradiography of acylated proteins*. Rat liver mitochondria were incubated with unlabelled free fatty acids with chain lengths from C$_8$ to C$_{22}$ (Control Con plus adjacent lanes from left to right). After 5 min of incubation the acids were chased with myristic-1-^{14}C acid. The SDS gel was exposed to X-ray film for 10 days. (Concentration of the unlabelled acids 75μM, labelled chase 25μM)

3) The half maximal concentration for the acylation and for the depression of the phosphate potential are both 30μM.

4) The nature of the acylated proteins is unknown thus far.

In the light of these results, we would like to put forward the following hypothesis for the regulation of the degree of coupling by fatty acids: The matrix contains acylase(s) and deacylase(s) specific for chain length in the range of C$_{12}$ to C$_{16}$. The modification of certain proteins catalyzed by these enzymes may lead to slips in pumps [17], activation of ATPase [18], and inhibition of carriers [19]. The concerted effects of these elementary processes lead to an overall change of the degree of coupling and the cellular phosphate potential such that the efficiency of oxidative phosphorylation is optimal under the given external constraints.

We have now solved the problems of why oxidative phosphorylation settles to operate at the state of optimal efficiency in the cell and how and why an appropriate degree of coupling is chosen. Inspection of (6), however, reveals a new problem. In this relation of conductance matching all parameters are assumed to be constant. This is certainly not true for L_1 since in a living cell the load fluctuates which would obviously compromise optimal efficiency of oxidative phosphorylation. The mechanisms by which the cell can minimize deviations from conductance matching in the presence of a fluctuating load conductance will be discussed in the next section.

5. Thermodynamic Buffering

Figure 5 shows an experiment where isolated rat liver mitochondria were incubated in the presence of glucose and varying concentrations of hexokinase [20]. This enzyme catalyzes the formation of glucose-6-phosphate and ADP from glucose and ATP and hence acts as a load. Increasing the hexokinase concentration leads to an increase in efficiency just up to the optimal value where a further increase of this load was not followed by a further shift of the phosphate potential than that corresponding to optimal efficiency. This effect is due to the presence of adenylate kinase in the intermembrane space of the mitochondria, which catalyzes

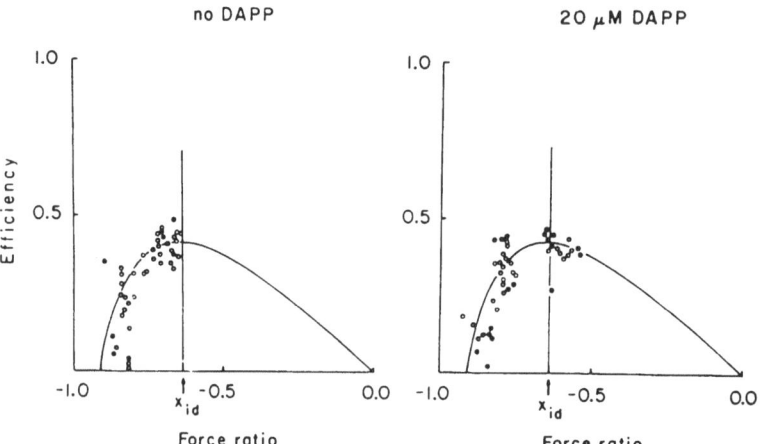

Fig. 5: *Thermodynamic buffering by adenylate kinase*. Mitochondria from rat liver incubated in the absence (panel A) and in the presence (panel B) of di-adenosine-pentaphosphate. Inhibition of the adenylate kinase allows driving the system past optimal efficiency. For experimental details see [20]

the reversible reaction ATP + AMP $\longleftrightarrow$ 2 ADP. Thus in the presence of too high loads part of the formed ADP is converted back into ATP and into AMP which accu-mulates. Therefore this reaction acts like a buffer which keeps the phosphate po-tential close to the value permitting optimal efficiency. In the presence of too low loads and provided that AMP be present in the system just the opposite oc-curs, but again with the result that oxidative phosphorylation operates close to optimal efficiency. If the experiment is repeated in the presence of di-adeno-sine-pentaphosphate, an inhibitor of the adenylate kinase, too high loads can drive the system to now operate at lower than optimal efficiency, as is illus-trated in panel B of Fig. 5.

A detailed analysis of the thermodynamic buffering in the presence of fluctu-ating load conductances has further revealed that the adenylate kinase reaction acts as a frequency filter with respect to perturbations of L_1 [21]. Very high frequency fluctuations of L_1 do not lead to deviations from optimal efficiency since the inertia of the adenine-nucleotide pool itself acts as a buffer. On the other hand, intermediary frequencies of load fluctuations were effectively damped by the adenylate kinase reaction, whereby for each frequency there exists an ac-tivity of this enzyme where buffering is maximal. At low frequencies the buffer-ing effect is virtually absent because here the adenylate kinase operates close to its own equilibrium and thus buffering no longer takes place because it is a transient phenomenon.

The analysis of the properties of another reversible ATP utilizing process, namely the one catalyzed by the creatine kinase: creatine + ATP $\longleftrightarrow$ creatine-phosphate + ADP, shows a similar buffering characterized by its own range of fre-quencies and values of q where buffering is maximal [22]. In summary, we can now classify ATP utilizing reactions into two categories: the irreversible ones which act as a load and the reversible ones which act as thermodynamic buffers. The so-phisticated interplay of these resistive and capacitive elements together with the regulatory mechanisms mentioned in the previous section allows oxidative phosphorylation to make optimal use of the exergy contained in the foodstuff.

6. Concluding Remarks

All the mechanisms discussed so far have not only the effect to optimize ATP production but, in addition, they also stabilize the steady state of oxidative phosphorylation against external fluctuations. In the light of the theoretical and also experimental results obtained during the last decades, the dogma has emerged, that stable systems notoriously fail to exhibit interesting dynamics or complexity [23,24]. Nonlinearity and far-from-equilibrium regimes of biochemical and biophysical processes are necessary prerequisites to destabilize steady states and it is only past such an instability that for example bifurcations can occur. These processes are, ultimately, driven by ATP and, as shown above, the production of ATP is globally asymptotically stable.

Does this now mean that the observed stability of ATP is ill chosen by nature to support all such processes as heart beat or brain waves which are characterized by an unstable steady state? I think not. One could perhaps classify biological processes into two categories: Tropotropic processes which have the goal to maintain the integrity of an organism, guarantee the constancy of the milieu interieur and mobilize energy in the form of ATP. The other category consists then of the ergotropic processes where functional aspects like pumping of blood, reaction towards stimuli from the outside world or processing of information are of major interest. Which processes should now be stable and which unstable? I think that ergotropic processes could profit from instabilities because richness of dynamic behavior and therefore functional adaptability is almost unthinkable in an asymptotically stable system. On the other hand, tropotropic processes should better be stable because, for example, wild fluctuations of let us say the pH of the blood or of the phosphate potential would constantly kick the steady states of the ergotropic reactions around, which could even have lethal consequences for the organism.

Perhaps the situation in the organism can be compared to a monetary society. One of the aims of a good and reliable banking system should be not only to provide and mobilize the currency when needed, but also to guarantee a reliable exchange rate [25]. Otherwise the development of trade, innovative technology and the like would be hampered. Therefore, on the basis of this admittedly far fetched analogy, one might say that the stabilization and optimization of ATP production is a necessary prerequisite to allow nonlinear processes to explore complex behavior far from equilibrium.

Acknowledgement

The acylation of mitochondrial proteins was studied in close collaboration with Erwin Sigel. This work was supported by grants from the Swiss National Science Foundation.

References

1. K.S. Spiegler: Principles of Energetics (Springer Verlag, Berlin, Heidelberg 1983)
2. P.W. Atkins: The Second Law (Scientific American Library, W.H. Freeman & Company, New York 1984)
3. O. Kedem, S.R. Caplan: Trans. Farad. Soc. 21, 1897 (1965)
4. J.W. Stucki: Eur. J. Biochem. 109, 269 (1980)
5. J.J. Lemasters, W.H. Billica: J. Biol. Chem. 256, 12949 (1981)
6. H.V. Westerhoff: Mosaic Nonequilibrium Thermodynamics, PhD Thesis, Amsterdam (Drukkerij Geria, Waarland 1983)
7. D. Pietrobon, S.R. Caplan: Biochemistry 24, 5764 (1985)
8. S. Soboll, J.W. Stucki: Biochim. Biophys. Acta 807, 245 (1985)
9. H.V. Westerhoff, K.J. Hellingwerf, K. VanDam: Proc. Natl. Acad. Sci. USA 80, 305 (1983)

10. J.W. Stucki, L.H. Lehmann, P. Mani: Biophys. Chem. 19, 131 (1984)
11. J.W. Stucki: Progr. Biophys. Mol. Biol. 33, 99 (1978)
12. J.W. Stucki: in preparation
13. J.W. Stucki: In Metabolic Compartmentation, ed. by H. Sies (Academic Press, New York 1982) p. 39
14. J. Lahav, A. Essig, S.R. Caplan: Biochim. Biophys. Acta 448, 389 (1976)
15. L. Wojtzak: J. Bioenerg. Biomembr. 8, 293 (1976)
16. F. Wold: TIBS 11, 58 (1986)
17. D. Pietrobon, M. Zoratti, G.F. Azzone, J.W. Stucki, D. Walz: Eur. J. Biochem. 127, 483 (1982)
18. B.C. Pressman, H.A. Lardy: Biochim. Biophys. Acta 18, 482 (1955)
19. F. Morel, G. Lauquin, J. Lunardi, J. Duszynski, P.V. Vignais: FEBS Lett. 39, 133 (1974)
20. J.W. Stucki: Eur. J. Biochem. 109, 257 (1980)
21. A.L. Veuthey, J.W. Stucki: Biophys. Chem. 26, 19 (1987)
22. A.L. Veuthey, J.W. Stucki: in preparation
23. G. Nicolis, I. Prigogine: Self-Organization in Nonequilibrium Systems (J. Wiley & Sons, New York 1977)
24. H. Haken: Synergetics. An Introduction (Springer-Verlag, Berlin, Heidelberg 1978)
25. P.A. Samuelson: Economics (McGraw-Hill, International Student Edition 1981)

Cell Lineage and Segmentation in Development

G.S. Stent

Department of Molecular Biology, University of California,
Berkeley, CA 94720, USA

1. Origins of Cell Lineage Studies

Studies of developmental cell lineage - that is of the fate of individual cells, or blastomeres, that arise in the early embryo - were begun in the 1870's, in the context of the controversy then raging about Ernst Haeckel's "biogenetic" law. The biogenetic law seemed to imply that cells of the metazoan blastula recapitulate the non-differentiated tissues of a remote sponge-like ancestor. Only after gastrulation would the germ layers - ectoderm, mesoderm, endoderm - be destined to form the highly differentiated tissues characteristic of more recent metazoan ancestors. This implication was tested by the founder of American experimental embryology, Charles O. WHITMAN (1887) [1]. By observing the cleavage pattern of early leech embryos - which is also the main experimental material of this brief review article - Whitman traced the fate of individual cells from the uncleaved egg to the germ-layer stage and concluded that, contrary to the implication of the biogenetic law, a characteristic postembryonic fate can be assigned to each identified blastomere and to the clone of its descendant cells. These findings suggested, moreover, that the differentiated properties that characterize a given cell of the mature animal are somehow determined by its genealogical line of descent from the egg.

Despite its highly promising beginnings, the study of developmental cell lineage went into decline after the turn of this century. It remained a biological backwater for the next 50 years, probably because the discovery of regulative and inductive phenomena in the development of echinoderms and chordates focussed the attention of the embryologists on cell interactions rather than on cell lineage as causal factors in cell differentiation. The first dramatic development in this direction was Hans Driesch's discovery in 1891, that upon separation of the two cells produced by the first cleavage of the sea urchin egg, each cell is capable of developing into a whole, albeit smaller embryo. This finding showed that, in accord with the implications of the biogenetic law and contrary to the view of cell lineage as a determinant of cell fate, individual blastomeres contain the entire developmental potential of the uncleaved egg - that is, they are totipotent.

When Hans Spemann and Hilde Mangold showed in 1924 that grafting an exogenous dorsal blastoporal lip on the ventral aspect of an amphibian gastrula induces the development of a second, supernumerary central nervous system, the attention of embryologists became focussed on the mechanism by which the cells in one part of the embryo induce the developmental fate of pluripotent cells in another part of the embryo.

For the next thirty years, the search for inducers formed the core project of experimental embryology. However, despite intensive efforts, not one substance was ever identified for which the role of a specific inducer could be convincingly demonstrated. Now, in retrospect, the reason for this failure is quite apparent: the experimental embryologists of the 1930's, 1940's and 1950's lacked the modern molecular biological knowledge which we now know to be necessary to account for the chemical basis of the determination of developmental cell fate.

2. Revival of Cell Lineage Studies

It may have been the disappointment over the lack of progress in uncovering the molecular basis of morphogenetic gradients and inducers that brought a revival of interest in the developmental role of cell lineage about 20 years ago. This revival was accompanied by the introduction of analytical techniques more precise and far-reaching than those available to Whitman and other 19th century pioneers. This recent renaissance of developmental cell lineage analysis reawoke interest in the study of embryos of protostomes, such as nematodes (SULSTON et al. [2]), leeches (STENT et al. [3]) and insects (GARCIA-BELLIDO and MERIAM [4]), in which the inductive aspects of development are much less prominent than in embryos of deuterostomes, such as echinoderms and chordates. And with this renewed interest in protostomal development Whitman's old idea of the determinative role of cell lineage came back into favor. Among the reasons which led to the renewed belief in a causal nexus between the line of descent of a cell and its fate is that in the embryogenesis of nematodes, leeches and insects serially and bilaterally homologous cell types are, on the whole, generated via homologous genealogical pathways. Indeed, as we shall see presently, this generative homology can be thought to account for the evolution of the segmented body plan of some metazoa in the first place. Another reason is that certain mutations or other manipulations which lead to changes in cell lineage also lead to changes in cell fate.

3. Typological and Topographic Commitment Hierarchies

It had been generally expected that commitment of embryonic cells to their developmental fate proceeds stepwise, according to a typologically hierarchic sequence. For instance, it was thought that in the developmental line of ancestry of a cholinergic motor neuron there would occur a commitment first to ectoderm rather than to mesoderm, then to nervous tissue rather than skin, then to neuron rather than glial cell, then to motor neuron rather than sensory neuron, and finally to synthesis of choline acetyltransferase rather than glutamic acid decarboxylase.

One important, albeit negative, insight brought by the modern cell lineage studies is that development does not generally proceed according to that expected typologically hierarchic commitment sequence. Instead it transpired that the commitment to differential cell fates, manifest in developmental cell lineage, appears to be typologically arbitrary. For instance, of two differentiated sister cells, one may be a neuron and the other an epidermal cell, whereas of two anatomically similar neurons, one may have arisen on the ectodermal branch and the other on the genealogically very remote mesodermal branch of the lineage tree. Instead of being typologically hierarchic, in nematodes and leeches, the commitment sequence turns out to be largely topographically hierarchic. That is to say, in these embryos, where cell migration plays a relatively minor (though definite) role, it is the position of two cells rather than their phenotype which tends to be correlated with the closeness of their genealogical relation. Or, in other words, the spatially ordered sequence of cell divisions represented by that genealogical relation is so arranged that most differentially committed postmitotic cells arise at, or very close to, the sites where they are actually needed (STENT [5]).

4. Segmentation

Another important insight brought by modern cell lineage studies pertains to the developmental origin of body segments which provide, in fact, a particularly important example of the topographically hierarchic character of the commitment sequence. Several protostomal as well as deuterostomal phyla, such as the annelids, arthropods and vertebrates, share the general structural feature of their bodies being composed of a periodic series of bilaterally symmetric segments. Each segment corresponds to a module, or metamere, of regularly iterated morphological elements, such as appendages, skin specializations, muscles or nerve cell ganglia. This basic morphological segmentation pattern is often obscured, however, because the metameres, rather than being exactly alike, usually differ at various positions along the longitudinal body axis, from head to tail (LANKESTER [6]).

An important concept relating to the embryonic origin of body segments dating back to William Bateson is that each segment is poised to generate the basic metamere pattern, and that the position-specific departures from the basic pattern, or segmental heterosis, arises from a specific deflection of the local tissues from their basic developmental pathway. In the fruit fly Drosophila a set of genes has been identified, and are presently under intensive molecular-biological study, whose products are necessary, and in some cases sufficient, for generating the position-specific heterotic departures from the basic metamere pattern. These genes are designated as homeotic genes, because, in line with the tradition of classical genetics, according to which genes are named, not after their normal function, but after their mutant phenotype, in homeotic Drosophila mutants segments manifest the basic metamere pattern rather than their position-specific heterotic departure from the basic pattern (OUWENEEL [7]; LEWIS [8]).

5. Leeches

Here I concentrate on the role of cell lineage in the development of the body segments of leeches. The bilaterally symmetric, tubular body of leeches, which form the class Hirudinea of the phylum Annelida, consists of 32 segments and a non-segmental prostomium. The metameric morphological elements include the segmental ganglion of the ventral nerve cord, a bilateral pair of excretory organs, or nephridia, and three subdivisions of the skin, or annuli. One of these three annuli lies in register with the segmental ganglion and includes circumferentially distributed sensory organs, or sensilla. The frontmost four and the rearmost seven body segments are fused, and, reflecting the position-dependent departures from the basic, midbody metamere pattern, i.e. heterosis, contain the specialized morphological features of the head and tail regions.

The location of the segment border, i.e. the interface between successive mid-body (i.e. unfused) metameres, had long been the subject of controversy. At the turn of the century, both William Castle and J.P. Moore claimed that the ganglion and its in-register sensillar annulus mark the middle of the segment. Accordingly, Castle and Moore chose the furrow separating two successive non-sensillar annuli as the location of the segment border. The Castle-Moore view of the centrality of the segment ganglion came to be generally accepted, not only for segmentation in leeches but also in insects. By contrast, Whitman had proposed earlier, on the basis of his embryological finding that the ganglion lies astride the septal margin of the somite, that the middle of the sensillar annulus marks the segment border rather than the middle of the segment, i.e. that the ganglion is an intersegmental rather than segmental structure. Now, with the wisdom of hindsight, it is obvious that this controversy was largely futile, since there is no objective way of fixing the beginnings and ends of the morphological modules of a periodically repeated structure. As we shall see, however, developmental cell lineage analyses do permit an objective definition of the spatial limits of individual generative (rather than morphological) metameres. These analyses showed that in this controversy, as in several others in which he had been involved, Whitman was closer to the truth than his adversaries.

6. Leech Embryogenesis

Development of the fertilized leech egg proceeds via a stereotyped sequence of (holoblastic) cleavages, giving rise to an embryo whose cells can be identified individually by various criteria, including size, position, birth order and cytoplasmic specializations (Fig. 1). The ectodermal and mesodermal segmental

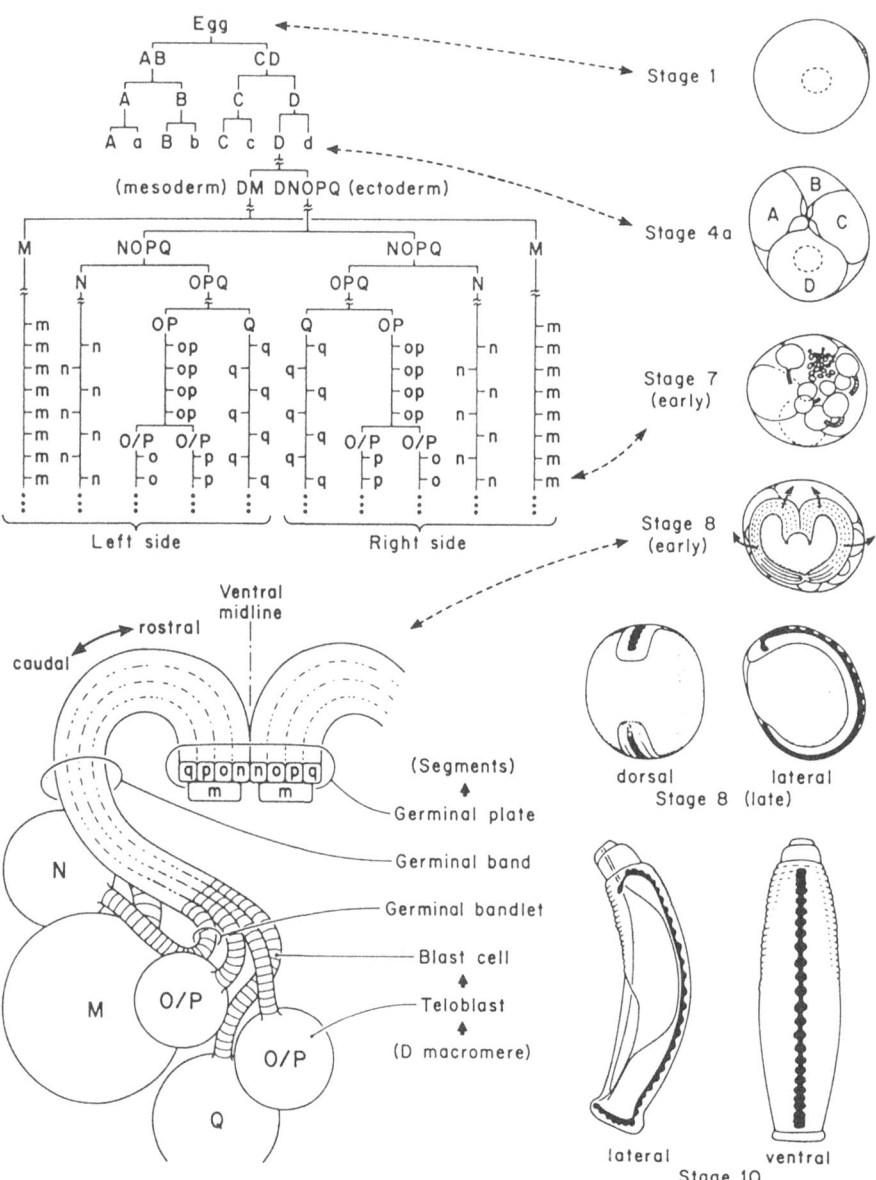

Fig. 1: Schematic summary of the development of the leech. Upper left: cell pedigree leading from the uncleaved egg to the teloblast pairs, M, N, O/P, O/P, and Q; and the paired primary blast cell bandlets. Lower left: hemilateral disposition of the teloblasts and their primary blast cell bandlets within the germinal band and germinal plate. Right margin: diagrammatic views of the embryo at various stages. In the stage 8 (early) embryo, the heart-shaped germinal bands migrate over the surface of the embryo in the directions indicated by the arrows. In the stage 8 (late) embryo the germinal plate is shown to lie on the ventral midline, with the nascent central nervous system and its ganglia indicated in black. In the stage 10 embryo shown, body closure is nearly complete (WEISBLAT et al. [11])

tissues arise from bilateral sets of five large blastomeres, designated teloblasts. By a series of several iterated, highly unequal divisions, each telo-blast generates a bandlet of several dozen much smaller primary blast cells. The bandlet designated m will give rise to the mesodermal and the bandlets designated n, o, p and q to the ectodermal components of the 32 segments. Ipsilateral bandlets merge to form left and right germinal bands and the two bands migrate over the surface of the embryo and eventually coalesce in rostrocaudal sequence to form a sheet of cells, the germinal plate, along the future ventral midline. At the time of its formation, the n, o, p and q bandlets lie in mediolateral alphabetical order on the superficial aspect of the germinal plate, while the m bandlet lies on its deep aspect.

Proliferation and differentiation of the m, n, o, p and q blast cell clones gives rise to a morphological periodicity of the germinal plate, reflecting formation of the 32 body segments. Within the germinal plate, the clones founded by the older blast cells of each bandlet lie anterior to the clones founded by the younger blast cells. Hence throughout most of its development, the germinal plate embodies a caudorostral developmental progression, according to which the development of a given segment is more advanced than that of its next posterior segment. Cell proliferation also causes the germinal plate to expand laterally around the circumference of the embryo, until the right and left leading edges of the plate meet along the future dorsal midline. At this point, formation of the body tube of the leech is complete.

The coordinates of the morphological periodicity pattern appear to be provided by an orthogonal network of longitudinal and circular muscle fibres, as visualized by immunofluorescent staining with a monoclonal anti-muscle antibody. In particular the rectangular territory of the ganglionic rudiment is enclosed by two serially successive circular muscle fibres at its anterior and posterior margins and by a bilateral pair of longitudinal muscle fibres at its lateral margins (TORRENCE and STUART [9]).

7. Cell Lineage Tracers

To refine and extend Whitman's century-old genealogical studies, we developed a novel lineage tracer technique. This technique consists of injecting a tracer molecule into an identified embryonic cell, which is passed on exclusively to the lineal descendants of that cell. These descendants can then be identified at a later developmental stage, by observing the distribution pattern of the tracer within the tissues. One type of tracer we have used is the enzyme horseradish peroxidase (HRP), which, upon a particular histochemical treatment of the embryo-nic tissues, causes formation of a black precipitate in the cells containing it (WEISBLAT et al. [10,11]) (Fig. 2). Another type of tracer of our design are ad-ducts of the fluorescent dyes fluorescein and rhodamine and carrier molecules, such as dextran (WEISBLAT et al. [12]). The distribution of these fluorescent tracers can be observed in living tissues under the fluorescence microscope.

The fluorescein labeled tracer has an additional, very useful property: it can serve also as a specific photosensitizer. Thus illuminating the embryonic tissues with light of a particular wavelength leads to death by photo-oxidation of all descendants of the tracer-injected precursor cell, but not of any other, genealo-gically unrelated cells with which the labeled cells may be intermingled (SHANKLAND [13]). This type of tracer makes it possible, therefore, to examine the developmental effects of the selective ablation of cells of particular lines of descent.

By use of the cell lineage tracer techniques to mark the differentiated post-mitotic descendants of individual blast cell bandlets we found that each bandlet gives rise to a particular hemilateral kinship group of segmentally iterated identified cells (WEISBLAT et al. [11]). Hence these findings extended Whitman's inference of a determinative role of genealogical relations for developmental

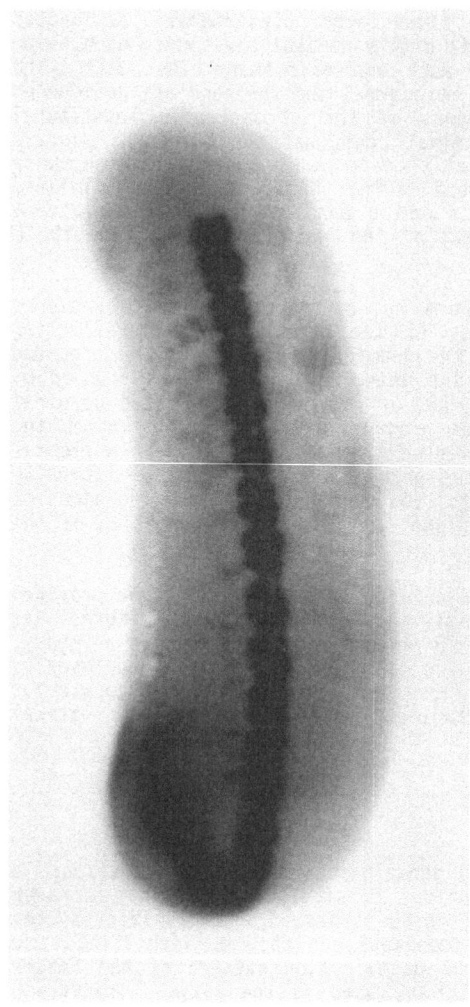

Fig. 2: Use of the horseradish peroxidase (HRP) lineage tracer technique to identify the developmental origin of embryonic tissues in the leech. Ventral view of a leech embryo in which HRP had been injected into the precursor teloblast of the n blast cell bandlet at an earlier stage of development. The blackened cells are present mainly in the segmental ganglia of the ventral nerve cord and represent the metameric kinship group of neurons derived from n_s and n_f primary blast cells that found each hemilateral segment

fate to the ultimate level of identified post-mitotic cells. They showed moreover, that there is little mixing of cells across the ventral and dorsal midlines of the embryo.

8. Segmental Founder Cells

The cell lineage tracer technique was adapted also for ascertaining the number of primary blast cells from each bandlet which found the kinship group of ectodermal and mesodermal cells of each morphological segment (WEISBLAT and SHANKLAND [14]). For this purpose, the teloblast of origin of a blast cell bandlet is injected with the lineage tracer after it has already produced a few unlabeled primary cells. In the resulting embryo, a boundary is then observed between anterior, unlabeled segments, derived from primary blast cells born prior to the injection, and posterior labeled segments born after the injection. At first, this boundary is rather sharp, suggesting that there is little longitudinal intermixing of blast cell clones. Now if each hemilateral segmental kinship group represents a clone descended from a single primary blast cell, then the antero-posterior label boundary should always coincide with the segment boundary. However, if each hemilateral segmental kinship group is founded by two successively born primary blast cells, then in about a half of the embryos, the label boundary should coincide with the segment boundary and in the other half of the embryos it should course midway between two segment boundaries.

The results of this experiment were as follows: In the case of the ectodermal n and q bandlets, the two types of label boundaries were observed with equal frequency. This showed that each hemilateral n kinship group is derived from two primary n blast cells (designated as n_s and n_f) which are serial successors in the blast cells bandlet, and the domains of the descendant clones which alternate rostrocaudally. The same is true for the two q segmental founder cells (designated q_f and q_s) and their descendant clones. However, in the case of the mesodermal m and ectodermal o and p bandlets, only a single label boundary was observed, and this single boundary coursed midway through the segment (as defined by Castle and Moore), i.e. midway through two successive segmental ganglia. This showed that each hemilateral m, o and p kinship group is derived from a single m, o and p primary blast cell and that, moreover, the generative metamere corresponds to Whitman's rather that to Castle and Moore's definition of the segment.

Thus we conclude that in the leech each hemilateral segment arises as seven distinct cell clones, six ectodermal clones and one mesodermal clone. Each of the seven heterologous, hemisegmental clones comprises a few dozen characteristic cells, which have arisen from their founder blast cell via a clone-specific, stereotyped cell lineage pattern. By contrast, bilaterally and serially homologous primary blast cells generate homologous cell lineage patterns. As specific ablation experiments using the photosensitizing tracer methods have shown, the periodicity of that pattern is (largely) an autonomous property of individual blast cells, rather than being the product of intercellular interactions. An interesting exception is formed by the o and p blast cells, of which one can take on the fate of the other and whose commitment to one of two alternative fates does depend on intercellular interactions (WEISBLAT and BLAIR [15]; SHANKLAND and WEISBLAT [16]).

In the course of growth and development of the hemisegmental founder blast cell sextet there occurs extensive (but stereotyped rather than random) interclonal cell mixing among the menbers of both heterologous clones of the same metamere and of homologous clones belonging to adjacent metameres. The latter kind of cell mixing eventually leads to a development interdigitation of serially successive generative metameres, and hence to a lack of isomorphism between generative metameres and morphological segments. For instance, the nephridial tissues and the muscle fibres derived from the same primary m blast cell clone must be assigned to different morphological segments of the postembryonic leech, no matter how its segment borders are defined. Moreover, the transegmental distribution of members of a single primary blast cell clone can be enhanced experimentally by photoablating an adjacent, serially homologous blast cell.

The differences in segment morphology, or heterosis, along the longitudinal axis of the leech body are reflected in corresponding modifications of the cell pattern to which some serially successive generative metameres give rise. These modifications do not seem to be attributable, however, to intrinsic differences between serially homologous primary blast cells - for instance to the rank of their birth - as can be inferred from experiments in which slippage of a blast cell bandlet has been displaced to an inappropriate segment. Under these conditions, the displaced primary blast cell will develop in accord with its ectopic position rather than its birth rank (SHANKLAND [13]).

9. Segmentation in Insects

Let us now briefly compare segmentation in insects, especially in the fruit fly Drosophila, with that in the phyletically related leeches. The body of the adult fly is obviously built of a dozen or more periodically iterated morphological modules, or metameres. But the rostrocaudal differentiation of these metameres, i.e. their heterosis, is much more pronounced in flies than in leeches, especially in the head and tail regions. For that reason it is difficult to make an unambiguous assignment of the total number of morphological metameres in the adult fly. However, the heterosis is less marked in the larva of the fly, which is generally considered to be composed of about 15 metameres, of which three are cephalic, three thoralic and nine or more abdominal.

From a morphological point of view the segmented structure of the fly larva bears considerable anatomical resemblance to the leech body. A periodically repeated circumferential furrow of the cuticle provides a convenient morphological landmark for the border separating two adjacent segments, in each of which a segmental ganglion of the ventral nerve cord lies in the middle. However, despite this morphological resemblance - in fact, undoubted anatomical homology - the early course of embryogenesis is radically different from the stereotyped sequence of holoblastic cleavages which we have previously considered for leech development. The fly egg begins its development with a series of synchronous mitotic divisions of the zygote nucleus, unaccompanied by cell division. In this way there arises an embryonic syncytium containing thousands of nuclei. Eventually most of these nuclei migrate to the periphery of the syncytium, where each nucleus becomes cellularized by an infolding of the embryonic cell membrane. In this way the fly embryo comes to consist of a uniform sheet of about 6000 cells, the cellular blastoderm, none of which has actually arisen by cell division. It is only in subsequent embryogenesis that cells of the blastoderm proceed to divide and generate cell clones.

Studies of the developmental genetics of Drosophila, coupled with post-cellular-blastoderm cell lineage studies, have also illuminated the role of generative metameres in the fly. Since the concept of cell lineage is not applicable to fly embryogenesis before formation of the approximately 6000 cells of the blastoderm, here the generative metameres cannot be traced back, as it can be in the leech, to the serial production of homologous sets of segmental founder blast cells by iterated divisions of a few paired teloblasts. Rather, it would appear that, in the fly, the generative metameres arise only upon subdivision of the two-dimensional sheet of blastoderm cells into a longitudinal series of 15 or more circumferential bands, of which each comprises a set of a few dozen founder cells committed to the formation of one generative metamere. It is noteworthy that a phase-shift between the boundary of the traditionally accepted morphological segment and the generative metamere has been recently inferred also to exist in Drosophila, where the generative metamere has been designated as "parasegment" (MARTINEZ-ARIAS and LAWRENCE [17]).

Individual circumferential bands, or para-segmental primordia, are three cells wide (in the longitudinal direction). There is good evidence that this periodicity of metameric cell commitment depends on the activity of a special set of a dozen or more genes (NÜSSLEIN-VOLHARD et al. [18]). The products of these genes divide up the sheet of blastoderm cells in a hierarchical manner, giving rise to a higher and higher spatial frequency of circumferential bands, of narrower and narrower width. On the top of that hierarchy appear to lie four so-called gap genes, which divide up the embryo into four large bands. Each of these gap genes appear to be activated within a certain, spatially defined range of a monotonic, caudorostral determinant gradient. This gradient is itself laid down under the influence of a few identified maternal genes during oogenesis (NÜSSLEIN-VOLHARD et al. [19]). A very complex theory which accounts for the diverse mutant phenotypes with alterations of the basic segmentation pattern and for the cell-specific expression of the segmentation-related genes in normal and mutant fly embryos, has been recently put forward by MEINHARDT [20].

Within each segmental primordium the cells are destined for various fates according to a circumferential pattern (POULSON [21]). An arc of founder cells straddling the ventral midline comprises the founder cells of the mesodermal metamere. The arcs lying to either side of the mesodermal founder arc and extending towards the future dorsal midline comprise the founder cells of the ectodermal metamere, with a mixed neural and epidermal fate. The ventral zone of the ectodermal arcs adjacent to the mesodermal arc provides ventral epidermis and neurons of the segmental ganglion. The dorsolateral zone of the ectodermal arcs provides dorsal epidermis and peripheral neurons (HARTENSTEIN and CAMPOS-ORTEGA [22]). Thus as regards their fate, the arcs of mesodermal and ectodermal segmental founder cells in the Drosophila blastoderm appear to correspond to the seven bilaterally paired metameric founder clones, or in the leech embryo derived from the m, n_s, n_f, o, p, q_s and q_f primary blast cells.

10. Polyclonal Commitment

How then is one to explain that the course of embryogenesis in these two phyletically related taxa, insects and leeches, is so radically different? How can it be that whereas in leeches determinate cell lineages can be traced back to the uncleaved egg, in insects cell lineage plays little or no role before the cellular blastoderm stage, at which a 6000-cell embryo suddenly springs forth, full-blown, like Athena from Zeus' head? This evolutionary divergence of the mechanism of embryogenesis can be accounted for by the increasing prevalence of polyclonal commitment in the phylogenetic line leading from an annelid-like ancestor to insects.

The notion of polyclonal commitment was first put forward by CRICK and LAWRENCE [23] in their explication of Garcia-Bellido's compartment concept designed to account for some findings related to the developmental genetics of the Drosophila wing (GARCIA-BELLIDO et al. [24]). As epitomized recently by MARTINEZ-ARIAS and LAWRENCE [17], its compartments make up what structuralist philosophers would call the deep structure of an insect, namely 'its (objective) internal representation', as distinct from 'our (subjective) external description'.

'Polyclone' refers to an ensemble of cells jointly committed to a common fate and representing all the clonal descendants of a small set of spatially contiguous founder cells located within a given domain of the embryo. For an ensemble to qualify as a polyclone, it is essential that its founder cells do not, in fact, constitute an ordinary (mono-)clone. The polyclone shares a common developmental fate, which, in the case of the Drosophila compartment, is reflected in its occupancy of a coherent body region with a topographically determinate boundary and in the topological connectedness of its cellular elements. The most generally applicable feature of the polyclone concept, however, is not the determinate boundary delimiting the area it may occupy, or the topological connectedness of its elements, but the joint commitment of its (non-clonal) set of pluripotent founder cells to a common fate. This fate may still include a set of subfates, with respect to which the cells have remained pluripotent. Thus, at a later stage two or more (non-clonal) sets of cells of the polyclone may become committed to take on different subfates. In this way, the polyclone would be split into two or more cellular subsets, or newly constituted sister polyclones.

It is in its invocation of a multicellular joint commitment process that polyclonal development in insects differs most significantly from the monoclonal development characteristic of embryogenesis in leeches where a single clonal founder cell is committed to a particular fate. Thus it is the topographic placement of a cell rather than its line of descent which governs its inclusion in the set of founder cells of a particular polyclone.

I conclude my remarks by recalling that Niels Bohr consigned true statements to two categories: ordinary truths, whose opposites are false, and deep truths, whose opposites are also deep truths. As we saw, statements about the role of cell lineage in development represent mainly deep truths, since, more often than not, their opposites are also true. Indeed, nearly 80 years ago E.B. WILSON [25] had already noted that a counter-example can usually be found for any generalization regarding the connection between cell lineage and developmental cell fate. As we now know, the role played by cell lineage varies greatly in the embryogenesis of different taxa, and even for different aspects of the embryogenesis within the same species. Cell lineage is evidently a more important developmental determinant in simple worms than in the much more complex insects and vertebrates. But even in simple worms, cell lineages share the governance of developmental cell fate with cell interactions.

This patchwork of developmental mechanisms, which achieves what appear to be essentially similar ends by a diversity of means, supports the notion set forth by JACOB [26] that ontogeny is related to phylogeny by 'tinkering' - that is, that evolution changed the course of embryogenesis by resort to any tool or trick that may have been handy when it was needed. In fact, the results of cell lineage studies suggest that by the time evolution put the pseudocoelomate nematode on the scene, it had already tried most of the items in its bag of tools and tricks for determining cell fate. Thus it does not seem very likely that during subsequent metazoan evolution there have emerged many novel developmental mechanisms at the cellular level. Rather, what does seem likely is that the insects and the vertebrates evolved from their humbler ancestors by opportunistic variations in the timing, iteration, and spatial localization of the cell commitment processes that were already at work in the embryos of worms.

References

1. C.O. Whitman: J. Morphol. 1, 105 (1887)
2. J.E. Sulston, E. Schierenberg, J.G. White, J.N. Thomson: Dev. Biol. 100, 64 (1983)
3. G.S. Stent, D.A. Weisblat, S.S. Blair, S.L. Zackson: In Neuronal Development, ed. by N. Spitzer (Plenum, New York 1982) p.1
4. A. Garcia-Bellido, J.R. Merriam: J. exp. Zool. 170, 61 (1969)
5. G.S. Stent: Phil. Trans. R. Soc. Lond. B312, 3 (1985)
6. E.R. Lankester: In Encyclopaedia Britannica, 13th ed., Vol. 18 (Encyclopaedia Britannica Co., London, New York 1926) p. 215
7. W.J. Ouweneel: Adv. Genet. 18, 179 (1976)
8. E.B. Lewis: Nature 276, 565 (1978)
9. S.A. Torrence, D.K. Stuart: J. Neurosci. 6, 2736 (1986)
10. D.A. Weisblat, R.T. Sawyer, G.S. Stent: Science 202, 1295 (1978)
11. D.A. Weisblat, S.Y. Kim, G.S. Stent: Dev. Biol. 104, 65 (1984)
12. D.A. Weisblat, S.L. Zackson, S.S. Blair, J.D. Young: Science 209, 1538 (1980)
13. M. Shankland: Nature 307, 541 (1984)
14. D.A. Weisblat, M. Shankland: Phil. Trans. R. Soc. Lond. B312, 39 (1985)
15. D.A. Weisblat, S.S. Blair: Dev. Biol. 101, 326 (1984)
16. M. Shankland, D.A. Weisblat: Dev. Biol. 106, 326 (1984)
17. A. Martinez-Arias, P.A. Lawrence: Nature 313, 639 (1985)
18. C. Nüsslein-Volhard, H. Kluding, G. Jürgens: Cold Spring Harbor Symp. quant. Biol. 50, 145 (1985)
19. C. Nüsslein-Volhard, E. Wieschaus, G. Jürgens: In Verhandlungen der Deutschen Zoologischen Gesellschaft (Gustav Fischer, Stuttgart 1982) p. 91
20. H. Meinhardt: J. Cell. Sci. Suppl. 4, 357 (1986)
21. D.F. Poulson: In The Biology of Drosophila, ed. by M. Demerec (Wiley, New York 1950) p. 168
22. V. Hartenstein, J.A. Campos-Ortega: Roux's Arch. Devl. Biol. 193, 308 (1984)
23. F.H.C. Crick, P.A. Lawrence: Science 189, 340 (1975)
24. A. Garcia-Bellido, P. Ripoll, G. Morata: Nature, new Biol. 245, 251 (1973)
25. E.B. Wilson: In Biological Lectures (Woods Hole: Marine Biological Lab. 1898), p.21 [Reprinted in Foundations of Experimental Biology, 2nd ed., ed. by B.H. Willier and J.M. Oppenheimer (Hafner Press, New York) p. 52]
26. F. Jacob: The Possible and the Actual (University of Washington Press, Seattle, London 1982) p. 71

Models of Cytoplasmic Motion

W. Alt

Abteilung Theoretische Biologie, Universität Bonn,
D-5300 Bonn 1, Fed. Rep. of Germany

1. Introduction

Since about two decades, protoplasmic streaming in various cell-types has been systematically investigated, for example in plant cells [1,2], in amoebae [3 - 5], see Fig.1, or in nerve axons [6].   Of particular interest are experiments with cytoplasma extracts or fragments, showing the same characteristic motile phenomena as *in situ*, but on much smaller spatial scales: uniform or circular streaming, migration (also in chemical or other gradients) and oscillatory contractions [7 - 12].

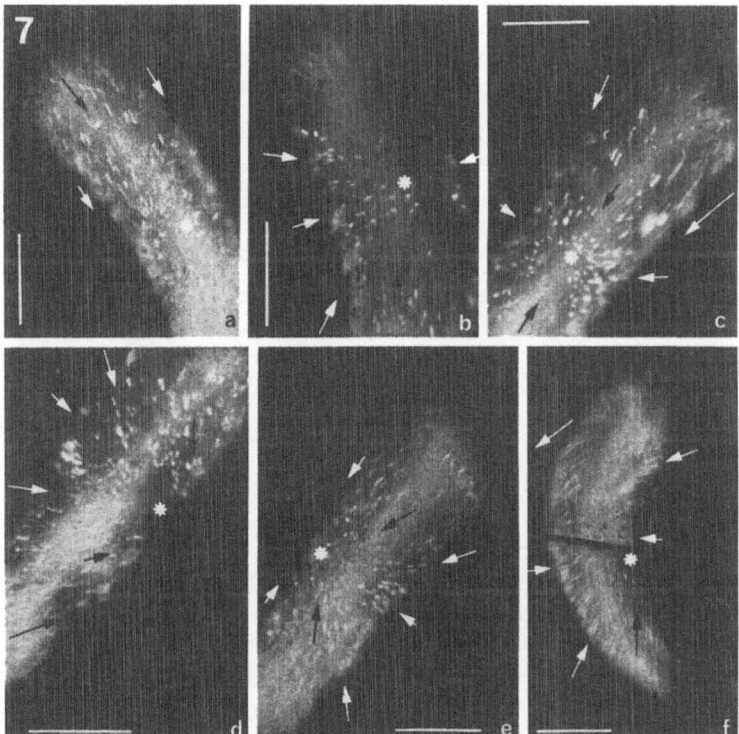

Fig. 1: Movements of *Amoeba proteus* cells showing transport of granules within the lower ectoplasma (outer cell cortex near the glass plate) reproduced from [5]. Streaming velocities (→) are mainly towards points of adhesion of the plasma membrane to the substratum (*). Bars = 100 μ

From the very beginning, theoretical models of these kinds of cytoplasmic motion were concentrated on two questions:

(1) What are the force generating elements leading to transport or contraction of the cytoplasma?

(2) How is their action regulated internally or externally?

In this contribution I would like to review some mathematical treatments of the problems above: two older models and two more recent mathematical approaches.

2. Viscous Fluid Model with Boundary Layer Force (Green Algae)

In an early model of cytoplasmic streaming within the cylindrical green alga *Nitella*, KAMIYA and KURODA [1] considered the "endoplasma" as a viscous fluid which is passively moved by some forces at the border to the fixed "ectoplasma", see Fig.2 . Since the viscosity coefficient of the endoplasma $\mu_{endo} \approx 1$ poise = dyn·sec/cm^2 had been determined from other experiments, the motive force F and the sliding resistance R at the boundary could be calculated by using the simple force balance equation

$$F = -R \cdot v + \mu_{endo} \cdot \partial_y v \; . \tag{1}$$

It turned out that this motive force with value $|F| \approx 1.6$ dyn/cm^2 was independent of different streaming velocities v and shearing gradients $\partial_y v$ achieved by varying the cell diameter.

In a mathematical description of non-Newtonian fluids, HAYASHI [13] calculated the velocity profile of such a cylindrical flow. Moreover, using the model that active filaments at the boundary exert a constant force onto the fluid in a small boundary layer, Hayashi could determine the value of this force from the velocity data to be $|F| \approx 1.2$ dyn/cm^2, which was in the same range as the older estimate.

In the mean time, cell biologists obtained more knowledge about the force generating filaments (actin, myosin, tubulin, dynein) and their interaction with

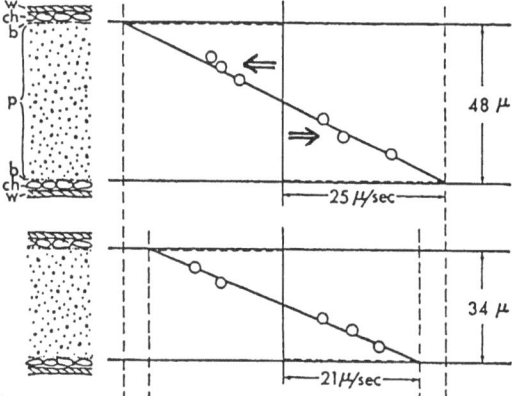

Fig. 2: Longitudinal sections through compressed *Nitella* cells, from [1], with cell wall (w), chloroplasts (ch), endoplasma (p) and the borderline (b) to the ectoplasma. The measured velocity profiles of the endoplasma (∘) are almost linear with maximal values at the borders, depending on the cell diameter

granules and organelles, whose movement has a considerable stochastic component in spite of the unifom mean speed v, see [2: Fig.2] . Therefore, more detailed stochastic models for the microscopic behavior of protoplasmic streaming in plant cells are desirable, which regard biochemical and -physical details and are able to reproduce the observed phenomena of cooperative motion.

3. Simple Two-Phase Continuum Model With Fiber Contraction   (Axoplasma)

In a pioneering, though more speculative model, ODELL [14] tried to explain the axoplasmic transport of particles along visible fibers in nerve cells by traveling waves of rhythmic fiber contractions. If θ denotes the local volume fraction of the fiber material and v its mean velocity in a cylindrical domain, see Fig.3, then the conservation law for the <u>fibrous phase</u>

$$\partial_t \theta + \text{div}(\theta \cdot v) = 0 \tag{2}$$

is accompanied by the conservation law for the <u>fluid phase</u> with local volume fraction $(1-\theta)$, namely

$$\partial_t(1-\theta) + \text{div}((1-\theta)\cdot w) = 0. \tag{3}$$

Here the mean fluid velocity w is coupled to the fiber velocity v by Darcy's law for porous media

$$w = v - \frac{1}{\phi} K(\theta) \cdot \nabla p , \tag{4}$$

with the hydraulic pressure p, the drag coefficient ϕ and the permeability tensor K, which regards the anisotropy of the fibrous material.

Clearly, when a moving "contraction wave", i.e. a fiber concentration function $\theta = \theta(y, x-ct)$ is prescribed, then a simple squeezing effect induces a corresponding net stream of the fluid. Assuming a particular rhythmic contraction profile across the axon, ODELL [14] even obtained the possibility of the observed bi-directional flows.

However, the question remained how the fiber contraction wave is induced. In an application of the same model to the <u>endoplasmic streaming in a moving amoeboid cell</u>, ODELL and FRISCH [15] postulated Ca-ions as regulators, which could be produced and diffuse from the tip of an extending pseudopod and alter the fiber contractivity by reversible binding.

Although modern video-micrographs clearly show that the axonal transport mechanism is quite different (namely a saltatory translocation of particles along single microtubules, see [6]) Odell's approach has recently been revived and generalized in two different directions. One of them develops a detailed model for

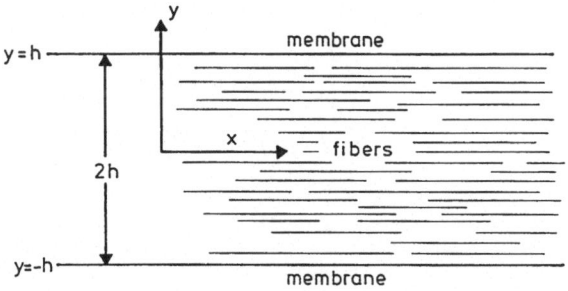

Fig. 3: Schematic representation of an axon segment in Odell's model [14]. Contractile fibers are aligned longitudinally and are thought to squeeze the viscous water-like fluid preferentially in the x-direction

Ca^{++}-regulations of contractions in ectoplasma gels (section 4), the other is based on a two-phase fluid model for the sol-gel dynamics in a protoplasma (section 5).

4. Visco-elastic Gel Model with Ca-ion Control   (Amoeboid Cells)

In the mean time many experiments revealed that the endoplasmic transport within amoeboid cells, as the fountain streaming in advancing pseudopods [5] or the shuttle streaming in *Physarum* strands [16] are induced by local contractions of the outer cell-cortex, the ectoplasmic tube, which is thicker than in plant cells (section 2). Such a developed ectoplasma is thought to be a meshwork of cross-linked filaments (F-actin) constituting a visco-elastic gel with the additional ability to contract itself (by mutual interaction with myosin filaments).

OSTER and ODELL [17] took the usual continuum model for visco-elastic materials and let the coefficients in the equation for the total stress

$$S = S_{elastic} + S_{contractive} + S_{osmotic} + S_{viscous}$$

depend on the concentration c of free intracellular Ca^{++}. Moreover, supposing an auto-catalytic release of Ca-ions from vesicular stores, they supplemented the model system by a cubic nonlinear diffusion equation for c.

Because of the analogy to systems as the nerve conduction equations, the simulated "model cytogel" behaves in a similar way as other "excitable media": Local release of Ca^{++} (by chemical stimulation or stretching) initiates local solation/gelation or contraction of the actomyosin gel which can

(a) propagate along an elongated strand as a solation- or contraction-wave following a traveling Ca-wave ---- a possible model for the movement of advancing pseudopods, see [18] and Fig.4 ----

(b) lead to compensating contractile oscillations when two ectoplasma-segments are allowed to shuttle forth and back the endoplasma between them ---- a possible model for shuttle streaming, see [19] ----.

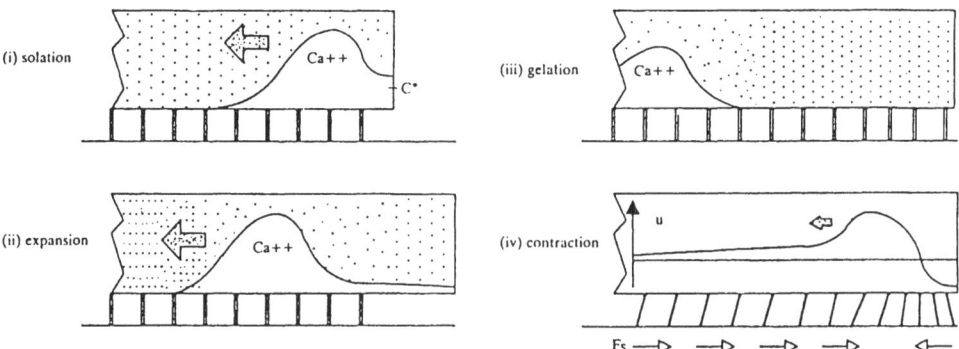

Fig. 4: Oster's model for the crawling of amoeboid cells, Fig. 9A in [18]. A membrane leak increasing the Ca-ion concentration at the tip of the pseudopod causes solation (i) and expansion (ii) of cytoplasma to the right, until re-gelation (iii) occurs which produces new attachment sites (|||) to the substratum and, simultaneously, contraction (iv) of the gel. An auto-catalytic Ca-wave traveling to the left then induces a corresponding wave of solation, gelation and contraction which finally yields a cell displacement to the right

In this context I should mention some other similar approaches trying to model

(a) cell movement as a free boundary problem, due to varying surface motion regulated by certain intracellular diffusible substances [20] or by the contractile system itself [21],

(b) rhythmic contraction waves in *Physarum* plasmodia [22], where the ectoplasma is also modelled as a visco-elastic medium, but with the active stress $S_{contractive}$ being induced purely by a mechanical feedback of the radial contraction after a time lag.

In both models (b) for shuttle streaming in *Physarum*, the coupling of the ectoplasmic contractivity to the hydrostatic pressure of the endoplasma plays a crucial role. This is different in a further model [23], where the spatial coordination of contractive oscillations is achieved only by tension and contraction forces within the ectoplasma itself. I should emphasize, however, that the true nature of these famous cytoplasma oscillations is far from being resolved, since some experiments even suggest the existence of a purely biochemical oscillator as a trigger.

These difficulties and uncertainties, based on an increasing complexity of both the biochemical interactions and the mathematical equations, have been one reason to concentrate the modelling effort on more simple experimentally verifiable systems, with the hope to describe the essential phenomena already by using easier systems of differential equations [24]. This is examplified for a one-dimensional case in the following section.

5. Two-Phase Viscous Fluid Model with Contraction and Reaction (Protoplasma)

Experiments with endoplasmic extracts of *Physarum* [11], amoebae [7] or amphibian eggs [12] suggest that there is a continuous transition from lower concentrated endoplasma or "sol"-state (containing only few actin filaments in a loose meshwork) to high concentrated ectoplasma or "gel"-state (constituting a crosslinked actomyosin network) and vice versa. Thus, in order to model the corresponding dynamics of gelation, solation and contraction let us consider, in the same spirit as in section 3, the continuum equations for two phases [25]:

θ : volume fraction of the network phase (polymeric F-actin)
with mean velocity v, and

$(1-\theta)$: volume fraction of the aqueous phase or solution phase
(containing monomeric G-actin and other proteins)
with mean velocity w.

Assuming simple linear assembly-disassembly kinetics for the F-actin, the mass balance equation (2) -- in space dimension one -- is generalized to

$$\partial_t \theta + \partial_x(\theta \cdot v) = \eta \cdot (\theta_{eq} - \theta) , \qquad (5)$$

where θ_{eq} denotes the equilibrium concentration and η is the sum of the association and the dissociation rate. Summation of (5) with the corresponding mass balance equation (3) gives

$$\partial_x \left[\theta \cdot v + (1-\theta) \cdot w \right] = 0 . \qquad (6)$$

The <u>force balance equation for the network phase</u> in the limit of low Reynolds number is

$$0 = F_{viscosity} + F_{contraction} + F_{swelling} + F_{drag} + F_{hydraulic} \quad ,$$

namely, according to the general model [25: (B5)] ,

$$0 = \partial_x \left[\mu\theta \cdot \partial_x v + \Psi(\theta) + \Sigma \cdot \ln(1-\theta) \right] - \phi \cdot \theta(1-\theta) \cdot (v-w) - \theta \cdot \partial_x P_F \quad , \tag{7}$$

where <u>viscosity</u> is assumed proportional to the network concentration θ. Furthermore, $\Psi(\theta)$ denotes the <u>contractile stress</u> due to mutual attraction of network filaments by binding to myosin, see [23], Σ is the coefficient of <u>solvation</u> for F-actin in the aqueous phase, leading to a swelling pressure $\Sigma \cdot |\ln(1-\theta)|$ in the network which increases to ∞ for $\theta \to 1$. Finally, the ϕ-term describes the drag of moving filaments and P_F denotes the so-called effective pressure in the aqueous phase. For further details and estimations of the parameters see [25]. Analogously, the <u>force balance equation for the aqueous phase</u> with negligible viscosity is, according to [25: (B4)] :

$$0 = -\phi \cdot (1-\theta)\theta \cdot (w-v) - (1-\theta) \cdot \partial_x P_F \quad . \tag{8}$$

Notice that this is a simple one-dimensional version of Darcy's law, namely equation (4) with $K(\theta) = 1/\theta$. By adding (7) and (8) and solving for the pressure gradient in (8) we obtain

$$\partial_x \left[\mu\theta \cdot \partial_x v + \Psi(\theta) + \Sigma \cdot \ln(1-\theta) \right] = \partial_x P_F = \phi \cdot \theta \cdot (v-w) \quad . \tag{9}$$

In the one-dimensional case we can easily solve for w in (6), if we impose a <u>no-flux boundary condition</u> at one border of the considered interval of length L, say at $x = 0$, yielding the simple "squeezing law"

$$w = -\frac{\theta}{(1-\theta)} \cdot v \quad , \tag{10}$$

meaning that network and solution flow are always antagonistic. Inserting this into (9) and rescaling space to the unit interval, we finally obtain the following linear second order elliptic equation for the mean network flux v:

$$\partial_x \left[\theta \cdot \partial_x v + \psi/2 \cdot \theta^2 + \sigma \cdot \ln(1-\theta) \right] = \phi \cdot \frac{\theta}{(1-\theta)} \cdot v \tag{11}$$

with the following time rates appearing in (5) and (11)

η : reaction rate of the F-actin network,

$\psi = \psi_0/\mu$: contraction rate, assuming the law $\Psi(\theta) = \psi_0/2 \cdot \theta^2$,

$\sigma = \Sigma/\mu$: solvation rate for potential swelling of the network,

and finally

$\phi = \phi \cdot L^2/\mu$: the dimensionless drag coefficient.

All parameters except η are related to the network viscosity coefficient μ.

Together with appropriate boundary conditions at $x=0$ and $x=1$ the differential equations (5) and (11) constitute a well-posed hyperbolic-elliptic system, namely -- in a more physical terminology -- the <u>generalized Stokes equations</u> for a reactive and contractive, highly viscous fluid with reaction kinetics

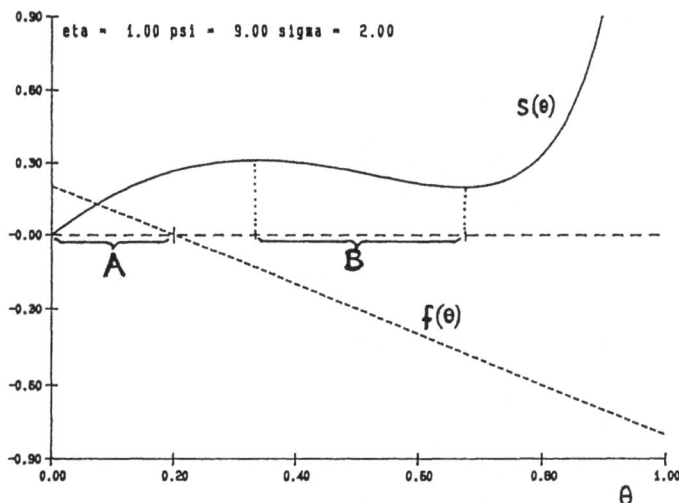

Fig. 5: Constitutive functions in the one-dimensional viscous fluid model for a protoplasma, with reaction kinetics f(θ) as in (12a) and swelling pressure s(θ) as in (12b) depending on the network concentration θ ≤ 1 . For further interpretations see the text

$$f(\theta) = \eta \cdot (\theta_{eq} - \theta) \tag{12a}$$

and "state function", i.e. net swelling pressure vs. concentration

$$s(\theta) = \sigma \cdot |\ln(1-\theta)| - \psi/2 \cdot \theta^2 , \tag{12b}$$

see Fig.5. The state function s(θ) is non-monotone iff ψ > 4σ. Then contraction dominates solvation for network concentrations in the interval B where s'(θ) is negative, resulting in a net contraction of the network in this regime. This is one regime of <u>chemical energy input</u> into the system, namely ATP needed for the <u>contractile activity</u> of myosin filaments. The other regime is the interval A where net assembly of actin filaments is achieved by <u>enzymatic activity</u> of certain proteins, see e.g.[26] .

We are going to show that these simple physico-chemical hypotheses already lead to some basic phenomena of biological motion as, for example, autonomous oscillatory contractions. Notice the analogy in the state function in Fig.5 to physical models of van der Waals fluids with moderate viscosity, where simulations sometimes also show oscillatory behavior [B. Nichols, personal communication].

In the situation of sticky boundaries (v = 0 at x=0 and x=1) former analysis and computer simulations for slightly modified equations already showed some interesting dynamical features: the formation of "contraction centers" at the boundaries or, with positive drag coefficient φ, also in the interior. These centers, representing gel formations, tend to compete with each other in an oscillatory or, for larger values of φ, even in a chaotic manner, see [24,27].

Here we will consider a different situation in order to model the experimental results of EZZELL et al.[12] showing <u>repeated waves of gelation and contraction</u> of protoplasma in a cuvette, see Fig.6. Starting with a uniform concentration of cytoplasma-extract from *Xenopus* eggs (containing both G- and F-actin at chemical equilibrium together with ATP and phosphorylated myosin) one first observes gelation and progressive contraction of the whole volume, away from the cuvette walls (A-B). After about 2.5 min the gel has contracted in the upper part of the cuvette at the water-air border (C). The process then repeats in a

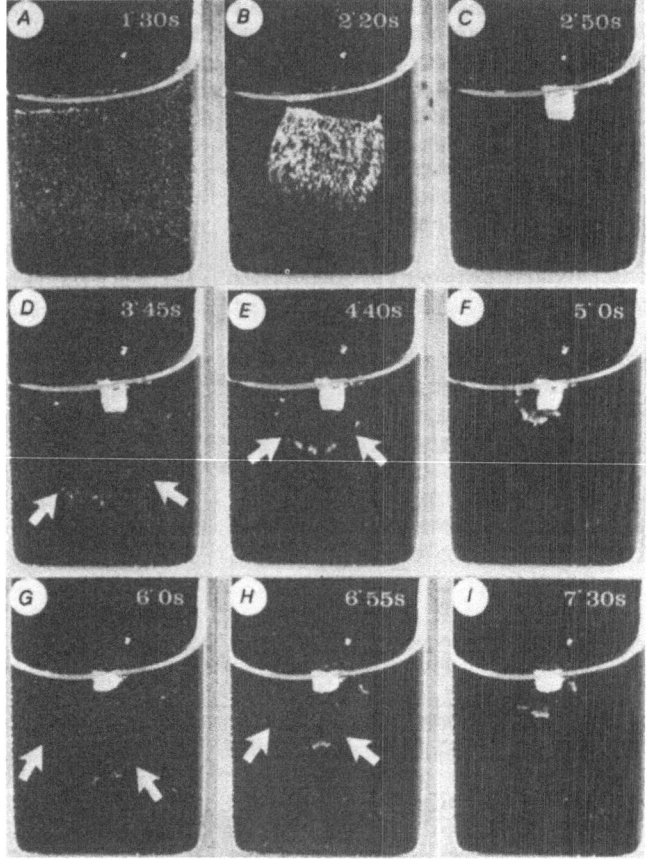

Fig. 6: Photographs of a cytoplasma extract from *Xenopus* eggs in a cuvette exhibiting three consecutive waves of gelation and contraction (A-C,D-F,G-I) with cycle period of about 2.5 min. Experiments by EZZELL et al. [12]

weakened form, whereby secondary waves of higher F-actin concentrations are attracted and finally absorbed into the first gel (D-F and G-I).

Since network proteins and water have similar specific mass [25], buoyancy forces should be neglected. Therefore, we might explain the observed top-bottom asymmetry by differences in the adhesion of the network filaments to the different surfaces. In an idealized one-dimensional model we assume a

<u>sticky boundary</u> at the top: $v = 0$ at $x = 0$ (13)

and

<u>non-sticky boundary</u> at the bottom: $v \leq 0$ at $x = 1$.   (14)

Because of no boundary flux ($\theta \cdot v = 0$) the latter condition requires

$$\theta = 0 \qquad\qquad \text{if} \quad v < 0 \qquad \text{at } x = 1 . \qquad (14a)$$

Moreover, differentiation in (11) yields the <u>Neumann boundary condition</u> in this case:

$$\partial_x v = s(\theta) \, / \, \theta \qquad\qquad \text{if} \quad v < 0 \qquad \text{at } x = 1 . \qquad (14b)$$

242

For <u>vanishing drag coefficient</u> $\varphi = 0$, this elliptic boundary value problem for v is amenable to a direct analysis: For each fixed time t equation (11) with boundary conditions (13) and (14) is equivalent to

$$\partial_x v(t,x) = \frac{s(\theta(t,x)) - S(t)^+}{\theta(t,x)} , \quad 0 \leq x \leq 1, \tag{15}$$

with the <u>averaged net swelling pressure</u>

$$S(t) = \int_0^1 s(\theta(t,x))/\theta(t,x) \, dx \Big/ \int_0^1 dx/\theta(t,x) . \tag{16}$$

Insertion of (15) into (5) gives the following ordinary differential equation for θ along characteristics with speed v

$$\partial_t \theta + v \cdot \partial_x \theta = f(\theta) - \left[s(\theta)-S(t)^+\right] = F(\theta) + S(t)^+ , \tag{17}$$

where $F(\theta) = f(\theta) - s(\theta)$ is the sum of the chemical (12a) and the contractive growth function (12b), see Fig.5 and the resulting plots of $F(\theta)$ in Figs. 7a and 8a.

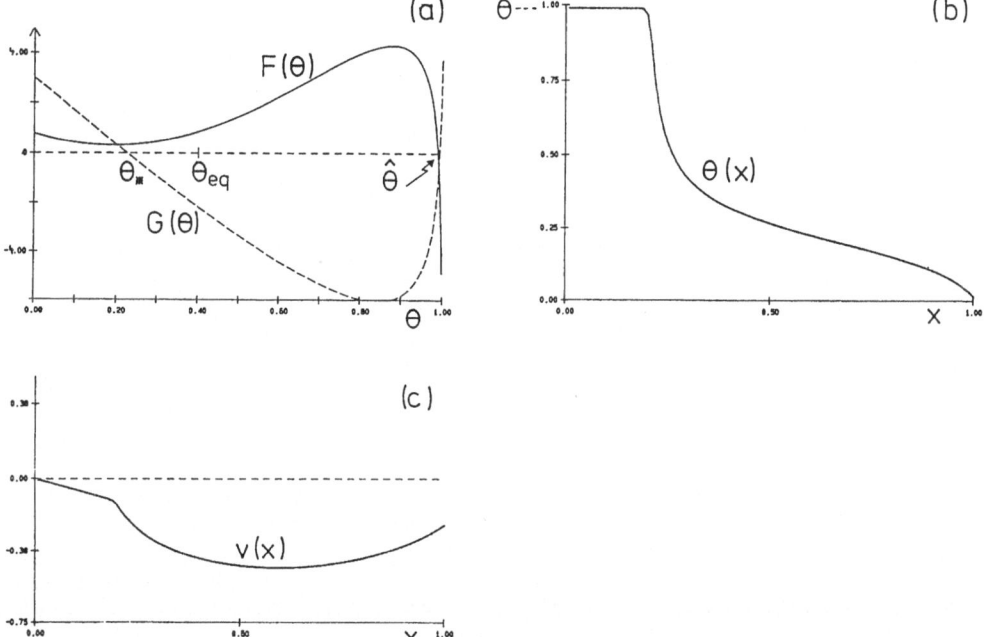

Fig. 7: (<u>a</u>) Plots of the "net growth function" $F(\theta)$ (——) appearing in (17), with zero $\hat{\theta}$, and the "relative swelling pressure" $G(\theta) := s(\theta)/\theta$ (-----) appearing in (15), with zero θ_*. Parameters are $\eta = 2$, $\psi = 30$ and $\sigma = 3$ min^{-1}

(<u>b</u>) Simulated asymptotic steady concentration profile $\theta(x)$, $0 \leq x \leq 1$, for vanishing drag $\varphi = 0$ modeling a contracted "gel" with plateau level $\hat{\theta}$

(<u>c</u>) Corresponding velocity distribution v(x) of actin filaments, being steadily attracted towards the "gel"

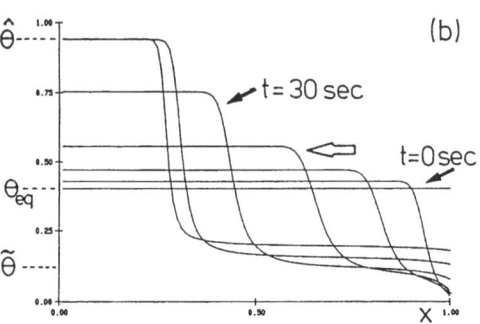

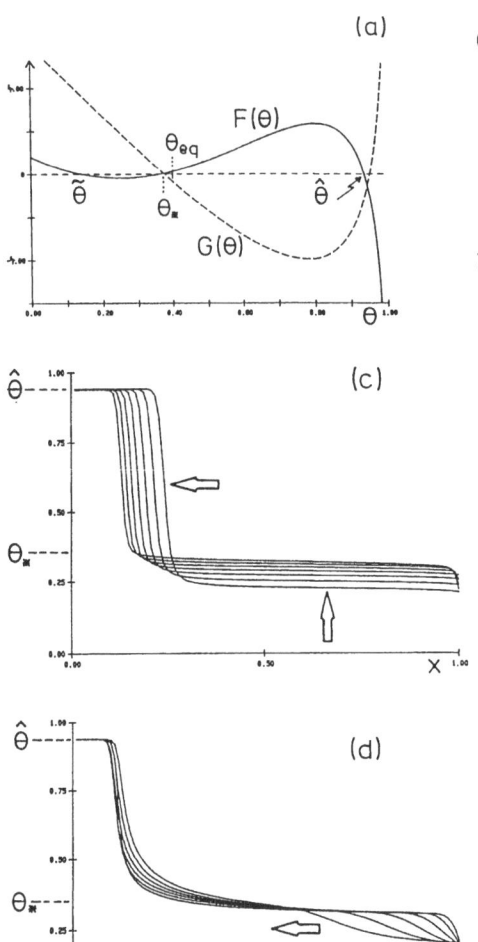

Fig. 8: (a) Same plot as in Fig. 7a with altered parameters $\eta = 2$, $\psi = 40$ and $\sigma = 6$ min^{-1}.
(b) Computer simulations of system (5),(11),(13),(14) for $\varphi = 0$, starting with $\theta_{eq} = 0.4$: Wavelike formation of a primary contracted "gel" within 45 sec,
(c) followed by a slow re-gelation in the depleted region within the next 45 sec.
(d) Formation of a further contraction wave towards the primary "gel" within 75 sec

As long as (the numerator of) the averaged swelling pressure $S(t)$ in (16) is negative, $S(t)^+$ is zero and (15) gives $v(t,1) < 0$ as well as the boundary condition (14b). Then, because of (14a) we have to set $\theta(t,1) = 0$. Thus, if on the average over the whole interval contraction dominates solvation, the actin network is pulled away from the boundary $x = 1$.

In the other case of a positive averaged swelling pressure $S(t)$ the network cannot cross this boundary, therefore we have $v = 0$ at $x=1$. Then the concentration $\theta(t,1)$ is free to increase or decrease there.

In particular, starting with <u>initial concentration</u> $\theta(0,x) = \theta_{eq}$ the condition (14b) for disruption from the boundary is $s(\theta_{eq}) < 0$. For parameter values as in Figs. 7 and 8 this criterium is fulfilled since the solvation-contraction equilibrium θ_* lies below θ_{eq}. Then the characteristics point into the interval at $x=1$ and the concentration θ develops along them according to the ODE $\theta' = F(\theta)$ in (17), thus approaching the lowest zero $\hat{\theta}$ of $F(\theta)$ for $t \to \infty$. In the case $\theta_{eq} < \hat{\theta}$ as in Fig.7a, one shows that the solution $\theta = \theta(t,x)$ converges to a steady state, which even can be described analytically, see Fig.7b.

In other cases, where the net growth function $F(\theta)$ becomes negative for values less than the chemical equilibrium with lowest zero $\tilde{\theta} < \theta_{eq}$ as in Fig.8a, the dynamical behavior of the model protoplasma is more complicated:

Figure 8b shows that after <u>disruption</u> of the network from the border x=1 the function $\theta(t,x)$ forms a <u>contraction plateau</u> near the other border with height $\hat{\theta}$, whereas in the remaining region it recovers from zero up to a lower level $\tilde{\theta}$. This value lying below the chemical equilibrium means that <u>newly assembled F-actin</u> is transported to the contracted "gel" on the left. However, due to network disassembly the width of this gel slowly shrinks, as shown in Fig.8c, thereby decreasing the contractive force and increasing the averaged <u>swelling pressure</u> S(t) above zero. This changes the dynamical behavior in equation (17) by lifting the growth function $F(\theta)$ and allowing the F-actin concentration on the right hand side to rise, slow <u>re-gelation</u>, until it passes the critical value θ_*. But now the swelling pressure S(t) has a chance to become negative again, so that a new disruption from the border x=1 can occur, see Fig.8c. This induces a new <u>contraction wave</u> traveling to the left as a "shoulder" with upper plateau level θ_* as shown in Fig.8d. Afterwards the process described above repeats in a weaker form. Longer time simulations ($\gtrsim 50$ min) yield a regular cycle of a contraction wave followed by re-gelation, with cycle period of about 1.5 min.

For <u>positive drag coefficient</u> φ these cyclic contraction waves are even more pronounced. In this case the network contractivity is more localized than in the case $\varphi = 0$, inducing the possibility that in regions of dynamic instability, where $F'(\theta) > 0$, local contraction centers appear and grow in size. The simulations in Fig.9, compared to those in Fig.8b-d, show that this occurs at the shoulders of the traveling contraction waves: The primary wave has a sharper "egde" (Fig.9a), while the secondary wave even develops a peak which is pulled to the left (Fig.9b).

Again, as in the case $\varphi = 0$, longer time simulations yield repetitive waves of contraction and re-gelation, but now with a larger cycle time of about 2.5 min, obviously due to the drag ($\varphi = 0.3$) acting against the network flow. The time scale above ($\eta = 2$) had been adjusted in order to describe the particular experimental situation in Fig.6 with similar cycle time. Then, supposing an actomyosin network viscosity of $\mu \approx 200$ poise, the simulations would correspond to a chemical equilibration time $T_{eq} = 1/\eta \approx 30$ sec and the following coefficients of drag $\phi \approx 60$ dyn·sec/cm^4, of solvation $\Sigma \approx 20$ dyn/cm^2 and of contraction $\psi_0 \approx 130$ dyn/cm^2. This would predict a contractive stress at equilibrium of $\Psi(\theta_{eq}) \approx 10$ dyn/cm^2, where $\theta_{eq} = 0.4$ was chosen unnaturally high only because of better graphical representation. A lower value would give also a lower contractive stress which then is comparable to other estimates, see section 2.

These properties of our "model protoplasma", which obviously are consistent with the experimental findings in [12], suggest that the basic mechanism of the observed oscillatory behavior of cytoplasma extracts is an <u>intrinsic property of the actomyosin network</u> itself, namely an autonomous interplay between the active processes

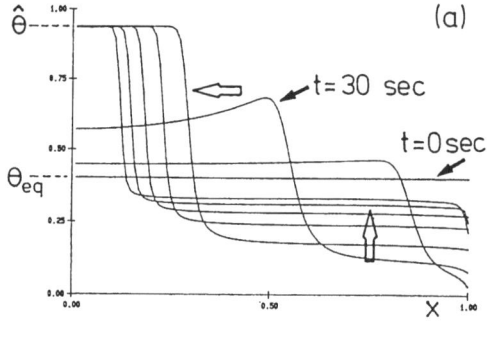

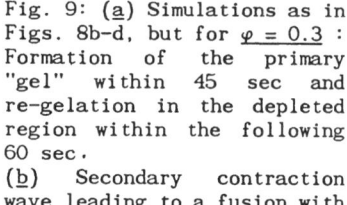

Fig. 9: (a) Simulations as in Figs. 8b-d, but for $\varphi = 0.3$: Formation of the primary "gel" within 45 sec and re-gelation in the depleted region within the following 60 sec.
(b) Secondary contraction wave leading to a fusion with the primary "gel" after about 1.5 min

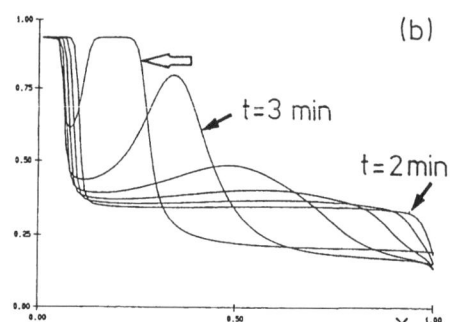

	(parameter)
(a) assembly-disassembly kinetics of actin filaments,	η
(b) contractive force of myosin filaments,	ψ
(c) solvation pressure by the aqueous phase,	σ

and finally the passive processes

(d) shear viscosity of the filament network	μ
(e) drag within the aqueous phase.	φ

Indeed, both the experiment and the model produce the recurrent waves of gelation and contraction <u>without any external regulation</u>. For example, the Ca^{++}-levels can be buffered experimentally, reflected by the assumption of constant parameters η and ψ in the mathematical system.

Similarly, the more extensive two-dimensional simulations for the same model by DEMBO and coworkers [28,29] explain the appearence of fountain streaming and other motile events in demembranated cytoplasma as autonomous processes, which in vivo might be additionally controlled and regulated by membrane proteins or diffusible substances (compare section 4). There is a hope that these "simple" physico-chemical hypotheses and the resulting systems of differential equations are able to model the basic mechanisms of cytoplasmic motion and can be extended or modified in order to describe several other phenomena of cellular motility.

Acknowledgement

I would like to thank Micah Dembo (Los Alamos) for stimulating collaboration in this field, Thomas Pohl (Bonn) for assistance in performing the computer simulations and Mario Markus (Dortmund) for the kind invitation to write a contribution to this volume.

246

References

1. N. Kamiya, K. Kuroda: Biorheology 10, 179 (1973)
2. D. Menzel, M. Schliwa: Eur. J. Cell Biol. 40, 275 (1986)
3. R.D. Allen, N.S. Allen: Ann. Rev. Biophys. Bioeng. 7, 469 (1978)
4. W. Stockem, H.-U. Hoffmann, B. Gruber: Cell Tissue Res. 232, 79 (1983)
5. A. Grebecki: Protoplasma 123, 116 (1984)
6. R.D. Allen, J. Metusals, I. Tasaki, S. Brady: Science 218, 1127 (1982)
7. D.L. Taylor, J.S. Condeelis, P.P. Moore, R.D. Allen: J. Cell Biol. 59, 378 (1973)
8. G. Albrecht-Bühler, Proc. Natl. Acad. Sci. USA 77, 6639 (1980)
9. K. Kuroda: In Cell Motility: Molecules and Organization, ed. by S.Hatano, H.Ishikawa, H.Sato (Univ. Tokyo Press 1979) p.347
10. W. Gawlitta, K.V. Wolf, H.-U. Hoffmann, W. Stockem: Cell Tissue Res. 209, 71 (1980)
11. F. Achenbach, K.E. Wohlfarth-Bottermann: Differentiation 19, 179 (1981)
12. R.M. Ezzell, A.J. Brothers, W.Z. Cande: Nature 306, 620 (1983)
13. Y. Hayashi: J. theor. Biol. 85, 451 and 469 (1980)
14. G.M. Odell: Lect. Math. in the Life Sci. 9, 141 (1977)
15. G.M. Odell, H.L. Frisch: J. theor. Biol. 50, 59 (1975)
16. K.E. Samans, I. Hinz, Z. Hejnowicz, K.E. Wohlfarth-Bottermann: J. Interdiscipl. Cycle Res. 15, 241 (1984)
17. G.F. Oster, G.M. Odell: Cell Motility 4, 469 (1984)
18. G.F. Oster: J. Embryol. exp. Morphol. 83, Suppl. 329 (1984)
19. G.M. Odell: J. Embryol. exp. Morphol. 83, Suppl. 261 (1984)
20. K. Tarumi, H. Schwegler: Bull. math. Biol. 49, 307 (1987)
21. W. Alt: In Temporal Order, ed. by L.Rensing, N.Jaeger, Springer Ser. Synerg. Vol.29 (Springer, Berlin 1985) p.163
22. S.I. Beilina, N.B. Matveeva, A.V. Priezhev, Yu.M. Romanenko, A.P. Sukhorov, V.A. Teplov: In Self-Organization, Autowaves and Structures far from Equilibrium, ed. by V.I.Krinski, Springer Ser. Synerg. Vol.28 (Springer, Berlin 1984) p.218
23. W. Alt: Progress in Zoology 34, 1 (1987)
24. M. Dembo, F. Harlow, W. Alt: In Cell Surface Dynamics. Concepts and Models, ed. by A.S.Perelson, Ch.DeLisi, F.W.Wiegel (Marcel Dekker, New York and Basel 1984) p.495
25. M. Dembo, F. Harlow: Biophys. J. 50, 109 (1986)
26. T.P. Stossel: Ann. Rev. Cell Biol. 1, 353 (1985)
27. W. Alt: In Modelling of Patterns in Space and Time, ed. by W.Jäger, J.Murray, Lect. Notes in Biomath. Vol.55 (Springer, Berlin 1984) p.1
28. M. Dembo, M. Maltrud, F. Harlow: Biophys. J. 50, 123 (1986)
29. M. Dembo: Biophys. J. 50, 1165 (1986)

Developmental Control of a Biological Rhythm: The Onset of Cyclic AMP Oscillations in *Dictyostelium* Cells

A. Goldbeter[1] *and J.-L. Martiel*[2]

[1]Faculté des Sciences, Université Libre de Bruxelles,
  Campus Plaine, C.P. 231, Bd. du Triomphe, B-1050 Bruxelles, Belgium
[2]Départment d'Informatique, Faculté de Médecine,
  Université de Grenoble, F-38700 La Tronche, France

1. Introduction

Biological rhythms are so ubiquitous and necessary to life that their occurrence is often taken for granted. The fact that troubles of these rhythms arise in association with physiological disorders underlies a conspicuous property of periodic behavior, namely, that it occurs only in well-defined conditions. Indeed, slight changes in conditions may lead to the suppression of a rhythm, or to a transition from periodic to chaotic behavior. Physiological disorders associated with such nonequilibrium transitions have been referred to as "dynamical diseases" (1,2).

Besides the breakdown of periodic behavior, a mirror question arises as to how biological rhythms are established in the course of development. Broadly speaking, one or more control parameters of a regulated biological system may change in time so that beyond some critical bifurcation value the behavior switches from stationary to periodic. Few biological rhythms are known for which the underlying changes that bring them about have been identified.

The purpose of this paper is to examine, on one specific example, the molecular basis of the transition to oscillatory behavior which takes place during development. The system considered is that of cyclic AMP (cAMP) oscillations which govern aggregation of the slime mold Dictyostelium discoideum after starvation. Among biological oscillations, this rhythm is one of the best understood at the molecular level. By analyzing a model for the periodic synthesis of cAMP, we show explicitly that the onset of oscillations can be comprehended in terms of a continuous increase in biochemical parameters such as the activity of key enzymes, and the total amount of cAMP receptor incorporated into the cell membrane.

2. Mechanism and Function of cAMP Oscillations in Dictyostelium

When deprived of nutrients, Dictyostelium discoideum amoebae aggregate by a chemotactic response to cAMP signals emitted with a periodicity close to 10 min by cells which behave as aggregation centers (3,4). The signals emitted by the centers are relayed toward the periphery of the aggregation field, so that as many as 10^5 amoebae can collect around any of the centers. The aggregates thus formed further develop into fruiting bodies composed of stalk cells and spores.

The periodic nature of cAMP signaling is responsible for the wavelike patterns of aggregation in D. discoideum, in contrast with the monotonous nature of the aggregation process in D. minutum which lacks at that stage the capability of relaying or synthesizing periodically the chemotactic factor (5). The wavelike patterns of aggregation of D. discoideum on agar present a striking resemblance to the spiral and concentric patterns observed in the Belousov-Zhabotinsky reaction (6,7) which is the prototype of chemical oscillatory system.

Both relay and oscillation of cAMP have also been demonstrated in cell suspension experiments, first by means of light scattering studies (8), and later by direct assay of cAMP (9,10). Besides their role in controlling aggregation, periodic cAMP pulses also govern differentiation. In the hours that follow starvation, such pulses - in contrast with constant cAMP stimuli - induce the synthesis of specific proteins needed for intercellular communication and aggregation (11,12).

The biochemical mechanism responsible for cAMP oscillations involves a self-amplification process and a limiting step. The former results from the stimulation of adenylate cyclase upon binding of extracellular cAMP to a cell surface receptor; the intracellular cAMP thus synthesized is transported into the extracellular medium: this creates a positive feedback in cAMP synthesis. Such autocatalysis would cause a runaway process, were it not for limiting factors, among which the most prominent is probably the cAMP-induced desensitization of the cAMP receptor through reversible phosphorylation (13,14).

It is of interest to note the relation of this mechanism to the one responsible for glycolytic oscillations in yeast, which are the prototype of metabolic periodicity (15-18). There, positive feedback is also implicated as the activation of phosphofructokinase by a reaction product gives rise to sustained oscillations of all intermediates of the glycolytic system, even though the input of hexose is constant (15,16).

3. Model for cAMP Signaling Based on Receptor Desensitization

On the basis of the above-described mechanism, a model has been developed (19-21), which is represented schematically in Fig.1. It can be viewed as an extension of previous models for cAMP oscillations (22,23). The latter models, closely related to a model proposed for glycolytic oscillations (24), did not take into account the recently observed process of receptor modification. A similar remark holds for an alternative model based on the putative inhibition of adenylate cyclase by calcium ions (25).

The model for cAMP signaling based on receptor desensitization consists of nine ordinary differential equations. These were initially reduced to seven (19) and, more recently, to three kinetic equations (21). We shall use the latter, simpler version to analyze here the development of the signaling system.

After reduction to three variables, the model for cAMP signaling is governed by the system of kinetic equations /1/.

$$\frac{d\rho_T}{dt} = -f_1(\gamma)\rho_T + f_2(\gamma)(1-\rho_T)$$

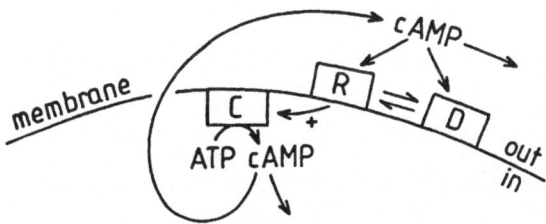

Fig.1: Model for cAMP signaling in Dictyostelium cells. Extracellular cAMP binds to the active (R) and desensitized (D) states of the cAMP receptor. Only binding to the active state elicits the activation of adenylate cyclase (C) which synthesizes cAMP from ATP. Arrows indicate synthesis of ATP, transport of cAMP into the extracellular medium, and hydrolysis of cAMP by phosphodiesterase

$$\frac{d\beta}{dt} = q\sigma\Phi(\rho_T,\gamma,\alpha) \quad - \quad (k_i+k_t)\beta$$

$$\frac{d\gamma}{dt} = (k_t\beta/h) \quad - \quad k_e\gamma \qquad\qquad /1/$$

with $\quad f_1(\gamma)=\dfrac{k_1+k_2\gamma}{1+\gamma}; \quad f_2(\gamma)=\dfrac{k_1L_1+k_2L_2c\gamma}{1+c\gamma}$

$$\Phi(\rho_T,\gamma,\alpha)=\frac{\alpha(\lambda\theta+\epsilon Y^2)}{1+\alpha\theta+\epsilon Y^2(1+\alpha)}; \quad Y=\frac{\rho_T\gamma}{1+\gamma}$$

In the above equations, ρ_T denotes the total fraction of receptor in active state, while β and γ denote the normalized concentrations of intracellular and extracellular cAMP. As to parameters, k_1 and k_2 measure the rate of receptor desensitization in the absence and presence of the ligand, respectively; k_t is the apparent first-order rate constant for cAMP transport into the extracellular medium; k_i and k_e are rate constants for hydrolysis by the intracellular and extracellular forms of phosphodiesterase; σ measures the normalized maximum activity of adenylate cyclase; h is a dilution factor; L_1, L_2 and c are ratios of constants related to receptor modification or to cAMP binding (see ref. 21 for further details on the equations and on the definition of the parameters).

A detailed analysis of eqs./1/ shows (21) that both types of dynamic behavior observed in the experiments can be accounted for by the model based on receptor desensitization. When the unique steady state admitted by eqs./1/ becomes unstable beyond a critical value of some control parameter (for example, k_t, k_i, k_e or the total receptor concentration R_T which appears in the definition of variable ρ_T), the system evolves to a stable limit cycle corresponding to sustained, autonomous oscillations of cAMP, accompanied by a periodic alternance of the receptor between its phosphorylated and dephosphorylated states.

For parameter values close to those which produce oscillations, the steady state is stable but excitable, as the system amplifies suprathreshold perturbations in a pulsatory manner. An absolute followed by a relative refractory period characterize this response; the recovery from refractoriness corresponds to the progressive return of the receptor from the desensitized to the active state (19-21). Such excitable behavior accounts for the relay of suprathreshold cAMP pulses observed in the course of aggregation (3,4,10).

Further reduction of eqs./1/ to a system of two kinetic equations allows to demonstrate by means of phase plane analysis the link between excitable and oscillatory behavior (21).

4. Development of the cAMP Signaling System

During the hours that follow starvation, D. discoideum amoebae undergo a sequence of developmental transitions as they are first unable to relay cAMP pulses, then acquire relay capability, before being able to produce cAMP pulses periodically in an autonomous manner (26). GOLDBETER and SEGEL (27) have proposed that such a sequence of developmental changes can be brought about by a continuous variation in key parameters such as the activity of adenylate cyclase and phosphodiesterase. The amoebae would follow a developmental path in this parameter space, which would bring them from a stable, nonexcitable steady state into a domain of excitability and, finally, into the domain of autonomous cAMP oscillations.

A similar explanation holds in the present model in the multiparameter space formed by the activity of adenylate cyclase (σ), intracellular (k_i) and extracellular (k_e) phosphodiesterase, and the concentration of cAMP receptor (R_T). All these parameters are known to increase in a sigmoidal manner over a 6-h period after starvation (28-31).

We may reproduce the experimentally observed sequence no relay-relay-oscillations when we incorporate into eqs./1/ the slow changes in the parameters. This is done by supplementing system /1/ with four differential equations for σ, R_T, k_i and k_e. In order to represent their sigmoidal increase in time, we assume that these parameters are proportional to a variable $X(t)$ whose time evolution is governed by the logistic eq./2/:

$$dX/dt = kX(X_{max} - X). \hspace{3cm} /2/$$

The parameters k and X_{max} are adjusted so that the evolution of X (and, hence, of the biochemical parameters) proceeds over slow characteristic times, X reaching its maximum X_{max} after 6 hours. Moreover, the initial value of X is taken as 1 and X_{max} as 50, so that X will change by a factor close to that observed experimentally for the activity of adenylate cyclase and phosphodiesterase during the interphase that follows starvation (28-31).

The parameters σ, k_e, k_i and R_T are taken as proportional to $X(t-\tau)$ where τ represents a time delay which may differ for each parameter (see legend to Fig.2) so that the continuous rise affects, successively, adenylate cyclase, the cAMP receptor and both forms of phosphodiesterase as observed in the experiments (30). The values of the delays and of the constants of proportionality to X are chosen so as to yield the profile and range observed for each of the evolving parameters after starvation.

The temporal evolution of the various parameters over a period of 6 h is shown in Fig.2a. There, the evolving receptor level is given as the fraction f_R equal to $R_T/(R_T)_f$, where $(R_T)_f$ denotes the receptor concentration reached after 6 h. The time evolution of intracellular cAMP (β) resulting from the above parameter changes is shown in Fig.2b. It indicates that 4 h after the beginning of starvation, the system becomes capable of autonomous oscillations of cAMP. Prior to the onset of oscillations, from time zero, we now perturb the signaling system every 10 min by an instantaneous increase in γ corresponding to a pulse of $3x10^{-8}$M extracellular cAMP. In these conditions, the continuous integration of eqs./1/ shows that the system becomes excitable after 3 h (Fig.2c); no significant amplification of the stimulus occurs before this time has elapsed.

When taking into account explicitly the continuous changes in the concentrations of enzymes and receptor which occur after starvation, the model thus reproduces the different modes of dynamic behavior in the time course observed during development, and thereby suggests a plausible molecular basis for the appearance of cAMP oscillations.

5. Discussion

In many instances in biology, rhythmic behavior begins in an abrupt manner, as a threshold is passed in development. Thus the heart tissue begins to contract rhythmically at a certain stage of embryonic growth (32). Likewise, spontaneous activity of the bursting neuron R15 in Aplysia only starts once a certain phase is reached in the development of this mollusc (33).

In an attempt to comprehend the onset of a biological rhythm at the molecular level, we have considered the mechanism that leads to the onset of cAMP oscillations in the slime mold Dictyostelium discoideum. During the developmental program initiated by starvation, D. discoideum amoebae at first do not respond to cAMP stimuli. Later they acquire the capability of relaying suprathreshold pulses of cAMP. This first transition is followed by a second one which brings the cells from the excitable state of relay into that of self-sustained oscillatory behavior. The latter state corresponds to the autonomous, rhythmic secretion of cAMP. The multiplicity of dynamic behavioral modes of the cAMP signaling system is particularly advantageous in that it poses constraints on theoretical explanations for the origin of rhythmic behavior. Any molecular mechanism for the onset of

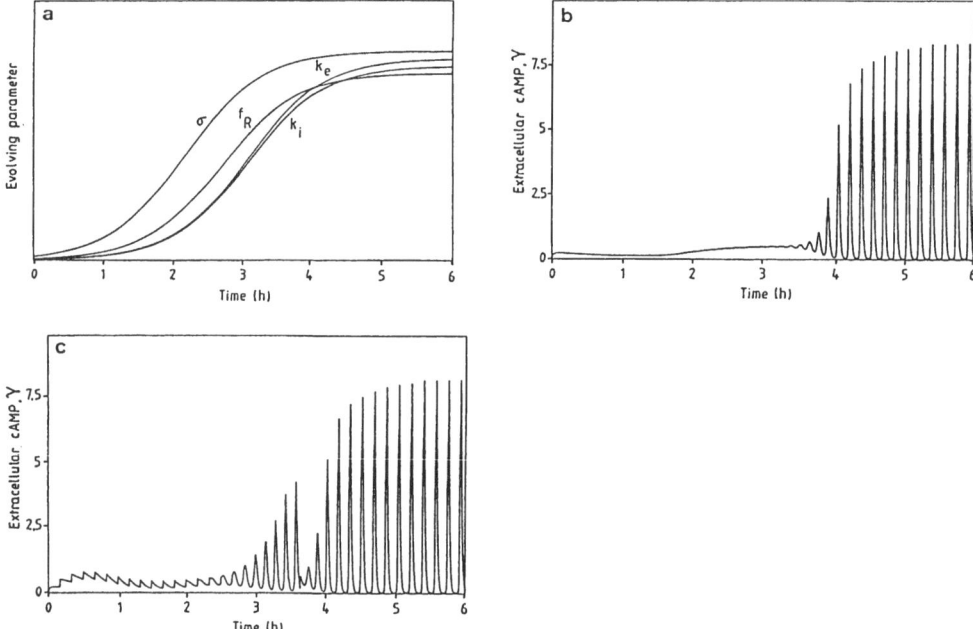

Fig.2: Possible molecular basis for developmental transitions leading to oscilla-tions in the cAMP signaling system of Dictyostelium cells. A continuous rise in adenylate cyclase (σ), extracellular (k_e) and intracellular (k_i) phosphodieste-rase, as well as in the number of receptor sites reflected by the fraction f_R which is proportional to R_T (panel a) may bring the system from a state of no re-lay to relay, and then to oscillations. In panel b is shown the onset of autono-mous oscillations of cAMP resulting from the changes in parameter values indica-ted in a. In panel c, the system is subjected to cAMP pulses at 10 min intervals until entrance in the oscillatory domain; the pulses are delivered in the form of additions of 0.3 units of γ. No amplification occurs at first, but the system be-gins to relay pulses before entering the oscillatory domain. The curves are ob-tained by numerical integration of eqs./1/ supplemented by evolution equations for σ, k_e, k_i, and f_R of the form $\sigma=0.012\,X(t)$, $f_R=X(t-\tau_1)$, $k_e=0.108\,X(t-\tau_2)$, $k_i=0.034\,X(t-\tau_2)$, where $X(t)$ is given by eq./2/ with $k=6\times10^{-4}\mathrm{min}^{-1}$. The delays τ_1, τ_2 are equal to 40 and 54 min, respectively; the initial value of X is taken as 1, with $X_{max}=50$ (see text). Other parameter values are $c=10$, $k_1=0.036$ min^{-1}, $k_2=0.666$ min^{-1}, $L_1=10$, $L_2=0.005$, $q=4\times10^3$, $k_t=0.9$ min^{-1}, $\alpha=3$, $h=5$, $\theta=0.01$, $\lambda=0.01$, $\varepsilon=1$. The fraction f_R is defined as the evolving, total receptor concentration R_T divided by the amount of receptor reached after 6h, $(R_T)_f$. Moreover, in eqs./1/, the factor 1 is replaced by f_R in the evolution equation for ρ_T as the total concentration of active receptor is normalized by division through $(R_T)_f$

cAMP oscillations should indeed account for the whole sequence of transitions ob-served during development.

We have shown that these sequential transitions can be brought about by the continuous increase in the activity of key enzymes of the cAMP signaling system, namely, adenylate cyclase and phosphodiesterase (see Fig.1), and in the total amount of cAMP receptor incorporated into the plasma membrane. These biochemical parameters are all known to increase sigmoidally in time, during a period of 6 h following the beginning of starvation.

The suggestion was previously made that a continuous rise in key enzymes of the cAMP signaling system is responsible for the sequential transitions in dynamic behavior: in parameter space, the amoebae would follow a developmental path which would cross, successively, a region of no relay, a region of excitability and, finally, a domain of sustained oscillations. Cells which would be the first to reach the latter domain would become the aggregation centers, at a time when other cells, less advanced on their developmental path, would reach the domain of relay. Such conjecture was based on a stability diagram showing the various behavioral domains in the adenylate cyclase-phosphodiesterase parameter space (27). Here, we have extended this analysis to the model for cAMP signaling based on receptor desensitization, and showed explicitly that the transitions occur, in the appropriate sequence, when the observed variation in the key parameters is built into the theoretical description.

The scenario that brings about cAMP oscillations in D. discoideum likely holds, mutatis mutandis, for other biological systems. Most biological rhythms correspond to limit cycle oscillations. For such rhythms, it is likely that in the course of development, the continuous variation in relevant control parameters brings about the discontinuous transition to self-sustained oscillatory behavior. While in the case of cAMP oscillations in Dictyostelium the evolving parameters are enzyme activities or a receptor concentration, in neurons the transition to spontaneous rhythmic behavior should rather originate from the slow appearance of specific ionic conductances controlling a pacemaker potential, or from the accumulation of an enzyme - such as a protein kinase - modulating the activity of an ionic channel (34). In neuronal networks (35), rhythms may originate, likewise, from the development of oscillatory properties of individual neurons or/and from the progressive establishment of inhibitory or excitatory connections between cells.

References

1. M.C. Mackey, L. Glass: Science, 197, 287 (1977)
2. L. Glass, M.C. Mackey: Ann. N.Y. Acad. Sci. 316, 214 (1979)
3. G. Gerisch: Ann. Rev. Physiol. 44, 535 (1982)
4. P. Devreotes: In The Development of Dictyostelium discoideum, ed. by W.F. Loomis (Academic Press, New York 1982) p.117
5. G. Gerisch: Curr. Top. Devel. Biol. 3, 157 (1968)
6. A.T. Winfree: Science 175, 634 (1972)
7. S.C. Müller, T. Plesser, B. Hess: Science 230, 661 (1985)
8. G. Gerisch, B. Hess: Proc. Nat. Acad. Sci. USA 71, 2118 (1974)
9. G. Gerisch, U. Wick: Biochem. Biophys. Res. Commun. 65, 364 (1975)
10. W. Roos, V. Nanjundiah, D. Malchow, G. Gerisch: FEBS Lett. 53, 139 (1975)
11. M. Darmon, P. Brachet, L.H. Pereira da Silva: Proc. Nat. Acad. Sci. USA 72, 3163 (1975)
12. G. Gerisch, H. Fromm, A. Huesgen, U. Wick: Nature 255, 547 (1975)
13. C. Klein, J. Lubs-Haukeness, S. Simons: J. Cell Biol. 100, 715 (1985)
14. P.N. Devreotes, J.A. Sherring: J. Biol. Chem. 260, 6378 (1985)
15. B. Hess, A. Boiteux, J. Krüger: Adv. Enzyme Regul. 7, 149 (1969)
16. B. Hess, A. Boiteux: Ann. Rev. Biochem. 40, 237 (1971)
17. A. Goldbeter, S.R. Caplan: Ann. Rev. Biophys. Bioeng. 5, 449 (1976)
18. M.J. Berridge, P.E. Rapp: J. Exp. Biol. 81, 217 (1979)
19. J.L. Martiel, A. Goldbeter: C.R. Acad. Sci. Sér.III 298, 549 (1984)
20. A. Goldbeter, J.L. Martiel: In Sensing and Response in Microorganisms, ed. by M. Eisenbach and M. Balaban (Elsevier Biomed. Div., Amsterdam 1985) p.185.
21. J.L. Martiel, A. Goldbeter: Biophys. J. in press (1987)
22. A. Goldbeter: Nature 253, 540 (1975)
23. A. Goldbeter, L.A. Segel: Proc. Nat. Acad. Sci. USA 74, 1543 (1977)
24. A. Boiteux, A. Goldbeter, B. Hess: Proc. Nat. Acad. Sci. USA 72, 3829 (1975)
25. P.E. Rapp, P.B. Monk, H.G. Othmer: Math. Biosci. 77, 35 (1985)
26. G. Gerisch, D. Malchow, W. Roos, U. Wick: J. Exp. Biol. 81, 33 (1979)
27. A. Goldbeter, L.A. Segel: Differentiation 17, 127 (1980)
28. C. Klein: FEBS Lett. 68, 125 (1976)
29. C. Klein, M. Darmon: Biochem. Biophys. Res. Commun. 67, 440 (1975)

30. C. Klein, M. Darmon: Nature 268, 76 (1977)
31. W.F. Loomis: Devel. Biol. 70, 1 (1979)
32. R.L. DeHaan: Curr. Top. Devel. Biol. 16, 117 (1980)
33. W.A. Adams, J.A. Benson: Progr. Biophys. Mol. Biol. 46, 1 (1985)
34. J.R. Lemos, W.B. Adams, I. Novak-Hofer, J.A. Benson, I.B. Levitan: In Neural Mechanisms of Conditioning, ed. by D.L. Alkon and C.D. Woody (Plenum, New York 1986) p.397
35. A.I. Selverston, M. Moulins: Ann. Rev. Physiol. 47, 29 (1985)

Periodic Cell Communication
in *Dictyostelium discoideum*

B. Wurster

Fakultät für Biologie, Universität Konstanz,
D-7750 Konstanz, Fed. Rep. of Germany

1. Introduction

The evolution of multicellular organisms required communication between cells. The cellular slime mould *Dictyostelium discoideum* is well suited for studies of intercellular communication at the transition from unicellular to multicellular organisms. Cells of *D. discoideum* feed and multiply as solitary amoebae. Upon depletion of the food source, growth phase cells differentiate into aggregative ones that assemble to form multicellular structures. These structures develop into fruiting bodies composed of stalk cells and spores. During differentiation to the aggregative state, cells acquire the capacity to communicate via chemical signals. In a population of cells, presumably those cells that first emit the signals become aggregation centers. Cells in the vicinity respond chemotactically to signal substances and they amplify and relay the signals. The centers emit the signals in a periodic manner, and aggregation occurs in steps of periodic movement [1-3].

An important signal substance of *D. discoideum* cells is cAMP [4-10]. Cyclic AMP is recognized by specific cell surface receptors and is destroyed by cAMP-phosphodiesterase [3, 11]. Stimulation of *D. discoideum* cells by cAMP causes synthesis and release of cAMP [5-8], as well as increases in the cellular contents of cGMP [12, 13] and inositol-1,4,5-trisphosphate [14], an influx of Ca^{++} ions [15, 16], an efflux of K^+ ions [17], and a decrease of the extracellular pH [18].

2. Spike-shaped and Sinusoidal Oscillations

Periodic activities of *D. discoideum* cells can be investigated in suspension. By means of an optical technique, two types of oscillations, spike-shaped and sinusoidal, were observed in the light-scattering properties of cell suspensions [19]. These oscillations reflect changes in cell shape and/or alterations in the size of cell aggregates [19]. During differentiation towards the aggregative state, spike-shaped oscillations occur before sinusoidal ones [19]. Usually, a series of 10 to 20 spike-shaped oscillations is followed by 10 to 20 sinusoidal cycles, and the first sinusoidal oscillations are observed between the last spikes. At 23°C, spike-shaped oscillations start with periods of about 9 min. The period length then steadily decreases so that after 10 spikes periods are about 7 min. Sinusoidal oscillations start with periods of about 7 min and end with periods of about 5.5 min.

Spike-shaped oscillations are accompanied by periodic synthesis and release of cAMP [9]. Exogenous cAMP causes phase shifts of these oscillations [19]. These results indicate that the communication substance cAMP can synchronize the periodic activities of cells during spike-shaped oscillations. An extracellular synchronizing substance may be merely coupled to an intracellular oscillator or it may represent a constituent of an oscillator with an extracellular loop. In the case of cAMP and spike-shaped oscillations, theoretical work [20, 21] suggests the latter possibility (Fig. 1).

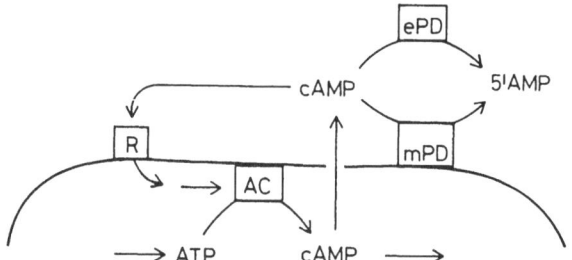

Fig. 1: Proposed mechanism for cAMP-dependent oscillations (adapted from [20,21]). Cyclic AMP is released by the cells. It binds to cAMP-receptors (R), and it is hydrolyzed by extracellular (ePD) and membrane-bound (mPD) cAMP-phosphodiesterase. The receptor-ligand complex causes activation of adenylate cyclase (AC), presumably via a G-protein and an additional soluble protein [22]. The receptor-ligand complex apparently is converted into a desensitized form [23,24]

Sinusoidal light-scattering oscillations occur in the absence of measurable cAMP oscillations [25,26]; exogenous cAMP does not significantly alter the phase [25]. The synchronizing compound as well as the reaction system underlying sinusoidal oscillations have not yet been identified.

Current approaches to the analysis of the reaction systems underlying periodic cell communication are i) identification of oscillating components, ii) isolation and characterization of oscillation mutants, and iii) isolation and identification of synchronizing substances.

3. Oscillating Components

In *D. discoideum* cells, oscillating components other than cAMP have been identified. These include cGMP [13], the redox state of cytochrome b [19], and the extracellular concentrations of Ca^{++} [26], K^+ [17], H^+ [27], and CO_2 [28]. Results are graphically presented in Fig. 2. Spike-shaped and sinusoidal oscillations in extracellular Ca^{++} were observed, and they were accompanied by spike-shaped and sinusoidal light-scattering oscillations, respectively. Oscillations in cAMP, which occur together with spike-shaped oscillations, disappear during early sinusoidal oscillations [25]. Oscillations in cAMP were only observed during spike-shaped oscillations. Periodic changes in extracellular K^+ started during late spike-shaped oscillations and continued during sinusoidal oscillations. Oscillations in external pH accompanied spike-shaped and sinusoidal oscillations, whereas changes in CO_2 and cytochrome b were only investigated with spike-shaped oscillations.

At the transition from spike-shaped to sinusoidal light-scattering oscillations a phase change occurs [25]. In contrast, no phase change was observed during Ca^{++} oscillations. Consequently, the phase relation between Ca^{++} and light scattering in spike-shaped oscillations differs from the phase relation in sinusoidal oscillations (Fig. 2). Also, the phase relation between Ca^{++} and H^+ changed slightly in going from spike-shaped to sinusoidal oscillations. These changes in phase and phase relationship cannot be readily explained, but they suggest the existence of complex reaction systems.

It will be interesting to learn which of the oscillating components, in addition to cAMP, are constituents of a reaction system that generates spike-shaped and/or sinusoidal oscillations. The finding that the Ca^{++} ionophore A23187 causes phase shifts of sinusoidal light-scattering oscillations [29] suggests that Ca^{++}

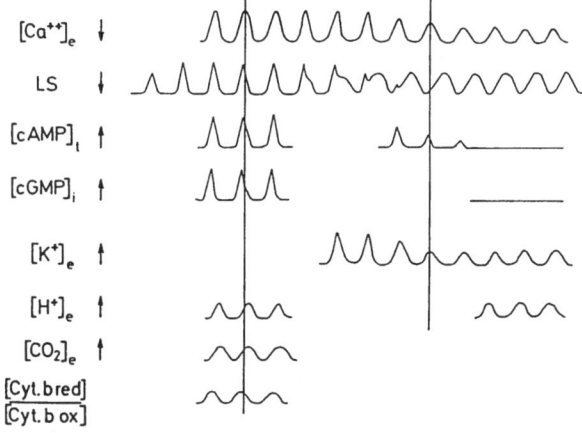

$[Ca^{++}]_e$ ↓

LS ↓

$[cAMP]_t$ ↑

$[cGMP]_i$ ↑

$[K^+]_e$ ↑

$[H^+]_e$ ↑

$[CO_2]_e$ ↑

$\dfrac{[Cyt.b\,red]}{[Cyt.b\,ox]}$

Fig. 2: Temporal correlation of various oscillating components in *D. discoideum*. For the purpose of comparison the period length has been normalized. Arrows indicate the direction of increase of the respective component. LS = light scattering. The subscripts e, i, t are abbreviations for extracellular, intracellular, and total, respectively. Taken from BUMANN et al. [26]

participates in a reaction system that controls sinusoidal oscillations. This reaction system may also function during spike-shaped oscillations [26]. The relevant component presumably is the intracellular Ca^{++} concentration. Exogenous Ca^{++} does not result in significant phase alterations. Reaction mechanisms involving the release of Ca^{++} and its uptake into intracellular compartments have been proposed [30,31].

4. Oscillation Mutants

Aggregation-defective mutants of *D. discoideum* often have aberrations in the oscillatory reaction systems, they either do not produce oscillations or oscillate with a period length different from that of the parent strain. The simultaneous defect in aggregation and oscillation appears to be due to the fact that periodic signals control differentiation towards the aggregative state [32,33] and aggregation of the cells [1-3]. Aggregation-defective mutants can be readily isolated. Examination of such mutants is still in the early stages but has already yielded interesting information.

Mutant HPX235 is deficient in intracellular, membrane-bound, and extracellular forms of cAMP-phosphodiesterase [34,35]. In suspension, HPX235 cells accumulate extracellular cAMP to concentrations of up to 30 nM; they neither differentiate nor display oscillations. However, when they are supplied with extracellular cAMP-phosphodiesterase, resulting in cAMP levels of about 1 nM, they differentiate towards the aggregative state and acquire the capacity to produce spike-shaped oscillations [34]. Obviously, degradation of extracellular cAMP is important for the occurrence of these oscillations. A deleterious effect of constantly elevated cAMP levels on oscillations was also observed with the normal strain Ax2. Continuous addition of cAMP resulting in steady-state concentrations of a few nM quenched light-scattering oscillations in suspensions of Ax2 cells [19, B.W. unpublished].

Cell suspensions of mutant Agip43 display spike-shaped oscillations in light-scattering with period lengths about 1.5 times larger than those of the parent strain [36]. These oscillations are not accompanied by measurable oscillations in

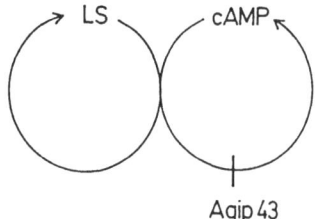

Fig. 3: Hypothetical scheme of coupled oscillators with proposed mutation in strain Agip43. LS = light scattering

cAMP, however, they can be shifted in phase by adding cAMP [36]. An intriguing interpretation of these results is that oscillations in cAMP levels and spike-shaped oscillations in light-scattering are produced by two distinct oscillatory reaction systems. It follows that in Agip43 the cAMP oscillator would be defective while the light-scattering oscillator would still be functioning. In the parent strain Ax2 the two oscillators should be coupled, since in these cells oscillations in cAMP and spike-shaped oscillations in light-scattering occur synchronously [9].

Cyclic AMP receptors at the cell surface presumably are a constituent of the cAMP oscillator (Fig. 1). In the case of coupled oscillators, cAMP receptors would also be connected to the light-scattering oscillator (Fig. 3). Such a connection could persist in the absence of an intact cAMP oscillator and would explain the phase shifts elicited by applied cAMP pulses in Agip43.

5. Isolation of Synchronizing Substances

Sustained rhythmic activity in a population requires a synchronizer [37]. If spike-shaped oscillations in Agip43 and sinusoidal oscillations in the parent strain Ax2 are not accompanied by the periodic synthesis and release of cAMP, what synchronizes these periodic activities? We are investigating the possibility that soluble synchronizing compounds exist in these cell suspensions, although we are aware that it could be a cell surface molecule that functions as a synchronizer.

A prerequisite for the isolation and subsequent identification of synchronizing substances from oscillating cell suspensions is an apparatus that allows recording of oscillations in large volumes. We have constructed a photometer with such properties [38]: The orifices of two optical glass fiber tubes are fixed at 5 mm distance. One tube is connected to the light source and the other tube guides the transmitted light to the photomultiplier. The rest is a conventional electronic set-up for photometry. The glass fiber tubes are immersed in the cell suspension which is supplied with oxygen and agitated by means of a magnetic stirrer. With this technique we could record synchronous oscillations in light-scattering properties from cell suspensions of 750 ml volume.

Oscillating cell suspensions were quenched at different phases by adding an equal volume of ethanol. Concentrated supernatants were added to an oscillating cell suspension to assay for the presence of synchronizing substances. We obtained evidence for the existence of a soluble synchronizing substance in oscillating suspensions of Agip43 cells. This compound is neither an anion nor a substrate for cAMP-phosphodiesterase [B. KANNAMÜLLER and B.W., unpublished]. Identification of this molecule is in progress.

6. Conclusions

Cells of *D. discoideum* communicate by means of periodic chemical signals. These signals control cell differentiation and aggregation. Periodic activities of *D. discoideum* cells can be observed in cell suspension as two types of oscillations in the light-scattering properties, spike-shaped and sinusoidal. In spike-shaped oscillations, cAMP is apparently a synchronizer and is presumably also a constituent of the oscillator. Results obtained with mutant Agip43 suggest that the reaction system producing spike-shaped oscillations is more complex and may consist of two coupled oscillators, only one of which involves cAMP. The reactions controlling sinusoidal oscillations are obscure, although there is an indication that intracellular Ca^{++} may participate in this process. Oscillating components, apart from cAMP and Ca^{++}, are cGMP, the redox state of cytochrome b, and the extracellular concentrations of K^+, H^+, and CO_2. It will be important to uncover whether one or more of these components are constituents of an oscillator or merely coupled to it. Other approaches to the molecular analysis of periodic signalling are identification of synchronizing substances and characterization of oscillation mutants. A favourable circumstance is that oscillation mutants can be obtained by selecting aggregation-defective strains.

Acknowledgements

I thank Drs. J. Bumann and G. Sweet for their criticism, and the Deutsche Forschungsgemeinschaft, Sonderforschungsbereich 156, for support.

References

1. J.T. Bonner: In The Development of Dictyostelium discoideum, ed. by W.F. Loomis (Academic Press, New York, London 1982) p. 1
2. G. Gerisch: Naturwissenschaften 58, 430 (1971)
3. P.N. Devreotes: In The Development of Dictyostelium discoideum, ed. by W.F. Loomis (Academic Press, New York, London 1982), p. 117
4. T. M. Konijn, J.G.C. van de Meene, J.T. Bonner, D.S. Barkley: Proc. Natl. Acad. Sci. U.S.A. 58, 1152 (1967)
5. B.M. Shaffer: Nature 255, 549 (1975)
6. W. Roos, V. Nanjundiah, D. Malchow, G. Gerisch: FEBS Lett. 53, 139 (1975)
7. A. Robertson, D. J. Drage, M.H. Cohen: Science 175, 333 (1972)
8. P.N. Devreotes, T.L. Steck: J. Cell Biol. 80, 300 (1979)
9. G. Gerisch, U. Wick: Biochem. Biophys. Res. Commun. 65, 364 (1975)
10. K.J. Tomchik, P.N. Devreotes: Science 212, 443 (1981)
11. G. Gerisch: Ann. Rev. Physiol. 44, 535 (1982)
12. J.M. Mato, F.A. Krens, P.J.M. van Haastert, T.M. Konijn: Proc. Natl. Acad. Sci. U.S.A. 74, 2348 (1977)
13. B. Wurster, K. Schubiger, U. Wick, G. Gerisch: FEBS Lett. 76, 141 (1977)
14. G.N. Europe-Finner, P.C. Newell: J. Cell Sci. 87, 221 (1987)
15. U. Wick, D. Malchow, G. Gerisch: Cell Biology International Reports 2, 71 (1978)
16. J. Bumann, B. Wurster, D. Malchow: J. Cell Biol. 98, 173 (1984)
17. S. Aeckerle, B. Wurster, D. Malchow: EMBO J. 4, 39 (1985)
18. D. Malchow, V. Nanjundiah, B. Wurster, F. Eckstein, G. Gerisch: Biochim. Biophys. Acta 538, 473 (1978)
19. G. Gerisch, B. Hess: Proc. Natl. Acad. Sci. U.S.A. 71, 2118 (1974)
20. A. Goldbeter, L.A. Segel: Proc. Natl. Acad. Sci. U.S.A. 74, 1543 (1977)
21. J.L. Martiel, A. Goldbeter: Nature 313, 590 (1985)
22. A. Theibert, P.N. Devreotes: J. Biol. Chem. 261, 5121 (1986)
23. P. Klein, A. Theibert, D. Fontana, P.N. Devreotes: J. Biol. Chem. 260, 1757 (1985)
24. P.J.M. van Haastert, R.J.W. de Wit, P.M.W. Janssens, F. Kesbeke, J. DeGoede: J. Biol. Chem. 261, 6904 (1986)
25. G. Gerisch, D. Malchow, W. Roos, U. Wick: J. Exp. Biol. 81, 33 (1979)
26. J. Bumann, D. Malchow, B. Wurster: Differentiation 31, 85 (1986)
27. D. Malchow, V. Nanjundiah, G. Gerisch: J. Cell Sci. 30, 319 (1978)
28. G. Gerisch, Y. Maeda, D. Malchow, W. Roos, U. Wick, B. Wurster:

In <u>Developments and Differentiation in the Cellular Slime Moulds</u>, ed. by P. Cappuccinelli, J.M. Ashworth (Elsevier/ North Holland Biomedical Press, Amsterdam 1977) p. 105

29. D. Malchow, R. Böhme, U. Gras: Biophys. Struct. Mech. <u>9</u>, 131 (1982)
30. M.J. Berridge, P.E. Rapp: J. Exp. Biol. <u>81</u>, 217 (1979)
31. P.E. Rapp, P.B. Monk, H.G. Othmer: Mathematical Biosciences <u>77</u>, 35 (1985)
32. M. Darmon, P. Brachet, L.H. Pereira da Silva: Proc. Natl. Acad. Sci. U.S.A. <u>72</u>, 3163 (1975)
33. G. Gerisch, H. Fromm, A. Huesgen, U. Wick: Nature <u>255</u>, 547 (1975)
34. P. Brachet, E.L. Dicou, C. Klein: Cell Differentiation <u>8</u>, 255 (1979)
35. J. Barra, P. Barrand, M. Blondelet, P. Brachet: Molec. Gen. Genet. <u>177</u>, 607 (1980)
36. B. Wurster, R. Mohn: J. Cell Sci. <u>87</u>, 723 (1987)
37. V. Nanjundiah: J. Theor. Biol. <u>121</u>, 375 (1986)
38. W. Kurzenberger, B. Wurster: in preparation.

Part V

From Complex Cellular Networks
to the Brain

Memory and Paralysis Phenomena in the Immune Response: Interpretation in Terms of Multiple Steady State Transitions

M. Kaufman

Faculté des Sciences, Université Libre de Bruxelles,
Service de Chimie Physique II, Campus Plaine, C.P. 231,
Bd. du Triomphe, B-1050 Bruxelles, Belgium

1. Introduction

An essential aspect of the immune response is the production by bone-marrow derived, B lymphocytes, of antibody molecules which specifically recognize the antigen that enters the organism and combine with it to cause its elimination.

For a long time, the immune system was conceived as an enormous library of precommitted and independent B lymphocytes, each lymphocyte bearing surface immunoglobulin receptors of a given specificity. Introduction of the antigen was believed to induce clonal expansion and differentiation of those lymphocytes displaying receptors able to bind the antigen. Although the basic premise of this theory, one lymphocyte - one antibody, is still considered to be correct, the recent immunological data suggest that the clonal selection theory [1] was an oversimplification. Several important findings such as for instance the cooperation between B lymphocytes and another group of thymic derived, T lymphocytes [2], the hapten-carrier phenomenon [3], the dual recognition system used by the T cells [4] and the involvment of idiotypes in clonal interactions [5,6], point out that the immune system cannot be simply considered as a large collection of independent clones.

It is now clear that the initiation and regulation of antibody production result from complex interactions between different sets and subsets of lymphocytes and their products, and a great deal of research is currently being done in an effort to elucidate the mechanisms of immunoregulation.

If one begins to know the details of some of the individual steps down to the molecular level, the understanding of the coordination and functioning of the system as a whole is still very partial. In the present stage of knowledge it becomes evident that intuition is unreliable, and that a language more precise than the verbal one is needed in order to describe and analyze this complex interactive network. Mathematical modelling, therefore, might help to clarify some of the complexities of the regulatory events and provide insight in the overall organization of the system.

During the last ten years, several mathematical models have been proposed in the framework of Jerne's idiotypic network concept [5,7] which implies the existence of self-regulatory interactions within the immune system even in the absence of foreign antigens. Richter's model [8], the first network model to be developed, describes an open-ended, linear net based on sequential activation-suppression interactions originating from the antigen. When an antigenic stimulus is applied, the linear arrangement of the units leads, however, to the perturbation of a very large number of clones which is unreasonable. This model has been modified by Hiernaux [9] in order to deal with the stability requirements of the immune system and the cyclic picture of the immune network [6]. Both versions do not introduce the distinction between B and T lymphocytes. Hoffmann's model [10,11] considers $T - B$ dichotomy and introduces the elegant concept of symmetrical idiotype - antiidiotype interactions [12]. No difference is made in his description between helper and suppressor T cells. The model accounts for multiple and specific stable steady states, in the absence of antigen, but its use for a dynamic analysis raises certain problems. Recently, Herzenberg et al. [14] have presented a circuit based regulatory system composed of interactions between T helper cells which amplify the immune response and T suppressor cells which damp the activity of the helpers. Their analysis mainly relies on intuitive arguments and a more careful study indicates that their cellular circuits fail to satisfy the requirement of a stable virgin state. Modifications of their schemes have been proposed by Eisenfeld [14].

Our purpose has been to investigate to what extent one may account for some important functional aspects of the immune system such as memory, paralysis and non-responsiveness, with simple models involving a small set of basic components and a small number of interactions. We did not try to describe every detail of an immune response, but rather our main goal has been to bring out some general property of the network that might explain the basic modes of behaviour upon antigenic challenge.

The approach that we have developed comprises two complementary methods of description: a logical or discrete formalisation in terms of discrete variables and functions, and a classical continuous analysis in terms of differential equation [15,16].

The logical formalisation has been essential to conceive and develop the model as it does not necessitate to specify the details of the cellular interactions or to introduce a great number of parameters. The essence of the method consists in encoding into logical relations the main regulatory interactions characterizing the system and deriving from these relations qualitative informations on the possible regime states and dynamic pathways [17]. The simplicity of such an analysis allows to extract the key elements responsible for a given behaviour and may even suggest minimum conditions that must be met for a better alternative. However, the logical analysis typically is qualitative and gives a somewhat "caricatural image" of the behaviour of the system.

The continuous description leads to a more refined analysis of the properties of the system. It requires the specification of the form of the mathematical functions describing the various cellular interactions, but on the other hand, it allows the relaxation of the strong nonlinearity inherent in the logical treatment. In addition, the neutralisation of the antigen is more naturally introduced in the continuous formalisation which leads to a more precise study of the behaviour of the system as a function of the characteristics of the antigenic stimulus.

2. The Model

The model that has emerged from this combined analysis [15,16] is presented on Fig. 1. It comprises the main components participating in the humoral immune response : the antigen or epitope (E), the antibody producing B cells, the regulatory T helper (T_H) and T suppressor (T_S) cells and the antibody molecules (Ab). A few but essential interactions are considered between these components.

The central part of the model consists in a triple loop involving the regulatory T cells. This core is formed by a negative feedback loop between helper and suppressor T lymphocytes [18] on which positive loops of the T_H and T_S populations on themselves are grafted. These positive loops take into account the occurrence of $T - T$ interactions both at the level of helper and suppressor circuits [12,18-20], and are here schematically represented by autocatalytic loops.

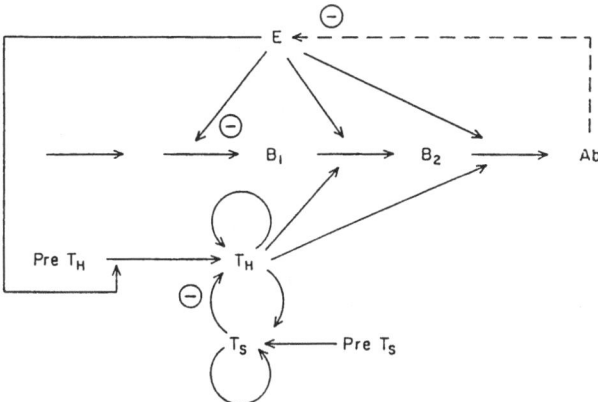

Figure 1 : Interaction diagram. Unless otherwise specified the regulations are positive

The complete network includes differentiation and maturation of the B cells. For the development of these B cells we retain two stages : virgin B_1 cells and more mature B_2 cells. Immature precursor cells are steadily produced from the bone-marrow but this flux is blocked at the stage of the precursors by the presence of antigen, even at low concentrations [21]. The cells which have escaped this negative signalling differentiate into B_1 lymphocytes. Provided the concerted action of antigen and a positive signal coming from activated T helper cells of the required specificity, these virgin lymphocytes further differentiate into more mature B_2 lymphocytes and finally reach the terminal maturation stage characterized by the secretion of antibody.

The antibodies synthetized during the response form an inactive complex with the antigen molecules. There is thus a negative feedback of antibody on the formation of new antibody secreting cells both by blocking the antigen from stimulating immunocompetent cells and by diminishing cellular collaboration between B and T_H lymphocytes.

The interaction scheme shown on Fig. 1 indicates which components influence each other but where two or more variables influence a same element, it does not specify how these interactions are connected with each other. In the absence of definite experimental data we have analyzed several possibilities. As discussed elsewhere [15,16,22], some choices appear to be crucial in order to observe an appropriate behaviour. The assumptions that we have made mainly consist in considering a distinct establishment and maintenance mechanism for the T cells, and in the role we ascribed to antigen at various levels of the network. In particular, for the cellular interactions, we consider antigen bridges [3] only where idiotype-antiidiotype recognition is excluded.

3. Analysis of the Model

A detailed presentation of the logical analysis of the model can be found in previous publications [15,22,23]. The main conclusions are that in the absence of antigen the system may persist, depending on its past history, in any one of three stable regime states : a "virgin" state in which only B_1 cells are present; a "suppressed" state where besides virgin B lymphocytes there is also a significant level of suppressor T lymphocytes; a "memory" state characterized by the presence of B_1, T suppressor and T helper cells. The logical analysis predicts, furthermore, a reasonable dynamics for primary and secondary responses upon single antigen injections, and the existence of paralysis phenomena when repeated antigenic stimuli are applied.

To illustrate the role of this multistationarity for the dynamic response of the system as a function of the amounts of antigen that are injected or on the way in which antigen is administered, we present here more quantitative results which were obtained with the continuous method [16,22].

3.1 Kinetic Equations

The following set of differential equations has been used to describe the model:

$$\frac{dx_1}{dt} = k_1 \, F_1^- \, (x_6) - k_2 \, x_1 \, x_3 \, F_2^+ \, (x_6) - d_1 \, x_1 \tag{1}$$

$$\frac{dx_2}{dt} = k_2 \, x_1 \, x_3 \, F_2^+ \, (x_6) + (m_2 - k_5) \, x_2 \, x_3 F^+ \, (x_6) - d_2 \, x_2 \tag{2}$$

$$\frac{dx_3}{dt} = k_3 \, F_3^+ \, (x_6) \, F_3^- \, (x_4) + m_3 \, F_3^+ \, (x_3) - d_3 \, x_3 \tag{3}$$

$$\frac{dx_4}{dt} = k_4 \, x_3 + m_4 \, F_4^+ \, (x_4) - d_4 \, x_4 \tag{4}$$

$$\frac{dx_5}{dt} = k_5 \, x_2 \, x_3 \, F^+ \, (x_6) - p \, k_6 \, x_5^p \, x_6^q - d_5 \, x_5 \tag{5}$$

$$\frac{dx_6}{dt} = -q \, k_6 \, x_5^p \, x_6^q - d_6 \, x_6 \tag{6}$$

where the x_i's, $i = 1$ to 6, are respectively the concentrations of the components B_1, B_2, T_H, T_S, Ab and E. The existence of a constant level of precursor cells of each required type is taken into account in the rate constants. p and q are the stochiometric coefficients of the antigen- antibody reaction. A term of spontaneous decay, $-d_i x_i$, is considered for each cellular or molecular species. The functions $F_i(x_j)$ are Hill functions defined by :

$$F_i^+ \, (x_j) = \frac{x_j^n}{\theta_{ij}^n + x_j^n} \qquad \text{for activation, and} \tag{7.a}$$

$$F_i^- \, (x_j) = 1 - F_i^+ \, (x_j) \tag{7.b}$$

for inhibition, with n the Hill coefficient and θ_{ij} a threshold parameter for the regulation of the $i - th$ variable by component j.

Compared to the discrete treatment [15,23] the nonlinearity is importantly reduced by considering a low degree of cooperativity ($n = 2$) and by retaining nonlinear sigmoid functions only for a small number of interactions.

3.2 Steady State Behaviour

The steady state behaviour of the system is essentially determined by equations (3) and (4) which correspond to the idiotypic regulatory core of the model. These equations describe the coupling between two positive feedback loops and the multiplicity of steady states depends on the number and on the strength of the constraints between these loops [16,24]. A typical situation is illustrated on Fig. 2 in the absence of antigen ($x_6 = 0$), for the parameter values listed in section 3.3.

266

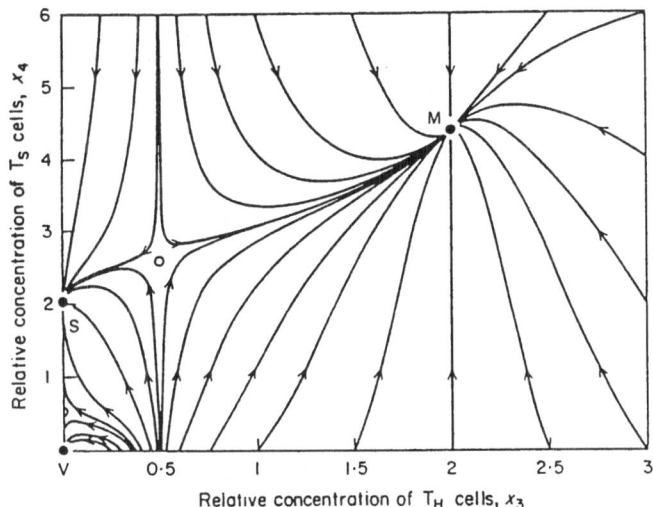

Figure 2 : Phase plane diagram corresponding to equations (3) and (4) in the absence of antigen

One observes the existence of three stable nodes separated by saddle points. The stable states account for biologically relevant virgin (V), suppressed (S) and memory (M) states. The correponding steady state values of the other components are : $x_1^0 = k_1/d_1, x_2^0 = x_5^0 = 0$.

Antigen introduces an additional coupling between the autocatalytic loops on x_3 and x_4 which tends to decrease the number of steady states as a function of the antigen level which is imposed in the system : for very low antigen concentrations there are still five steady states, but this number lowers to one when the antigen concentration is increased. Figures 3a and 3b illutrate how the domains of attraction vary with the antigen level.

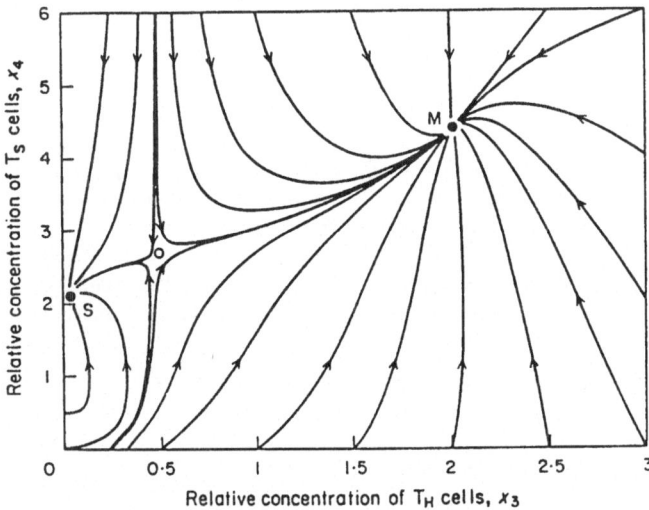

Figure 3a : Phase plane diagram corresponding to equations (3) and (4) in the presence of a constant level of antigen $\bar{x}_6 = .03$ The arrows on the trajectories show the evolution in time

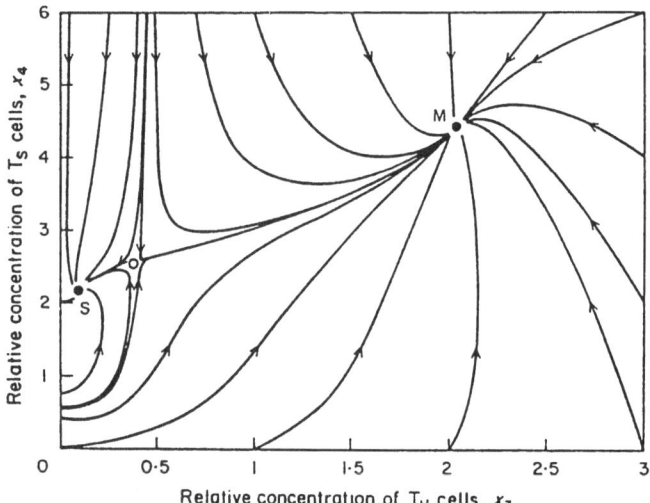

Figure 3b : Phase plane diagram corresponding to equations (3) and (4) in the presence of a constant level of antigen $\bar{x}_6 = .05$. The arrows on the trajectories show the evolution in time

Starting from the virgin state only the suppressed state S can be reached on Fig. 3a, whereas on Fig. 3b it is the memory state M which will be attained. The corresponding steady state levels of x_1, x_2 and x_5 are now dependent on x_3^0 and $\bar{x}_6$ and may be straigth-forwardly deduced.

The steady state which subsists at high antigen concentrations corresponds to high amounts of helper and suppressor cells. However, the corresponding levels of virgin and more mature B cells and of antibody are negligible, despite the presence of antigen. Here, inhibition of the influx of B lymphocytes prevails and this steady state may be characterized as a "paralyzed" state.

3.3 Dynamic Behaviour

The high multiplicity of steady states that arises in our model, both in the absence and presence of antigen, is essential for the system response to antigenic stimulation : depending on its past history, or on how much and in what way antigen is administered, the system stabilizes in a different steady state. This in turn determines the dynamic behaviour upon a next antigenic challenge. In the following the basic modes of response that are derived from the model are illustrated by considering two typical situations [16,22]. The dynamic behaviours have been obtained by numerical integration of equations (1) to (7) together with initial conditions and for the parameter values listed below:

$$k_1 = .6 \qquad k_2 = .1 \qquad k_3 = k_5 = m_2 = .4 \qquad k_4 = k_6 = .2 \quad m_3 = m_4 = .5$$
$$d_1 = d_2 = .1 \qquad d_3 = d_4 = d_5 = .2 \qquad d_6 = .05$$
$$\theta_{44}/\theta_{34} = 1. \qquad \theta_{16}/\theta_{26} = \theta_{36}/\theta_{26} = \theta/\theta_{26} = .1$$

$n = 2$ for all the sigmoid functions. x_6, x_3 and x_4 are reduced to their threshold value θ_{26}, θ_{33} and θ_{44} respectively.

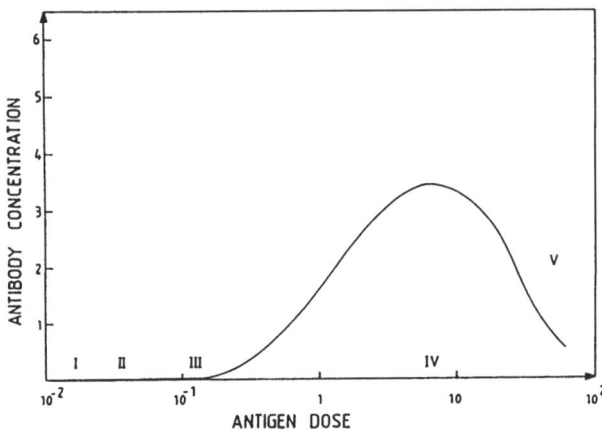

Figure 4 : Antibody response as a function of the antigen dose

The results which are presented are qualitative results since the parameter values have not been adjusted in order to take quantitative data into account.

DYNAMIC BEHAVIOUR UPON SINGLE ANTIGEN PULSES : Figure 4 summarizes the response of the system on a single antigen injection as a function of the size of the antigen dose. The initial state is the virgin state. The curve that is obtained has the characteristic bell shape of the experimentally observed dose - response curves [25].

As a function of the antigen dose, five different behaviours may be distinguished :

1. The system remains insensitive to a very low antigen dose corresponding to region I. A next stronger antigen injection will provoke a normal primary response.

2. A low antigen dose corresponding to region II induces suppression : antigen brings the system, without antibody production, towards the suppressed state from where it responds poorly to a second, optimal dose.

3. An antigen dose situated in region III is still unable to stimulate the differentiation of the B cells. However this time, the state of immune memory is reached and, as shown in Fig. 5, the system responds more rapidly and stronger to a next optimal injection than the virgin system. This extreme difference between primary and secondary response is known as the phenomenon of "priming without antibody secretion".

4. Antigenic stimulation with a dose corresponding to region IV leads to a significant primary response. Here again the memory state is established. The response to a next challenge will be enhanced by the presence of the T helper cells which are required at several stages of the differentiation and maturation of the antibody producing cells, and a secondary response follows (Fig. 6).

5. When the antigen dose is further increased (region V) the amplitude of the antibody response declines and becomes negligible. Both suppressor and helper T cells are activated in this region but rapid depletion of the B cells prevails.

DYNAMIC BEHAVIOUR AFTER PRETREATMENT: The procedure that is used consists in maintaining a given antigen level in the system during some time interval. Then, the response to an antigen challenge which normally evokes a response in a non-pretreated system is recorded. The behaviour of the system after pretreatment is summarized on Fig.

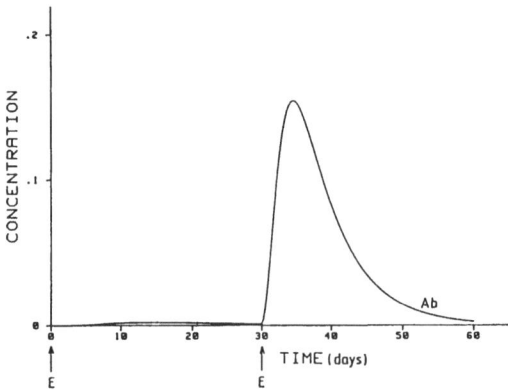

Figure 5 : Priming without antibody secretion (reduced concentrations)

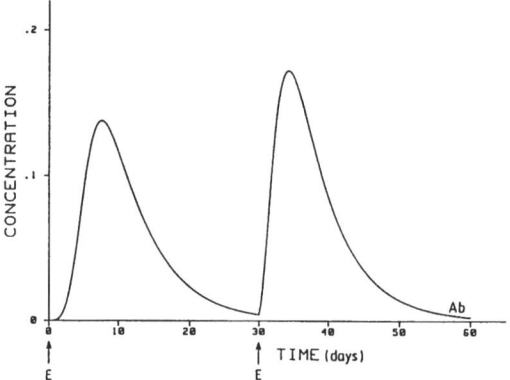

Figure 6 : Primary and secondary response (reduced concentrations)

7 which shows the size of the response to an optimal dose as a function of the antigen dose that has been used for pretreatment. The results are in very good qualitative agreement with the experimental data, e.g. Shellam et al. [26].

Here again one distinguishes four regions :

1. A "virgin" region, I, where pretreatment leaves the system in the vicinity of the virgin state.

2. A "low dose paralysis" region, II, where pretreatment drives the system into the suppressed state (see Fig. 3a). A subsequent normal stimulation results in low amounts of antibody.

3. An "immunity" region, III, where pretreatment induces the memory state (see Fig. 3b) without producing antibodies. The next antigenic challenge provokes a secondary type of response.

4. A "high dose paralysis" region, IV, where pretreatment now leads the system towards a paralyzed state characterized by high levels of T_H and T_S cells but strong depletion of the B cells. A next antigen injection fails to evoke a response. This paralysis is gradually lost when antigen disappears by spontaneous decay.

270

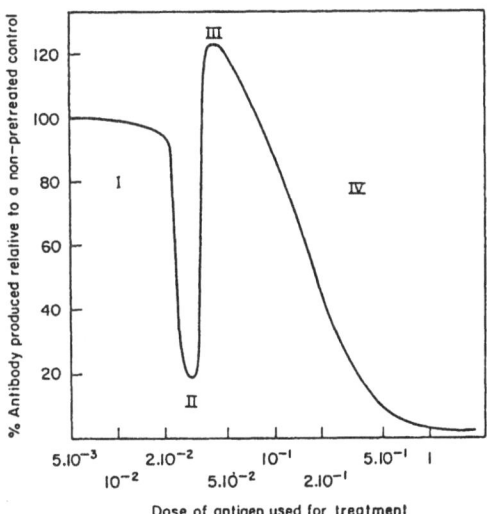

Figure 7 : The effect of pretreatment with different antigen doses is reflected in the size of the secondary response relative to a non-pretreated control

High and low dose paralysis are thus associated in our model with different steady state branches which implies that the underlying mechanisms are different. Low dose paralysis is related to an inefficient activation of the B cells : the suppressor T cells inhibit the development of the T helper cells which are needed for the differentiation of the antibody producing cells. High dose paralysis mainly results from clonal depletion of the B_1 cells in the continuous presence of antigen. The two regions of paralysis are separated by an immunity hump.

4. Conclusion

Our work represents an attempt at developing a theoretical framework for the study of complex cellular networks and at integrating various processes known to regulate antibody production into a coherent ensemble.

The model that is presented has been established on the basis of well-documented facts about lymphocyte interactions and some reasonable assumptions. It comprises several interconnected positive and negative feedback loops. The positive loops are responsible for the occurrence of a high multiplicity of steady states, whereas the negative loops imply the possibility of oscillatory behaviour.

This minimal scheme accounts for aspects as essential as primary vs. secondary response, low and high dose paralysis and low responsiveness. These different response modes are interpreted in terms of dynamic transitions among the various steady states and depend on how much and in what way antigen is injected, or on the previous history of the system. Some of these transitions are irreversible : once the system has encountered the antigen it never returns to the virgin state; others, such as high dose paralysis, are reversed when the antigen disappears. Suppression, on the other hand, may be lifted by applying appropriate antigenic stimuli.

Our approach certainly is oversimplified and concerns interactions within an antigen-defined set of lymphocytes, uncoupled to the remainder of the immune system. A next step will be to incorporate this cellular " unit" into a broader network.

5. References

1. F.M. Burnet : In *The Clonal Selection Theory of Acquired Immunity* (Cambridge University Press, Cambridge 1959).
2. J.F.A.P. Miller : *Thymus* 1, 3 (1979).
3. N.A. Mitchison : *Eur. J. Immunol.* 1, 18 (1971).
4. R.M. Zinkernagel, P. Doherty : *Adv. Immunol.* 27, 51 (1979).
5. N.K. Jerne : *Ann. Immunol. Inst. Pasteur* (Paris), 125C, 373 (1974).
6. J. Urbain, C. Wuilmart, P.A. Cazenave : *Contemp. Top. Mol. Immunol.* 8, 113 (1981).
7. N.K. Jerne : *Harvey Lect.* 70, 93 (1976).
8. P.H. Richter : *Eur. J. Immunol.* 5, 350 (1975).
9. J. Hiernaux : *Immunochemistry* 14, 733 (1977).
10. G.W. Hoffmann : *Eur. J. Immunol.* 5, 638 (1975).
11. N. Gunther, G.W. Hoffmann : *J. Theor. Biol.* 94, 815 (1982).
12. G.W. Hoffmann : *Contemp. Top. Mol. Immunobiol.* 11, 185 (1980).
13. L.A. Herzenberg, S.J. Black, L.A. Herzenberg : *Eur. J. Immunol.* 10, 1 (1980).
14. J. Eisenfeld : *Ann. N.Y. Acad. Sci.* , in press (1987).
15. M. Kaufman, J. Urbain, R. Thomas : *J. Theor. Biol.* 114, 527 (1985).
16. M. Kaufman, R. Thomas : *J. Theor. Biol.*, in press (1987).
17. R. Thomas : *Adv. Chem. Phys.* 55, 247 (1984).
18. D.R. Green, P.M. Flood, R.K. Gershon : *Ann. Rev. Immunol.* 1, 439 (1983).
19. T. Tada, K. Okumura : *Adv. Immunol.* 28, 1 (1980).
20. M.E. Dorf, B. Benacerraf : *Ann. Rev. Immunol.* 2, 127 (1984).
21. J. Lederberg : *Science* 129, 1649 (1959).
22. M. Kaufman : In *Theor. Immunol. Workshop, Santa Fe*, ed. by A. Perelson (Addison-Wesley, Reading, Massachusetts 1988).
23. M. Kaufman : In *Dynamical Systems and Cellular Automata*, eds. J. Demongeot, T. Goles, M. Tchuente (Academic Press, New York 1985) p.207.
24. R. Thomas : *Discr. Appli. Math.*, in press (1987).
25. J.G. Howard, N.A. Mitchison : *Progr. Allergy* 18, 43 (1975).
26. G.R. Shellam, J.V. Nossal : *Immunol.* 14, 273 (1968).

Spatiotemporal Patterns of Block in an Ionic Model of Cardiac Purkinje Fibre

M.R. Guevara

Department of Physiology, McGill University,
3655 Drummond Street, Montreal, Quebec, Canada H3G 1Y6

1. Introduction

The electrical activity of cardiac muscle serves as the trigger for the mechanical contraction of the heart that is essential for the preservation of life. In the mammalian heart, there is a well-organized sequence of activation in both space and time. This synchronized activity hinges upon the existence of a specialized electrical conduction system. The sinoatrial node, a specialized structure sitting high up in the right atrium, fires first, sending an electrical signal to the right and left atria (initiating contraction of the muscle in both atria). From the right atrium, the signal proceeds to another specialized structure, the atrioventricular node, which then relays the signal on to the bundle of His, the bundle branches, and the Purkinje network. This network is composed of fine strands of Purkinje cells which ramify over the inner surface of the ventricles, delivering the cardiac impulse to the working myocardium of both ventricles. Thus, the entire specialized conduction system serves to generate and conduct the cardiac impulse in a highly coordinated manner, with one activation of any heart cell for each activation of any other heart cell. In addition, the difference in activation times of any two given cells remains fixed from beat to beat.

A disease process can lead to a disturbance in the above orderly spatiotemporal sequence, leading to the appearance of a cardiac arrhythmia. One leading mechanism for the production of many arrhythmias is block of conduction of the cardiac impulse. This block can be complete, with no impulses traversing the region showing a depressed ability to conduct. The block can also be partial, with some, but not all, impulses propagating through the region of block.

Block not uncommonly occurs in the His-Purkinje system, especially following a myocardial infarction. In contrast to block occurring at higher levels ("supranodal block"), this infranodal form of block can have serious consequences, since it sometimes leads to sudden complete block of propagation of the cardiac impulse through to the ventricles. We investigate an ionic model of conduction in a strand of Purkinje fibre, and demonstrate that many patterns of conduction similar to those seen clinically and experimentally are observed in the model. While previous modelling studies have been carried out on propagation in Purkinje fibre, there has not been a systematic study on the effects of periodic stimulation. Some of the results mentioned below are mentioned in passing in GUEVARA [1].

2. Numerical Methods

When block occurs in cardiac tissue, slowly rising action potentials with rather low conduction velocities are commonly found in the region of block [2]. Since high conduction velocities are usually associated with the presence of the fast inward sodium current, I_{Na}, we implement a region of impaired conduction in the mid-section of a Purkinje strand by simply removing the fast sodium current I_{Na} in that part of the strand [3,4,5]. A strand of length 0.625 cm is discretized into 100 segments, with the cellular membrane in segments 1-40 (the proximal end) and

61-100 (the distal end) being represented by the MNT equations of McALLISTER et al. [6], while that in segments 41-60 (the region of block) is also represented by the MNT equations, but with I_{Na} = 0. We have used the equations as they appear in the text, and not in Table 1 of [6]. In particular, we have used equation (16) of [6] rather than the equation given in Table 1, for reasons previously given [7]. In addition, equations (25) and (26) were used rather than (27) and (28) for the current I_{qr}. L'Hôpital's rule was applied when necessary in computing the rate constants α_m, α_d, α_q, and α_s, as well as the current I_{K1}.

We have numerically integrated the one-dimensional cable equation

$$\frac{a}{2R_i} \frac{\partial^2 V}{\partial x^2} = C \frac{\partial V}{\partial t} + I \quad , \tag{1}$$

where t is the time (ms), x is the distance along the strand (cm), V(x,t) is the transmembrane potential difference (mV) at (x,t), I(x,t) is the total transmembrane current (μA/cm^2) at (x,t), a is the radius of the cable (0.00025 cm), R_i is the specific resistivity of the intracellular medium (0.02 KΩ-cm), and C is the specific membrane capacitance (10 μF/cm^2). For purposes of comparison, we have chosen the values of a, R_i, C, and the total cable length to be equivalent to those used in previous studies using the MNT model [3,5,8,9]. Note that for historical reasons, the value of C used in the MNT model [6] and in this article is tenfold higher than the currently accepted experimental value. To compensate for this, the value of R_i used here is set at one-tenth of the experimental value usually quoted. In obtaining (1), the specific external resistivity was set to zero. Since the input resistance R_m in the MNT model is about 3 KΩ-cm^2 [5], the time constant $\tau = R_m C$ is about 30 ms and the space constant $\lambda = (R_m a/2R_i)^{\frac{1}{2}}$ is about 0.14 cm.

Equation (1) was discretized with a spatial step (Δx) of 0.00625 cm and a temporal step (Δt) of 0.1 ms. We denote the transmembrane potential in segment i (1 < i < 100) by V_i. The central-difference approximation was taken for both the spatial and temporal derivatives, yielding the Crank-Nicholson integration scheme [3,9], an implicit method which is stable even for large values of $\Delta t/(\Delta x)^2$. We used the FORTRAN implementation given in ROACHE [10] of the Thomas algorithm to solve the resulting tridiagonal system. The values of the gating variables at time t + Δt were calculated from those at time t using an analytic solution to the equations governing them (eqn. (5) of [11] or eqn. (16b) of [9]). The contribution of the total current to the change in voltage was calculated, not using the formula appearing in eqn. (16a) of [9], but rather the formula appearing in footnote (2) of [9]. Dirichlet boundary conditions are assumed - i.e. "sealed-end" at both ends of cable with $\partial V/\partial x$ = 0 (V_1 = V_2 and V_{99} = V_{100} for all t). The initial condition in all segments is V_i = -70 mV, with all activation and inactivation variables set to the asymptotic or steady-state values appropriate to that voltage. All computations were carried out on a Hewlett-Packard minicomputer (model 1000F) in FORTRAN using single precision arithmetic (6.4-6.9 significant decimal digits).

3. Results

Figure 1 shows propagation in a cable without a region of impaired conduction in its interior. Note that there is a uniform spread of the impulse down the central part of the cable, with the speed of conduction being about 0.52 m/s at mid-cable. This speed is computed from the difference in times at which the action potential upstrokes cross -25 mV in segments 40 and 60, using linear interpolation between data points. This value is less than that found previously by other authors: 0.73 m/s [3], 0.72 - 0.73 m/sec [8], 0.658 m/sec [5], 0.659 - 0.667 m/s [12], and 0.68 - 0.73 m/s [9]. Note the edge effect in Fig. 1 near the proximal end of the cable, where there is an increase of conduction velocity due to boundary conditions. A similar effect occurs near the distal end. For a "more realistic"

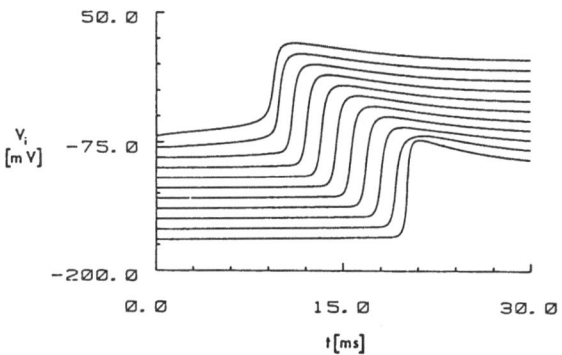

Fig. 1. Transmembrane potential in different segments of the MNT cable. Top to bottom: segments 1, 10, 20, 30, 40, 50, 60, 70, 80, 90, 100. A depolarizing stimulus current pulse of amplitude 100 $\mu A/cm^2$ and duration 10 ms is applied to segment 2 at t=0 ms. The true membrane potential is shown for segment 1; the voltages of other segments are displayed downwards successively in steps of 10 mV

[3] fibre diameter of 50 μm, the conduction velocity increases to 4.54 m/s, which is considerably larger than the value found by another author (2.31 m/s [3]).

The action potential waveform in Fig. 1 compares well with that seen in other studies. However, the maximal rate of rise of the upstroke of the action potential at mid-cable is only about 129 V/s, in comparison with 304-320 V/s [8], 388 V/s [5], 340-384 V/s [12], and 286-320 V/s [9]. The differences in these values, and in the conduction velocities, are due to the fact that different integration algorithms and stimulation protocols are employed in these various studies.

We have not investigated systematically the effects of reducing the sizes of the spatial and temporal steps used above. However, the traces shown in Fig. 1 superimpose almost exactly with those obtained when Δt is reduced tenfold to a value of 0.01 ms. For example, the conduction velocity increases marginally to 0.54 m/s, while the maximal upstroke velocity decreases marginally to 126 V/s. Similarly, if Δx is halved to 31.25 μm, with the cable length being maintained at 0.625 cm, the conduction speed increases slightly to 0.54 m/s, and the upstroke velocity decreases to 118 V/sec. We therefore use the values Δt = 0.1 ms and Δx = 0.00625 cm.

Figure 2A shows the effect of periodic stimulation in a length of intact cable which does not contain a region of impaired conduction. A periodic train of depolarizing current pulses of duration 20 ms and amplitude 500 $\mu A/cm^2$ is injected into segment 2 of the cable. The time between stimuli (t_s) is 320 msec. A 1:1 pattern of conduction occurs, with all action potentials generated proximally traversing the entire length of the cable. In addition, following a transient due to initial conditions, the latency between the onset of the stimulus pulse and the upstroke phase of the following action potential in any given segment remains fixed from stimulus to stimulus. As t_s is decreased to t_s = 319 ms, there is an abrupt transition to a 2:1 pattern (Fig. 2B). We cannot rule out the possibility that this pattern would eventually convert into a 1:1 pattern should the simulation time be extended sufficiently beyond 6 seconds. However, the shape of the last few action potentials seems to have converged.

Thus, in the intact cable, as t_s is decreased, there is a direct transition from a 1:1 to a 2:1 pattern of conduction. Note that in this instance the existence of a zone of impaired conduction is not needed to generate the pattern of block. However, a similar transition from a 1:1 to a 2:1 pattern can be seen

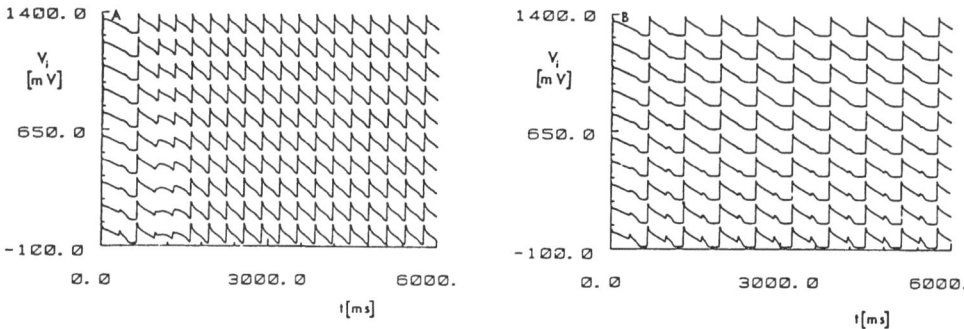

Fig. 2.   Periodic stimulation (pulse amplitude = 500 µA/cm^2; pulse duration = 20 ms; segment stimulated = 2) of a standard MNT cable.   (A) t_s = 320 ms: 1:1 conduction.   (B) t_s = 319 ms: 2:1 conduction.   Bottom to top: transmembrane potential in segments 10, 20, 30, 40, 50, 60, 70, 80, 90, 100.   The true transmembrane potential is shown only for segment 10; other tracings are displaced from one another by 150 mV

in a cable with a region of block at $t_s \cong 360$ ms.   Thus, in the case when an impaired region of conduction exists, the 1:1 pattern cannot be maintained to as low a value of t_s as in the intact cable.   In both cases the 2:1 pattern has its origin in the proximal end of the cable, where the stimulation frequency is high enough to allow refractoriness to be encountered.   The local response produced in proximal segments gradually decrements in amplitude as the mid-section of the cable is approached.

If R_i is increased tenfold, a very different sequence of conduction patterns is seen in the model as t_s is decreased.   This increase in longitudinal resistance causes the conduction velocity to fall to about 0.16 m/s in a cable without a region of block.   In addition to increasing R_i tenfold, we also incorporate a region of impaired conduction.   As in the case considered above, 1:1 conduction is seen if t_s is not too small (Fig. 3A).   In contrast, however, there is not a direct transition to a 2:1 conduction pattern as t_s is decreased further. Instead, one sees Wenckebach cycles [13] containing skipped beats: Fig. 3B shows a 4:3 Wenckebach rhythm, in which every fourth beat does not propagate through to the distal end of the cable.   Note that conduction slows as the mid-section of the cable (where I_{Na} = 0) is approached, and that the action potential upstroke velocity, amplitude, and overshoot potential in this region of impaired conduction are all decreased ("decremental" conduction).   As the action potential exits this region, the normal value of these and other parameters are gradually recovered ("incremental" conduction).   The conduction time across the cable gradually increases from beat to beat as the Wenckebach cycle develops, due to a progressive decrease in conduction velocity during the course of the Wenckebach cycle.   This decrease in velocity is associated with a progressive decrease in the rate of rise of the action potential upstroke (especially apparent in segment 50).   This decrease is in turn connected with a decrease in recovery time due to a progressive increase in action potential duration as the cycle progresses.   In addition, two-component upstrokes and humps on the repolarization phase of the action potential of the type commonly seen in experimental recordings of block [14] can also be seen.   The first component of the two-component upstroke is associated with activation of the proximal end of the cable, the second component with activation of the distal end.   In a similar vein, the humps on the repolarization phase of action potentials in proximal segments are electrotonic reflections of delayed activation of distal segments.

As t_s is decreased in the Wenckebach zone, one sees different patterns containing Wenckebach cycles, with the conduction ratio (the number of conducted beats divided by the number of stimuli) decreasing until a 3:2 pattern of

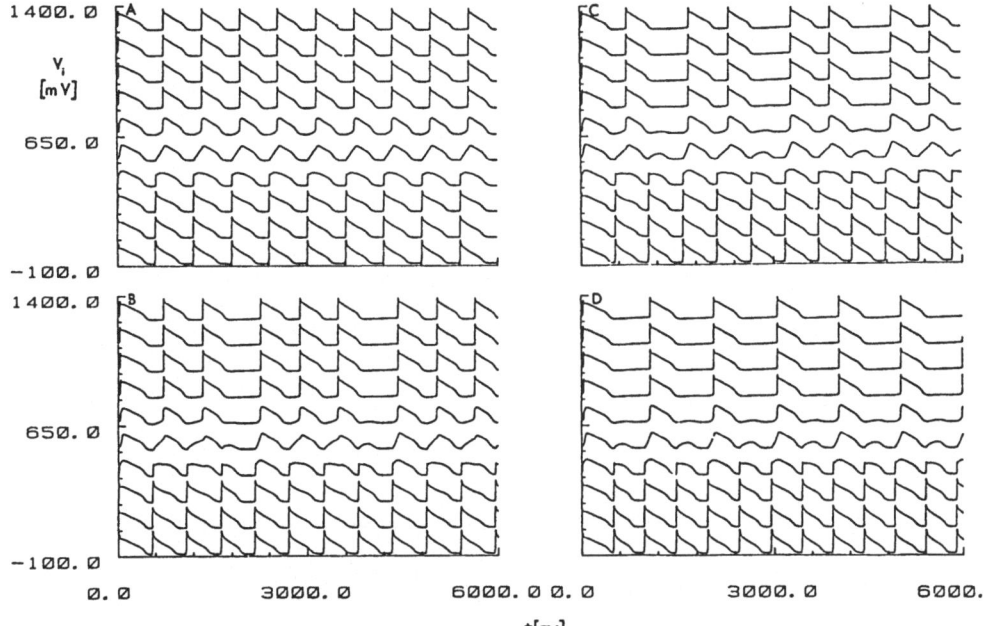

Fig. 3. Periodic stimulation (pulse amplitude = 200 μA/cm^2; pulse duration = 20 ms; segment stimulated = 10) of an MNT cable with R_i = 0.2 KΩ-cm and a region of impaired conduction in its interior. (A) t_s = 600 ms: 1:1 conduction. (B) t_s = 540 ms: 4:3 conduction. (C) t_s = 530 ms: 3:2 conduction. (D) t_s = 490 ms: 2:1 conduction. Transmembrane potentials displayed in same manner as in Fig. 2

conduction is found (Fig. 3C). Again, as the Wenckebach cycle progresses, there is a progressive increase in action potential duration, decrease in recovery time, and decrease in upstroke velocity and conduction velocity, all of which operate in a positive feedback manner ⌊15,16⌋ to generate the Wenckebach periodicity.

For t_s = 490 ms, a 2:1 pattern is seen, with every second stimulus provoking an action potential that decrements in amplitude as it traverses the region of block, eventually extinguishing (Fig. 3D). Note the alternation of action potential morphology in the proximal end of the cable. As the cable is traversed, the relative magnitude of the alternans becomes more pronounced. Simply from the simulation of Fig. 3D, one cannot say whether the alternation in very proximal segments is primary (i.e. largely due to the high frequency of stimulation) or secondary (i.e. largely attributable to electrotonic injection of current from segments in the distal end). However, simulation of the space-clamped MNT membrane with $I_{Na} \neq 0$ shows that stimulation at t_s = 490 ms over a large range of current pulse amplitudes does not result in the 2:2 pattern of alternans, but rather in a 1:1 pattern. Thus, the alternans is secondary, being due to electrotonic effects. Similarly, close inspection of the action potentials in all segments of Fig. 3B and Fig. 3C shows a period 4 cycle in the former and a period 3 cycle in the latter.

4. Discussion

There are many simplifications inherent in the approach to modelling block that we have taken above. Use of the one-dimensional cable equation is not strictly correct for several reasons. First, Purkinje cells have a cylindrical shape, with their longitudinal axis lying parallel to the axis of the strand. Two adjacent cells are joined together at the intercalated disc, a specialized region of close opposition of the adjoining membranes of the two cells. Low-resistance gap-junctional channels pierce through these two membranes, allowing electrical communication between the two adjacent cells. Thus, a more refined model should take into account these specialized junctional areas, and not just lump their resistances together with that of the cytoplasm ⌊17,18⌋. Secondly, Purkinje fibres often are bound together in bundles. In that case, one should take into account the effect on one strand of local current flow produced by action potential propagation in a neighbouring strand. For example, one might expect synchronization of conduction velocities in non-identical neighbouring strands, as occurs in a model of a pair of nerve axons running side-by-side ⌊19⌋. The effect of branching and changing fibre diameter should also be considered ⌊20⌋. Thirdly, in the decade since the formulation of the MNT model, experimental evidence has accumulated showing that this model is incorrect in several respects. Thus, it would be advisable for future studies to employ a more up-to-date model of Purkinje fibre (e.g. ⌊22⌋). Finally, models of mechanisms of propagation completely different from the cable model - such as percolation ⌊22⌋ - should be kept in mind, especially when considering propagation in tissues such as the atrioventricular node that show evidence of inhomogeneous conduction and dead-end pathways ⌊23⌋.

Despite the above reservations, the one-dimensional cable model does produce behaviour similar to that seen experimentally in Purkinje fibre. For example, Fig. 2 shows an abrupt transition from a 1:1 to a 2:1 pattern of conduction as t_s is decreased by the very small amount of 1 ms. This is due to effectively "all-or-none" propagation, in which a decrease in recovery time of 1 ms can cause block of conduction. A similar all-or-none characteristic of the propagating impulse has been described in experimental work on Purkinje fibre ⌈24⌋. Note that effectively all-or-none propagation is not the same as effectively all-or-none excitation, which has been described recently in modelling work on the space-clamped MNT model ⌊7⌋. While the latter might imply the former, the converse is not necessarily true.

The direct transition from a 1:1 to a 2:1 pattern shown in Fig. 2 is reminiscent of clinically observed Mobitz type II block ⌊25⌋, in that Wenckebach patterns are not seen during the transition from the 1:1 to the 2:1 rhythm. The paroxysmal nature of Mobitz II block is also reflected in these simulations, since, at fixed t_s, any small change in the equations describing the system when the 1:1 ↔ 2:1 border is close could cause an abrupt conversion of a 1:1 into a 2:1 pattern or vice versa. Our modelling work also agrees with the clinical observation that Mobitz II block is most commonly seen in areas of the heart, such as the His-Purkinje system, where action potentials with fast upstroke phases are observed (Fig. 2); in contrast, Wenckebach block is usually seen in areas, such as the atrioventricular node, where slow action potentials are present (Fig. 3).

While Wenckebach (also called Mobitz type I) block is classically generated in the atrioventricular node, it has also been identified in the His-Purkinje system and in many other areas of the heart (see reference in ⌊1⌋). To the best of our knowledge, Wenckebach rhythms have not been described in experimental work on healthy Purkinje fibres. However, they have been observed in experimental work on isolated Purkinje fibres in which conduction has been depressed by locally raising the external potassium concentration, decreasing the temperature of the bathing solution, injecting a constant bias current, or simply crushing the tissue (for references see [1]). These experimental observations again agree with our modelling work, which shows that Wenckebach rhythms are only seen if something is done to decrease the ability of Purkinje fibre to conduct (Figs. 2,3).

At least two-score mechanisms have been proposed to explain Wenckebach rhythmicity in the heart since its first clinical description in 1899 ⌊13⌋. Our modelling work shows that a wide variety of interventions, one of which is shown in Fig. 3, can result in Wenckebach rhythms in a simple cable model. These simulations demonstrate unequivocally that mechanisms such as inhomogeneous conduction are not necessary for the production of Wenckebach rhythmicity - the electrical properties of the cell membrane and the cytoplasm are sufficient. Wenckebach patterns of conduction have been described in many biological systems other than the heart. For example, they can be seen in nerve and in smooth muscle as well as in electronic analogues of nerve, smooth muscle and cardiac muscle [1]. Is there a factor common to these various systems that explains why Wenckebach sequence of patterns occurs in such diverse systems? The answer to this question is "Yes": in all these systems, a prematurely elicited action potential travels more slowly than one elicited after a long period of rest. In fact, the degree of slowing generally increases monotonically with the degree of prematurity, until the point is reached at which conduction blocks. Recent theoretical work has shown that this fact alone is sufficient to predict that Wenckebach patterns of block must occur as the stimulation frequency increases [16,26]. This analysis involves deriving a one-dimensional finite-difference equation and considering its qualitative dynamics.

Other forms of block associated with the name Wenckebach, such as reverse Wenckebach, alternating Wenckebach, and millisecond Wenckebach can also be seen in simulations with the MNT cable. These rhythms have been described in the His-Purkinje system as well as in other areas of the heart (see references in [1]). The electrophysiological mechanisms underlying these more esoteric variants of Wenckebach block are largely unknown. In fact, different mechanisms are usually invoked to explain each class of rhythms on a case-by-case basis. However, the fact that all of these rhythms can be seen in a simple cable model shows that membrane properties are enough to account for their existence. In fact, the simple fact that conduction slows and then eventually blocks with increasing prematurity of stimulation is sufficient to account for the existence of all of these rhythms [1,16,25-27].

In the proximal segments of Fig. 3D, one sees one alternation of action potential morphology, with the alternation perhaps being most marked in the action potential duration. Recordings possessing a similar appearance have been made in experimental work during fast pacing of Purkinje fibre and many other cardiac tissues (see references in [1]). However, in those cases, 2:1 block did not occur. Thus, we call the alternans in Fig. 3D "secondary", since it is mainly due to electrotonic effects and is not identical with the alternans seen in an isopotential system, which we call "primary".

Alternation in the morphology of electrocardiographic complexes can be seen clinically in many diseases affecting the heart. Alternation of the ventricular complexes frequently occurs clinically and experimentally following myocardial ischaemia. In these cases the phase of alternation often immediately precedes the phase of induction of malignant tachyarrhythmias such as ventricular tachycardia and fibrillation. This induction is usually associated temporally with the occurrence of a premature ventricular beat. Should the value of t_s used in Fig. 3D be increased slightly, one would obtain a rhythm consisting of many successive 2:1 cycles interspersed with occasional 1:1 cycles, with the conduction ratio being just larger than one-half. This would result in an occasional extra beat that would slip through to the distal end of the cable, corresponding to the intermittent occurrence of a premature ventricular beat on the electrocardiogram. A similar behaviour would also be expected at fixed t_s with a change in any parameter that would tend to improve conduction. This production of a premature beat when conduction is improved is reminiscent of the production of premature beats and the induction of ventricular tachyarrhythmias during experimental or clinical reperfusion of a blocked coronary bed [28]. This would also explain the close temporal relationship between the phase of alternans and the phase of induction of tachyarrhythmias.

Patterns of activity resembling those seen in any one segment of the cable of Figs. 2 and 3 can also be generated in the MNT model by stimulation of a patch of space-clamped membrane with a periodic train of constant current pulses [29]. Note that this situation is quite different from the case considered above, where an isopotential segment is not subjected to a periodic train of constant current pulses. For example, during 1:1 conduction, when the input is of period t_s, the waveform of the current input does not have the square shape of a current pulse, since this input stems largely from the propagating cardiac impulse in most segments. To make matters even worse, during non-1:1 rhythms such as 3:2 block (Fig. 3C), a segment is faced with an input of period nt_s (with $n \geqslant 2$), not t_s, since the input current to a segment varies from beat-to-beat as the Wenckebach cycle progresses.

Many different patterns of conduction have been described in both clinical and experimental cardiac electrophysiology. These include the various forms of Wenckebach patterns mentioned above, Mobitz II block, and patterns showing some form of alternation in the morphology or timing of the electrocardiographic complexes. Recent experimental work in an isopotential system shows the existence of patterns analogous to all of these patterns [1]. In that case, the response of the preparation to premature stimulation allows one to predict the existence of all of these patterns and to thus attain a unified perspective. This ability to predict hinges upon reduction of the dynamics to consideration of that of a one-dimensional finite-difference equation. Mathematically, the Wenckebach types of patterns arise out of tangent bifurcations, while the various types of alternans patterns arise out of period-doubling bifurcations. In contrast, the full extent of the bifurcations taking place in the modelling work shown above is not clear to us at the present time. For example, one might expect that the alternans seen in segment 10 of Fig. 3D should be the result of a period-doubling bifurcation. However, as t_s is decreased (Fig. 3A,B,C,D), examination of the voltage waveform in that segment alone leads to the conclusion that the alternating waveform is attained as the limiting process of an infinite number of tangent bifurcations.

The situation when conduction is present is thus much more difficult to analyze than the isopotential situation. Patterns can be described in the input/output way that we have used above, relating the number of action potentials propagating through to the distal end of the cable to the number of stimuli. Using such a description, we have recently been able to obtain a unified perspective in clinical situations where a sequence of Wenckebach rhythms, analogous to those occurring in the simulations of Fig. 3, are seen [16]. Unlike the isopotential case [1], we have not yet extended this scheme to include the situations when a direct 1:1 ↔ 2:1 transition (Fig. 2) or a beat-to-beat alternation of conduction time occurs. However, we expect that further bifurcation analysis of the modelling results outlined above will eventually lead to a unified theory of block of conduction in cardiac tissue.

5. Acknowledgements

The author thanks Drs. R. Siegel, B. Victorri, and A. Vinet for helpful discussions, Dr. J.S. Outerbridge and P. Krnjevic for help with computers, Christine Pamplin and Sandra James for typing the manuscript, and Robert Lamarche for photographing the figures. This work was supported by grants from the Medical Research Council, the Canadian Heart Foundation (CHF), and the Natural Sciences and Engineering Research Council (NSERC). The author would also like to thank the CHF and NSERC for pre- and post-doctoral fellowship support (1981-86).

References

1. M.R. Guevara: Chaotic Cardiac Dynamics, Doctoral Thesis (McGill University, Montreal 1984)
2. D.P. Zipes, J.C. Bailey, V. Elharrar (eds.): The Slow Inward Current and Cardiac Arrhythmias (Martinus Nijhoff, The Hague 1980)
3. G.H. Sharp, R.W. Joyner: Biophys. J. 31, 403 (1980)
4. R.W. Joyner: Biophys. J. 35, 113 (1981)

5. J. Henry: In Computing Methods in Applied Sciences and Engineering, V, ed. by R. Glowinski, J.L. Lions (North-Holland, Amsterdam 1982) p. 621
6. R.E. McAllister, D. Noble, R.W. Tsien: J. Physiol. (Lond.) 251, 1 (1975).
7. M.R. Guevara, A. Shrier: Biophys. J. 52, 165 (1987)
8. B. Victorri: Simulation Numérique de Potentiels d'Action Cardiaque, Rapport Technique EP82-R-23 (Ecole Polytechnique de Montréal, Montreal 1982)
9. B. Victorri, A. Vinet, F.A. Roberge, J.-P. Drouhard: Comp. Biomed. Res. 18, 10 (1985)
10. P.J. Roache: Computational Fluid Dynamics (Hermosa, Albuquerque 1976)
11. R.E. McAllister: Biophys. J. 8, 951 (1968)
12. J. Henry: In Mathematical Methods in Immunology and Medicine, ed. by G.I. Marchuk, L.N. Belykh (North-Holland, Amsterdam 1983) p. 285
13. K.F. Wenckebach: Zeitschr. f. Klin. Med. 37, 475 (1899)
14. F.J.L. van Capelle: Slow Conduction and Cardiac Arrhythmias, Doctoral Thesis (Universiteit van Amsterdam, Amsterdam 1983)
15. M.N. Levy, P.J. Martin, J. Edelstein, L.B. Goldberg: Prog. Cardiovasc. Dis. 16, 601 (1974)
16. A. Shrier, H. Dubarsky, M. Rosengarten, M.R. Guevara, S. Nattel, L. Glass: Circ. (in press)
17. M.S. Spach: In Normal and Abnormal Conduction in the Heart, ed. by A. Paes de Carvalho, B.F. Hoffman, M. Lieberman (Futura, Mount Kisco 1982) p. 145
18. M.S. Spach. J.M. Kootsey: Am. J. Physiol. 244, H3 (1983)
19. K. Maeda, T. Yagi, A. Noguchi: IEEE Trans. Biomed. Eng. BME-27, 139 (1980)
20. M.B. Berkinblit, N.D. Vvedenskaya, I. Dudzyavichus, S.A. Kovalev, S.V. Fomin, A.V. Kholopov, L.M. Chailakhyan: Biophysics 15, 545 (1970)
21. D. DiFrancesco, D. Noble: Phil. Trans. Roy. Soc. Lond. B 307, 353 (1985)
22. J.M. Smith, R.J. Cohen: Proc. Natl. Acad. Sci. USA 81, 233 (1984)
23. Y. Watanabe, L.S. Dreifus: Cardiovasc. Res. 1, 150 (1967)
24. B.I. Sasyniuk, C. Mendez: Circ. Res. 28, 3 (1971)
25. W. Mobitz: Zeitsch. f. d. ges. exp. Med. 41, 180 (1924)
26. L. Glass, M.R. Guevara, A. Shrier: Ann. N.Y. Acad. Sci. 504, 168 (1987)
27. D.M. Decherd, A. Ruskin: Brit. Heart J. 8, 6 (1946)
28. E. Downar, M.J. Janse, D. Durrer: Circ. 55, 455 (1977)
29. M.R. Guevara, A. Shrier: unpublished

Perturbations of Next-Period Functions: Applications to Circadian Rhythms

H. Degn

Institute of Biochemistry, Odense University,
Campusvej 55, DK-5230 Odense M, Denmark

1. Introduction

A circadian rhythm is an oscillation within a living organism with a period of about 24 hours. When the organism is kept isolated under constant conditions the circadian oscillator works with its own free-running period. When the organism is exposed to physical clues of the time of day the circadian rhythm is gradually entrained and assumes a precise 24 hour period. If the entraining stimuli are withdrawn, the circadian oscillator reverts gradually to its free-running period. A single stimulus under otherwise constant conditions elicits a transient of gradually changing period. A phase shift of an entraining oscillation is followed by a gradual synchronization to the new phase [1,2].

In the above summary of the properties of circadian oscillators the word gradual expresses a functional relationship between the lengths of adjacent periods. The purpose of the present work is to study the consequences of assuming the simplest possible functional relationship between adjacent periods, namely a linear one

$$p_{n+1} = Ap_n + B, \tag{1}$$

where p_n is the length of the n'th period and A and B are constants. Equation (1) is a good approximation for all limit cycle oscillators, presumably including circadian oscillators, near the limit cycle.

2. Properties of Linear Period Transfer Function

Figure 1 shows a graphical representation of (1) for two different values of the slope, A. The point of intersection between the line of (1) and the line $p_{n+1} = p_n$ is the fixed point where the free-running period prevails. The fixed point is stable if $0 < A < 1$. After a perturbation the oscillator will return to the free-running cycle through a transient whose length depends on A. At $A = 0$ the length of the transient is zero, i.e. it takes only one cycle to revert to the free-running period. The length of the transient goes to infinity as A approaches 1. When A equals 1 all periods are stable, corresponding to a conservative oscillator. The trajectories marked with arrows in Fig. 1 represent geometrical iterations of (1). It is noted that for $0 < A < 1$ the fixed point can be approached from two sides corresponding to the inside and the outside of a two-dimensional limit cycle. If $-1 < A < 0$ the fixed point is also stable. However in this case the transient alternates between the two sides of the fixed point. This behaviour corresponds to an unstable period-two in a limit cycle oscillator of at least three dimensions.

3. Linear Period Transfer Function as a Model of Circadian Rhythms

Equation (1) can be considered a model of a circadian oscillator where all details are omitted except the stability properties of the free-running period. Since these properties are decisive for the response to small perturbations, the basic phenomena of small perturbations and entrainment to periods not too far from the free-running

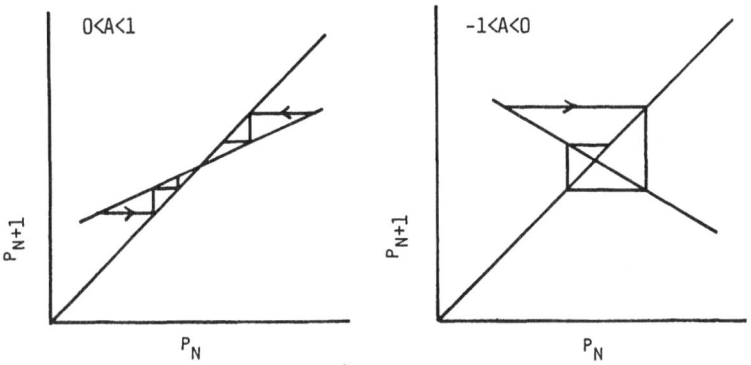

Fig. 1. Graphical iterations of linear period transfer function (1)

period should be reproduced by (1). In order to test this hypothesis we introduce time as the sum of periods in a second equation and add a periodic perturbation term $F(t_n)$ to (1). Finally we replace the constant B in (1) by $p_{fr}(A-1)$ where p_{fr} is the length of the free-running period in units of 24 hours. We then obtain the following minimal model of circadian rhythms

$$p_{n+1} = Ap_n + p_{fr}(A-1) + RF(t_n) ,$$

$$t_{n+1} = t_n + p_{n+1} ,$$

(2)

where R is the strength factor for the periodic perturbation and t_n is the time at the end of the n'th period. Clearly the properties of circadian oscillators call for positive values of A. A similar model has been studied previously for negative values of A [3].

4. Response to Single Perturbation

Experiments with circadian rhythms often consist in keeping the organism under constant conditions where the free-running cycle is realized. A short stimulus, usually light, is applied in order to perturb the circadian oscillator, and the return to the free-running cycle is recorded. When the stimulus is applied at different phases of the free-running cycle a phase response curve can be determined. A single perturbation of the model is studied by iterating with R = 0 and the initial period equal to $p_{fr} \pm d$, where d is the perturbation. The result is shown in Fig. 2. It is seen that the time but not the number of periods taken for the transient to vanish depends on the magnitude of d. The length of the transient increases with increasing A.

In the present model the phase response curve is embodied in the periodic function $F(t_n)$. The magnitude of the perturbation of p_{n+1} is determined by the phase of $F(t_n)$ at the end of p_n. This is the other way around compared to a real circadian oscillator but the effect is the same. We shall call $F(t_n)$ the virtual phase response function.

5. Steady State Entrainment

A circadian oscillator with a free-running period longer than 24 hours needs period shortening responses to periodic stimuli in order to become synchronized with a 24 hour period. Period lengthening responses are not required. The opposite is true for a circadian oscillator with a free-running period of less than 24 hours. It follows that the simplest possible virtual phase response function for the former case is a straight line through the origin with a negative slope. However, the one-

283

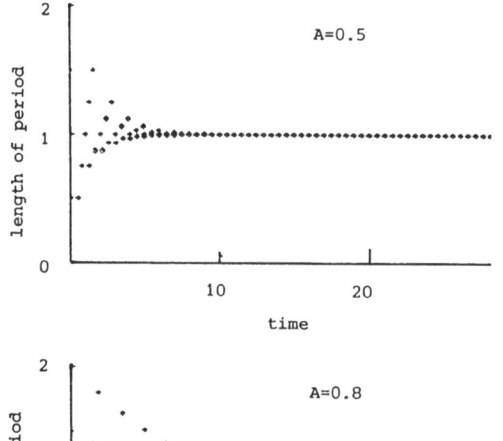

A=0.5

Fig. 2. Transients of period length
after single perturbations of (2)
(R = 0)

A=0.8

sided response is not safe if the free-running period is close to the synchronizing
period, in which case both positive and negative corrections are required for stable
entrainment. Fig. 3 shows the result of iterating (2) with a discontinuous linear
phase response function with a negative slope and intersecting the abscissa at t =
0.5. Random values for p_0 were chosen between 0 and 2, and p_{100} was plotted at different values of the strength factor R. The synchronizing period was 1. It is seen
that there is a lower critical value of R which must be exceeded before 1:1 entrainment can take place. The 1:1 entrainment is stable over a wide range of values of R.
However, other attractors also exist and become prominent as A is increased. In
other words, the model is multistable.

The virtual phase response function used in the above example was discontinuous.
We shall now consider continuous virtual phase response functions beginning with
the sine function. Figure 4 shows the effect of sinusoidal perturbation in (2). It

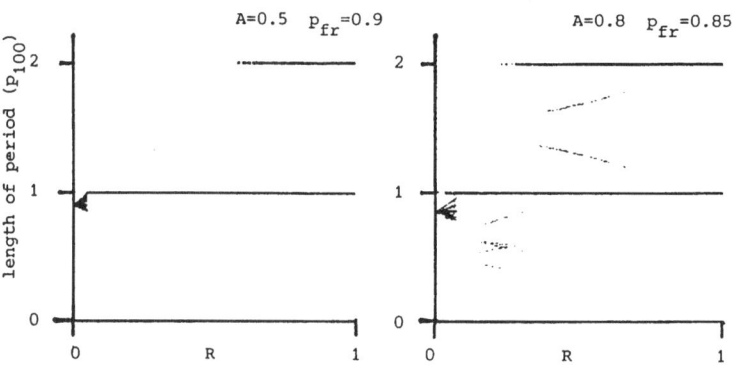

Fig. 3. Steady state entrainment with linear virtual phase response function in (2)

284

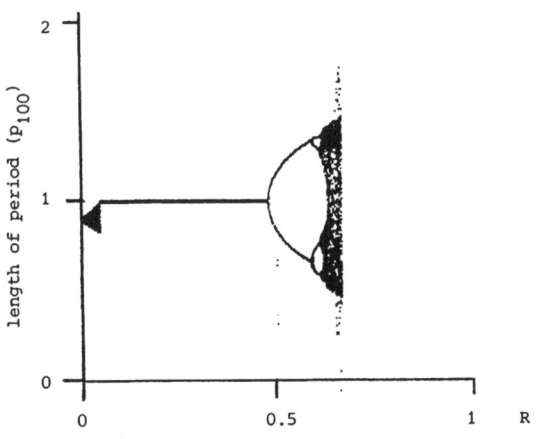

Fig. 4. Steady state entrainment with sinusoidal virtual phase response function

is seen that the 1:1 entrainment is stable over a wide range of values of R. At increasing R, eventually period doubling bifurcations and chaos occur. It is only the descending phase of the sine wave which can cause entrainment. The chaos observed in Fig. 4 occurs when the ascending phase of the sine wave takes part in the perturbation.

Neither the linear nor the sinusoidal virtual phase response curve are particularly realistic. The following expression was invented in order to approximate the most common type of phase response curve found in the literature:

$$F(t_n) = 0.32\sin(2\pi t_n)/(1.05-\cos(2\pi t_n)) \,. \tag{3}$$

The curve produced by (3) is shown in Fig. 5. It has a long descending phase and a short ascending phase. When (3) was inserted in (2) the entrainment properties were not very different from those produced by the linear virtual phase response curve. This is illustrated in Fig. 6 where the length of the period at steady state entrainment (p_{100}) is plotted against the synchronizing period at a fixed value of the strength factor, R, for the two cases.

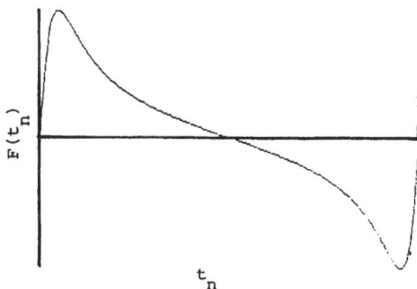

Fig. 5. Graph of equation (3) approximating real phase response curve

6. Phase Shift

In nature phase shift of the entraining oscillation occurs when the organism travels to another longitude. Numerous experiments have been done with phase shifts in the laboratory. The response of the model (2) to phase shifts in the periodic perturbation is shown in Fig. 7. The length of the period was plotted at the end of each

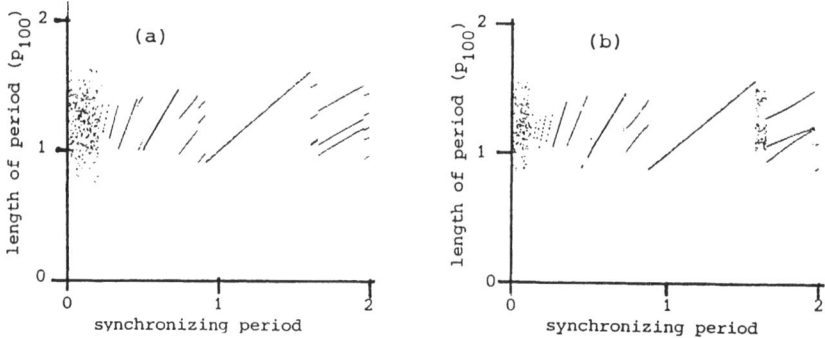

Fig. 6. Entrained period as a function of entraining period in the model (2) with (a) linear and (b) approximated real (Eq. 3) phase response function, A = 0.5, R = 0.3.

period as a function of time. It is seen that the phase shift is followed by a damped oscillation in the length of the period. Resynchronization after a phase shift lasts at least two weeks for A = 0.8. When the effect of phase shifts was studied with different values of the parameters it was found that there is no simple rule that determines whether the oscillatory transient begins with a minimum or a maximum.

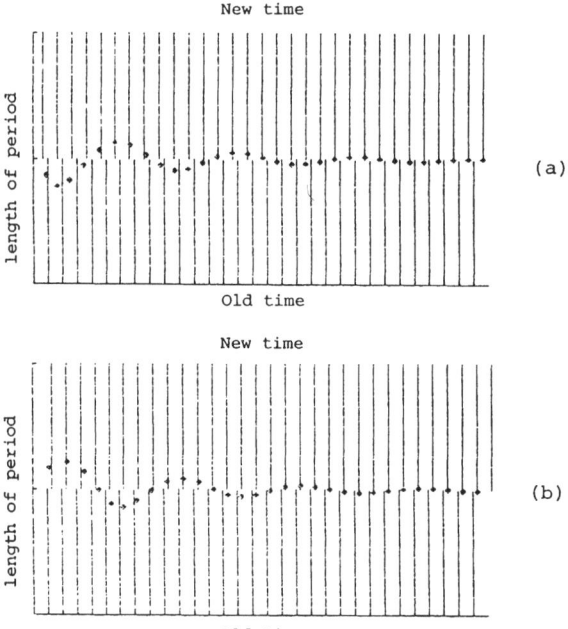

Fig. 7. Transient following a phase shift of the entraining oscillator. The phase shift was (a) 0.33 and (b) -0.33. A = 0.8, R = 0.2, p_{fr} = 1.2

7. Conclusion

The minimal model of circadian rhythms contains no information on the details of the mechanism of the circadian oscillator. The model therefore can be used to distinguish the types of experiments which cannot be used to elucidate the mechanism of the circadian oscillator. The vast amounts of observational data on circadian rhythms in the literature were collected and presented under the influence of theoretical concepts different from the ones presented here. Therefore data suited for comparisons with the present results are not easily extracted from the literature. For example the finding of an oscillatory transient in the length of the period following a phase shift would be hard to verify from published data. Since such a phenomenon would have far-reaching implications for the wellbeing of travelers it would be worth looking for it.

8. References

1. E. Bünning: The Physiological Clock, (Springer, New York 1973)
2. J. Aschoff (editor): Biological Rhythms, Handbook of Behavioral Neurobiology, Vol. 4, (Plenum, New York 1981)
3. H. Degn: In Chemical Applications of Topology and Graph Theory, ed. by R.B. King, (Elsevier 1983)

Brain Development and Self-Organization

W. Singer[1] *and Chr. v.d. Malsburg*[2]

[1]Max-Planck-Institut für Hirnforschung,
 Deutschordenstraße 46, D-6000 Frankfurt 71, Fed. Rep. of Germany
[2]Max-Planck-Institut für Biophysikalische Chemie,
 Am Faßberg, D-3400 Göttingen-Nikolausberg, Fed. Rep. of Germany

1. The Problem

The encephalization which accompanies the phylogenetic emergence of mammals has led to a dramatic increase of nerve cells and nervous connections that exceeds by far the disproportionally small increase of the genom. This implies that the information stored in the genom cannot alone suffice to specify the connectivity of higher nervous systems. The following numbers illustrate the magnitude of the specification problems that have to be solved during brain development.

One mm^3 of cerebral cortex contains approximately 40,000 nerve cells. Each of these contacts with its efferents at least 10,000 other nerve cells both within cerebral cortex and other brain structures and is in turn contacted by the axons of a similar number of other neurones. The neuronal cables which establish these connections attain several km of total length within 1 mm^3 of cortical tissue. The whole brain contains somewhere between 10^{12} and 10^{14} nerve cells. If these would form a chain, touching each other with their dendritic processes, this chain could be wrapped around the earth's equator about 20,000 times. These numbers exemplify the combinatory complexity of possible relations that can be established by interneuronal connections and they also indicate the enormous amount of information that can be stored by specifying these connections and reducing the degrees of freedom inherent in the architecture of neuronal networks. Since neuronal connections are highly selective in the mature brain and since it is this selectivity which determines the specificity of neuronal functions, it has become of primordial interest to ask how such a complex system can be developed from a set of genetic instructions that is necessarily less complex than the ensuing system.

2. General Principles of Organogenesis

The development of organs and organisms is based on a permanent, interactive exchange of information stored in the genom and information available in the cellular "environment" of the genom. The latter determines which of the genetic commands are expressed at particular developmental stages. As the developmental process proceeds and organs differentiate the microenvironment of the genom changes and hence modifies continuously the further expression of the genes. This reflexive process is commonly addressed as "self-organization". During early phases of brain development gene expression and posttranslational modifications of gene products are controlled as in any other organ by the biochemical signals which are produced in the cellular micro environment. Later, however, brain development starts to differ radically from the development of other organs because electrical activity is added to the biochemical messengers as a further signaling system in the self-organizing dialogue between the genes and their respective environment. Mechanisms are implemented which enable these electrical signals to influence gene expression and posttranslational modifications.

The fact that electrical signals influence brain development has a number of extremely important implications. Electrical signals are transported by neuronal processes over great distances and with high topological selectivity. This enlarges dramatically the range and the complexity of the "environment" that is available to the self-organization process. From a certain developmental stage onwards the brain possesses functioning sense organs which convert signals from within the organism and even from extracorporal space into electrical messages. Thus, the "environment" relevant for brain self-organization ultimately includes all domains with which the evolving brain is capable to interact and from which it can receive messages. Another important aspect is, that the very same electrical signals which convey sensory messages are used by the brain as information carriers for computational processes. Hence, the powerful capacities of nerve nets to perform complex logic operations on large parameter sets become also available to the self-organization process. As the complexity of the brain increases, the computational power and the complexity of its interactions with its environment increase as well. As a consequence, the set of parameters determining further brain development becomes also more complex and capable of supporting self-organization towards even more differentiated structures. It is because of this spiral of reciprocal interactions between the genom and its increasingly complex environment that a rather small set of genetic rules suffices to promote the development of such highly differentiated structures as the human brain.

It follows from these considerations that self-organization must rely on rather general rules and simple principles of error correction. This, in turn, has as a consequence that structures and patterns emerging from self-organizing processes possess a high degree of regularity, e.g. homogeneity, repetitivity or continuity. This conclusion is fundamental to the venture of doing research on structures as complex as the nervous system. Knowing the mechanism of ontogeny is of extreme importance: One cannot understand the function of nerve nets without knowing their structure, and one cannot know their structure without knowing the principles of their ontogeny.

3. General Principles of Pattern Formation

There are well-studied paradigms of pattern formation in non-biological self-organizing systems such as e.g. convection and crystallization. We will attempt to give here a general description of the rules of pattern-formation in self-organizing systems. It has to be verbal since there is yet no canonical mathematical description of organization.

To make things more concrete, we consider as an example convective pattern formation as it occurs with the Bénard phenomenon. From this and many other organizing systems the following three principles may be abstracted:

(1) *Fluctuations self-amplify*. This self-amplification is analogous to the reproduction in Darwinian evolution. In the Bénard system, fluctuations are created by thermal motion.

(2) *Limitation of some resource leads to competition among fluctuations and to the selection of the most vigorously growing (the "fittest") at the expense of others*. In the Bénard system, upward movement in one place requires downward movement in other places. The columns with least density will win and rise.

(3) *Fluctuations cooperate. The presence of a fluctuation can enhance the fitness of some of the others, in spite of the overall competition in the field. (In many systems the "fitness" of a fluctuation is identical with the degree of cooperativity with other fluctuations.)* The liquid near a column of rising liquid is dragged up by viscosity.

A fundamental and very important observation on organizing systems is the fact that global order can arise from local interactions. The intermolecular forces acting within a volume of liquid are of extremely short range, yet the patterns of convective movement they give rise to may be coherent and ordered on a large scale. This fact will be of extreme importance to the brain in which local interactions between neighboring cellular elements are to create states of global order, ultimately leading to coherent behavior.

In the nervous system, the stage for the generation of connection patterns is ultimately set by prespecified rules for the interaction of cellular processes and signals, and by the environment. Since nerve cells are connected by long axons, there is an important and exciting difference between the nervous system and most other examples studied so far. Neural interactions are not necessarily topographic in form, connected cells being "neighbors" although they may be located at different ends of the brain. This gives rise to genuinely new phenomena. Some of the ordered structures within the nervous system may not "look" ordered to our eye, which relies essentially on spatial continuity.

4. Neural Network Organization in General

Two types of variables are relevant to network organization: signals and interconnections. Signals are action potentials propagated along axons. Connections are characterized by a weight variable, which measures the size of the effect that a nervous impulse arriving at a nerve terminal exerts on the postsynaptic element.

A given network creates certain activity patterns, which are determined by the structure of the network and by input activity. Due to synaptic plasticity connections are modified in their strength in response to cellular signals in the activity patterns. Modifications in synaptic strengths in turn lead to modified activity patterns. In order to obtain reorganization (instead of stabilization) of a network the feed-back loop between changes in synaptic strengths and changes in activity patterns has to be positive, so that coherent deviations from an undifferentiated state self-amplify (according to the first of the principles formulated above). The process is constrained by the requirement that modifications in a synaptic connection have to be based on signals which are locally available. These are the presynaptic signal, the postsynaptic signal, and possibly modulatory signals which are broadcast by central structures.

These requirements, self-reinforcement and locality, suffice to specify the mechanism of synaptic plasticity in excitatory synapses: A strong synapse leads to coincidences of pre- and postsynaptic signals, and in turn the synapse is increased in strength by such coincidences. Donald HEBB [1] gave this formulation:

> "When an axon of a cell A is near enough to excite cell B and repeatedly or persistently takes part in firing it, some growth process or metabolic change takes place in one or both cells such that A's efficiency, as one of the cells firing B, is increased."

Hebb's rule corresponds to the "self-reproduction" of the general scheme of organization. In order to stabilize the system, some competition for a limited "resource" has to be introduced. Most likely, there is a mechanism of isostasy, by which each cell keeps the temporal average of its activity (taken over the span of some hours) constant. As a consequence the increase in strength in some synapses has to be compensated by a decrease in others. Only the more successful synapses can grow, the less successful ones have to get weaker and, eventually, to disappear. For technical reasons, some models discuss a simpler competition rule for synapses, in which the sum of the synaptic weights of all synapses converging on a cell is kept constant. This rule leads to functional deficits and is probably not realistic. Synaptic plasticity, constrained by competition, implements the principles 2 and 3 of organizing systems, as listed above.

One synapse on its own is not efficient in producing favorable events. Only sets of synapses which converge on a postsynaptic neurone and carry coincident signals can effectively activate that neurone and can thereby cooperate in creating favorable events. In order for such coincidences to appear consistently there must be a causal connection between presynaptic cells. Coincidences may be due to excitatory links between sets of presynaptic neurones. They may, however, also be due to simultaneous stimulation of different sensory cells. In this case they point to the existence of causal connections in the external world. The Hebbian modification algorithm is thus a means for the development of neuronal representations of causal relations.

The rules of cooperation and competition act on a local scale. The phenomenon of self-organization is the emergence of globally ordered states, as discussed with the example of global convection patterns in the Bénard phenomenon. These states are ordered in the sense of optimal mutual consistency of all local rules with each other. The fact that the external world takes part in the game leads to adaptation of the nervous system to it.

The rules for adjustment of synaptic weights we have introduced so far are able to produce ordered connection patterns. They do, however, not necessarily organize the nervous system to optimal biological utility. For this, two types of control are necessary: 1) Genetic control of boundary conditions and interaction rules in order to favor certain useful connection patterns. 2) Control by central structures which are able to evaluate the degree of biological desirability of activity states. If a state proves to be useful, a gating signal is sent to all of the brain or to an appropriate part of it to authorize synaptic plasticity, which then stabilizes the state just reached or contributes to the likelihood of it happening again.

5. Self-Organization in the Visual System

After this brief introduction of principles we now enter a more detailed discussion of a few paradigmatic cases of network organization. Most of our knowledge on activity-dependent specification of neuronal connections comes from the visual system. "Seeing" has to be learned during a critical period of early postnatal development and this "learning" process consists mainly if not exclusively of experience-dependent selective stabilization of neuronal connections. The most dramatic evidence for this comes from patients who suffered from congenital opacities of the eyes during early childhood and therefore were unable to perceive contours. With the development of lens transplants the optical media of these patients eyes could be restored but unexpectedly, these patients were unable to recover visual functions when operated as juveniles or adults. Experiments in visually deprived animals have revealed that these functional deficits are due to abnormalities in visual centres of the cerebral cortex, indicating that certain cortical functions can only be developed if visual experience is available.

The following example illustrates this. Higher mammals and man who have frontally positioned eyes with overlapping visual fields can compute from the differences between the images in the two eyes the distance of objects in space. This ability has two obvious advantages: first, the distance of an object in space can be assessed with great precision even if object and observer are stationary and do not produce any motion parallaxes. Second, the separation of figures from ground, a primordial prerequisite for pattern recognition, is greatly facilitated by evaluating spatial distance. The basis for this function are neurones in the visual cortex which possess two receptive fields, one in each eye, that are precisely superimposed in visual space and have the same internal structure. Thus, during development the one million afferents arriving from each eye have to be arranged so that only those pairs of afferents converge onto cortical cells which originate from precisely corresponding retinal loci. The problem is that there is no way to predict with any great precision which retinal loci will actually be corresponding in the mature visual system. Retinal correspondence depends on parameters such as the size of the eye balls, the position of the eye balls in the orbit and the interocular distance. Clearly, these parameters are strongly influenced by epigenetic factors. It follows that genetic instructions alone, even if they were quantitatively sufficient, cannot in principle suffice to determine with the required precision the pattern of interocular connections.

An elegant possibility exists, however, to identify fibers as coming from corresponding retinal loci by evaluating their electrical activity. When a target is fixated with both eyes, corresponding retinal loci are stimulated by the same contours. Therefore neuronal responses in afferents from corresponding retinal loci are more correlated than those in afferents from non-corresponding loci. As the following observations suggest there is a developmental mechanism capable of consolidating selectively those retinal afferents which convey correlated activation patterns.

By the time of natural eye opening most neurones in the visual cortex have established connections with both eyes but - and this is typical for the development of most nerve connections - these are formed with rather low topological precision and quantitatively outnumber by far the connections which are actually preserved in the mature system. It is only during a second, activity-dependent pruning process that the connections attain the selectivity characteristic of the adult system. This adaptive process has first been documented by the pioneering experiments of Hubel and Wiesel and is now a much investigated model of neuronal plasticity (for review of the extensive literature see [2]).

When the signals from the two eyes are incongruent, either because one eye is occluded or because the images in the two eyes are not in register - as is the case with squint - cortical cells lose their binocular connections. In the first case they stop responding to the deprived eye with the consequence that this eye becomes functionally blind. In the other cases cortical cells segregate into two approximately equally large groups, one responding exclusively to the ipsilateral and the other exclusively to the contralateral eye. Fusion of the images in the two eyes is then no longer possible. These experience-dependent modifications of excitatory connections depend on the statistical correlation between the activation patterns in the respective afferents and their common target neurones (for review see [3]). Connections stabilize if the probability is high that pre- and postjunctional elements are active at the same time and they destabilize when the postsynaptic cell is active while the afferent pathway is silent. Thus, only those converging pathways are selectively stabilized and permanently associated with each other that convey correlated activity. Conversely, converging pathways compete with each other if these convey uncorrelated activity. One subset consolidates on the expense of the respective other and eventually those afferents win which have the highest probability of being active in temporal contiguity with the postsynaptic target cell.

It is obvious that such activity-dependent selection can solve our specification problem and stabilize selectively those afferents from the two eyes which originate from corresponding retinal loci. It is furthermore clear that such experience-dependent pruning of neuronal circuits is a very powerful and versatile mechanism to establish associations and hence has all the potential to serve as a basis for learning. Before discussing this fascinating aspect, however, further constraints of the experience-dependent developmental process have to be considered. In our special case selection of afferent connections from the two eyes can only be successful if it occurs while the animal is actually fixating a non-ambiguous target with both eyes. Pruning must not take place when the two eyes are moving in an uncoordinated way. In this latter case, the images processed by the two eyes are different and hence all retinal signals, even those originating from corresponding retinal loci are uncorrelated. All afferents from the two eyes would therefore compete with each other and the consequence would be complete disruption of binocular connections. The selection process must therefore by non-retinal control systems capable of determining the instances at which retinal activity may induce changes in circuitry. The evidence reviewed below suggests that the mammalian brain possesses such gating systems.

6. Central Gating of Plasticity

Vision-dependent modifications of neuronal connections fail to occur when the animals do not pay attention to available visual stimuli or when they cannot use visual stimuli in the appropriate behavioral context. Such is the case when the animals cannot actively explore the visual surrounding either because they are restrained or anesthetized or because the visual signals are manipulated in a way which makes them behaviorally meaningless [4,5]. Such occurs e.g. when the pathways are disrupted which come from the eye muscles and inform the brain about the actual position and movement of the eyes [6]. It is natural that this information is required to determine when retinal signals are appropriate to guide circuit selection, - especially during development when both sensory and motor processes are still unprecise and uncalibrated. Experiments involving unilateral lesions in one brain hemisphere suggest that experience-dependent modifications require the presence of gating signals that are generated by the brain itself and are closely related to arousal and attention. Surgical disruption of the projections which control the arousal state of the brain and its level of attention leads, if performed unilaterally, to a sensory hemineglect. The animal tends to pay no attention to stimuli presented in the sensory space that corresponds to the hemisphere with the lesion. In vision this is the hemifield contralateral to the side of the lesion. In such cases it was demonstrated that visual signals that were made available simultaneously to both the normal and the lesioned, less "attentive" hemisphere, led to experience-dependent modifications of circuitry only in the former and not in the latter [7]. Conversely, direct electrical activation of these arousing systems greatly facilitates vision-dependent modifications of neuronal transmission in cerebral cortex. With appropriate stimulation long-lasting modifications of cortical transmission can manifest themselves within as little as 30 min [8,9]. This corroborates the notion that the experience-dependent selection of neuronal connections does not solely depend on the local interactions between the involved sensory afferents and cortical target cells but is in addition controlled by gating signals which are produced by the brain itself and mediate information about more global behavioral conditions.

Recent evidence suggest the interesting possibility that the experience-dependent modifications of cortical transmission are gated by more than one of the systems implicated in the control of global brain states [10]. Disruption of any of these systems alone is not sufficient to arrest plasticity. This is particularly interesting, because the various systems complement each other in a variety of functions but they differ with respect to the inputs they receive from other brain structures. While one of these - the noradrenergic projection - is activated very effectively by virtually all sensory inputs that enter the brain the other, - the cholinergic projection - seems to be mainly under the control of limbic structures which have to do with the control of motivation and selective attention. Both systems originate from structures of the central core of the brain which are phylogenetically much older than the cerebral cortex, both influence through widely distributed axonal arbors large areas of the brain simultaneously, and both do not seem to activate neurones in the direct way that is characteristic of the specific sensory afferents but they rather modulate the excitability of nerve cells. Thus, the first, the noradrenergic system, could serve as an alerting or activating system that is dominated by external stimuli while the second, the cholinergic system, might regulate brain activity more as a function of internally generated drives. At present, it is still an open question how these modulatory systems actually gate use-dependent long-term modifications of neuronal transmission and ultimately of neuronal connectivity. Indications are emerging that the molecular mechanisms of the use-dependent long-term changes in the mammalian brain actually be similar to those that underlay certain learning processes in the simple nervous systems of molluscs. In all cases it appears necessary that electrical activity opens membrane channels for Ca^{2+}-ions. These, after having entered the neuronal elements, appear to serve as one of the messengers that convert electrical events into the biochemical signals which are ultimately required for long-term modifications of transmission (for review and citations see [11]).

7. Consequences of Self-Organization at Higher Levels of Cortical Organization

So far we have dealt with a self-organization process that serves essentially to match ordered topographic maps. We shall now take this approach one step further and investigate the consequences of self-organization in the feature space in which "conceptual" rather than topological vicinity is the relevant selection criterion. It has been known for a long time that one of the prominent features of cortical organization is the presence of an extremely dense network of far-reaching connections which are tangential to the cortical lamination. However, it is only during the last few years that this important feature of cortical organization has received the attention of experimentalists. It is now known that these connections consist mainly of axon collaterals of pyramidal cells, are excitatory, and contact preferentially the apical dendrites of other pyramidal cells [12-14]. These pathways are thus capable of mediating reciprocal excitatory interactions between cortical neurones that are non-adjacent and process signals from non-neighboring loci in the visual field.

There are indications that these connections are organized in a selective way linking neurone clusters that tend to be spaced periodically [15] and share certain functional properties [16]. Developmental studies in the cat have shown that these tangential connections appear essentially postnatally [17], go through a phase of exuberant proliferation during which they are extremely numerous and far-reaching, and subsequently become pruned under the influence of visual experience. If visual experience is unrestricted, subpopulations of these pathways are stabilized, if visual experience is not available, only a rudimentary network of horizontal connections is maintained [18].

This intrinsic tangential network thus develops in very much the same way as the long-range connections between the eyes and their target structures (see above) and as the association projections between different cortical areas [19,20]. The latter is not surprising since many of the corticocortical association fibers are actually collaterals of the intrinsic tangential projections or vice versa. We propose, therefore, that the system of horizontal connections self-organizes according to the same principles as the thalamocortical connections. We assume that out of the initial exuberant Anlage only those connections are stabilized by coherency matching which span between neurone clusters whose activation patterns show some statistical correlation. For neurone clusters that are too remote from each other to share a common input from retinocortical afferents, the degree of correlated activation does no longer depend upon particular neighborhood relations but will be determined essentially by coherencies between particular features of the visual scene.

Selective stabilization of tangential intrinsic connections could thus generate a non-topographically organized map that matches the coherent properties of "feature constellations" in physical reality rather than topographic coherencies. Just as binocular cells become "detectors" of coherency between pairs of corresponding retinal ganglion cells in the two eyes the selectively coupled distributed clusters of feature detectors become as an ensemble detectors of coherencies within elementary constellations of features.

As in the formation of binocular connections there will be competition between many possible constellations of features which all show some coherency, and only the most consistent and most frequently occurring constellations will win. Those which match best the already established connectivity pattern will have a competitive advantage. This then leads to the development of a map which represents not only coherent relations in particular feature constellations, but also the statistical probability with which these constellations occur in the physical world. It is furthermore likely that the selection process at this level of cortical organization is also dependent on the activity of central gating systems as it is the case already for selection processes at lower levels. This then would provide the additional option to override the probability functions inherent in the structure of the physical world and to represent preferentially constellations that are relevant to the system in a behavioral context.

Thus, by simple iteration of the very same processes of self-organization which at peripheral levels of the visual system lead to map matching, it is possible to generate non-topographic maps which represent dimensional neighborhood relations in feature space.

Once such maps are established, they serve as detectors of coherencies in visual scenes. This, in turn, is the basis of any preattentive segmentation of scenes into "figures" and a necessary prerequisite for any subsequent identification of patterns. If such a cooperating cluster of neurones is presented with a pattern that contains the appropriate coherent property, the detector ensemble gets into a resonant state and the corresponding neurones distinguish themselves as members of a resonant ensemble because of coherent reverberation.

8. Self-Organization and Learning

Finally, we wish to draw attention to the possibility that the ontogenetic self-organizing processes discussed in this chapter may share more than only formal resemblance with the mature cortex. First, there is evidence that experience-dependent long-term changes of neuronal response properties can occur also in the striate cortex of adult cats [5]. Both weakening of previously functional connections and selective strengthening of inactivated connections have been observed. Second, the modifications in the adult appear also to be based on selective strengthening of connections between coherently activated groups of neurones, Thus, contingency matching seems to be the basic algorithm also in adult plasticity. This suggests the possibility that some of the neuronal mechanisms which subserve ontogenetic self-organization actually persist into adulthood and then serve to mediate adaptive changes such as underly learning and memory. Most of the prerequisites postulated for learning mechanisms would be fulfilled: The modifications of the coupling strength of neuronal connections are activity-dependent; the modification rules are based on local contingency matching and hence have the ability to associate events that the contiguous in time and space; the occurrence of modifications is gated by central core systems and hence can, in principle, be made dependent on global states such as arousal, attention and motivation; finally, modifications are long-lasting and thus can serve to establish permanent representations. The essential features of learning, - the evaluation of relations between events by selective correlations and the internalization of these relations by circuit modification, - are thus shared by developmental self-organization.

The equivalence of experience-dependent self-organization and learning is made explicit in recent computer simulation experiments. Multilayer systems with the above-described properties self-organize and express maps and columnar organizations [21-23]. These very same systems, if exposed to patterns, such as e.g. the letters of the alphabet, establish representations of these patterns. After termination of the learning process, these representations can be used for classification and recognition of the learnt patterns [24,25]. In conclusion then it appears as if a rather restricted set of self-organizing principles is sufficient to account not only for the development but also for the maintenance of a variety of the characteristic structural and functional features of cortical organization. The main differences between the developmental and the mature state appear to concern the turnover of connections rather than processing and modification algorithms. During development growth processes continuously supply new connections, thus maintaining a large repertoire for use-dependent selection. Moreover, during development connections not selected for consolidation are removed physically. Once growth processes are terminated the effects of developmental pruning become irreversible. In adulthood, by contrast, the repertoire of modifiable connections is much more restricted and probably fixed. However, when connections become weakened they do not seem to disappear physically and hence may remain reactivatable. What seems to change then during the transition from the developmental to the mature state are the constraints of self-organization rather than the basic principles.

References

1. D.O. Hebb: The Organization of Behavior (Wiley, New York 1949)
2. Y. Frégnac, M. Imbert: Physiol. Rev. 64, 325 (1984)
3. W. Singer: Vision Res. 25 (3), 389 (1985)
4. P. Buisseret, E. Gary-Bobo, M. Imbert: Nature 272, 816 (1978)
5. W. Singer, F. Tretter, U. Yinon: J. Physiol. 324, 221 (1982)
6. P. Buisseret, W. Singer: Exp. Brain Res. 51, 443 (1983)
7. W. Singer: Exp. Brain Res. 47, 209 (1982)
8. W. Singer, J.P. Rauschecker: Exp. Brain Res. 47, 223 (1982)
9. J.M. Greuel, H.J. Luhmann, W. Singer, in preparation
10. M.F. Bear, W. Singer: Nature 320, 172 (1986)
11. W. Singer: In The Neural and Molecular Bases of Learning,
    ed. by J.-P. Changeux, M. Konishi (John Wiley, New York 1987) p. 301
12. C.D. Gilbert, T.N. Wiesel: Nature 280, 120 (1979)
13. C.D. Gilbert, T.N. Wiesel: J. Neurosci. 3, 1116 (1983)
14. Z.F. Kisvárdy, K.A.C. Martin, T.F. Freund, Z. Maglóczky, D. Whitteridge,
    P. Somogyi: Exp. Brain Res. 64, 541 (1986)
15. K.S. Rockland, J.S. Lund: Science 215, 1532 (1982)
16. D.Y. T'so, C.D. Gilbert, T.N. Wiesel: J. Neurosci. 6, 1160 (1986)
17. D.J. Price, C. Blakemore: Nature 316, 721 (1985)
18. H.J. Luhmann, L. Martinez-Millán, W. Singer: Exp. Brain Res. 63, 443 (1986)
19. G.M. Innocenti, D.O. Frost, J. Illes: J. Neurosci. 5, 255 (1985)
20. D.J. Price, C. Blakemore: J. Neurosci. 5, 2443 (1985)
21. C. v.d. Malsburg: Kybernetik 14, 8 (1973)
22. C. v.d. Malsburg: Biol. Cybern. 32, 49 (1979)
23. R. Linsker: Proc. Natl. Acad. Sci. USA 83, 8390 (1986)
24. G.M. Edelman, G.N. Reeke Jr.: Proc. Natl. Acad. Sci. USA 79, 2091 (1982)
25. H. Frohn, H. Geiger, W. Singer: Biol. Cybern. 55, 333 (1987)

Do Coherent Patterns of the Strange Attractor EEG Reflect Deterministic Sensory-Cognitive States of the Brain?

E. Başar, C. Başar-Eroglu, and J. Röschke

Institute of Physiology, Medical University Lübeck,
Ratzeburger Allee 160, D-2400 Lübeck, Fed. Rep. of Germany

1. Introductory Remarks on EEG, Evoked Potentials, Strange Attractors, and Deterministic Processes of the Brain

One of the main concerns of brain research is to measure the brain's electrical activity and, in this way, to try to detect the coding of behaviorally relevant information in the central nervous system (CNS). It is usually assumed that there is no uniform code for behaviorally relevant information in the neuronal networks which constitute the CNS. There are also no standard methods for clearly describing the functional and behavioral components of the brain's electrical activity. Analysis of the electroencephalogram (EEG), of evoked potentials (EPs), and of endogenous potentials (P300 family) are among the most fundamental research tools for understanding the sensory and cognitive information processing in the brain. Since Berger's discovery of the EEG and Adrian's measuring of cortical field potentials, the important applications of these powerful neurological techniques have been described in several books [1-3].

Some authors take the view that the spontaneous EEG activity is an expression of the incessant, irregular background neural firing. Do we have the right to consider the spontaneous EEG activity of the brain as a background noise in the sense of ideal communication theory? Or rather, is the EEG a most important fluctuation, which controls the sensory-evoked and event-related potentials?

We have written elsewhere that the EEG plays an active role in the signals transmitted through various structures and recorded at various sites in the brain and that the EEG should not be considered as a noisy signal. Especially, we have assumed that regular patterns of the EEG reflect coherent states of the brain during which cognitive and sensory inputs are processed [3-5]. A preliminary phase portrait analysis of the spontaneous and evoked alpha activity of the brain indicated that the EEG might reflect properties of a strange attractor and therefore an analysis of EEG was undertaken in order to compare the EEG with the so-called "Rössler-Attractor" and attractors similar to the attractors of Navier-Stokes perturbation system [6]. By comparing the EEG and evoked potentials together, the following tentative formulation has been used in order to show the transition of the EEG to a stimulus-induced change in the EEG which is called "evoked potential": If a brain structure under study is in a desynchronized (or chaotic?) state, then the excitation (sensory stimulation) would put its activity in a temporary attractor. This attractor was called "instantaneous attracting cycle", rather than a "limit cycle". Several assumptions were also undertaken to describe the EEG activities in order to classify with terms of attractors [4]. Furthermore, the EEG was considered as deterministic signal and the expression "internal evoked potential" was introduced to describe that the EEG was considered as deterministic electrical activity stemming from hidden sources to sensory and cognitive inputs into the CNS.

In our newest approaches we used the algorithm of GRASSBERGER and PROCACCIA [7], similar to the pioneering analysis of BABLOYANTZ and NICOLIS [8]. The SWS-sleep EEG of the intracranial structures of the cat brain was embedded into phase space and dimensions of the attractors of the cat cortex, hippocampus and reticular formation were computed. The results confirmed the findings of BABLOYANTZ [9] for special structures of the cat brain and also our long-standing assumptions that the EEG represents an integrative signal stemming from deterministic processes and that it represents a strange attractor [10,11]. If the EEG is a signal stemming from deterministic neuronal processes it should be possible to find replicable EEG-states if repeatable initial conditions for a sensory-cognitive input to CNS can be experimentally established. In the part 2 of the present study our analysis of the cat EEG will be presented in order to show that EEG is a deterministic signal, whereas in part 3 experiments will be described in order to demonstrate that the EEG can reach repeatable patterns and confirms properties of a strange attractor.

2. Is the EEG a Strange Attractor?

2.1. Nonlinear Dynamics in Neurophysiology

For many physiological systems small changes in the initial conditions lead to small changes in the systems outcome. For example, electrical excitation of a motoneuron leads to a contraction of the muscle. This means, similar initial conditions lead to similar effects. We have described elsewhere that the amplitude and phases of evoked potentials depend strongly on the spontaneous electrical activity of the brain which immediately precedes the stimulation. The theory of nonlinear dynamical systems states that nonlinear systems are able to generate deterministic chaos under selected conditions. Chaos in the sense of nonlinear dynamics means that the behavior of a system is not predictable over long times; but nevertheless there exists a prescription (i.e. in terms of differential equations) for calculating the future behavior from given initial conditions.

2.2. The Concept of "Attractor"

According to the considerations above, the description of systems behavior (in our case the EEG from different brain structures) must be analyzed not only in the time domain or frequency domain, but also in the __phase space__. In general, a phase space is identified with a topological manifold. A n-dimensional phase space is spanned by a set of n independent linear vectors. This requirement is generally sufficient. There are several possibilities for defining a phase space. We consider a proposal of TAKENS [12] and span a 10-dimensional phase space by $x(t)$, $x(t+\tau)$, ..., $x(t+9\tau)$ where τ means a fixed time increment. Every instantaneous state of a system is therefore represented by a set $(x_1, ..., x_n)$, which defines a point in the phase space. The sequence of such states (or points) over the time scale defines a curve in the phase space, called a __trajectory__. As time increases, the trajectories either penetrate the entire phase space or they converge to a lower-dimensional subset. In this latter case, the set to which the trajectories converge is called an __attractor__.

If the dimension of an __attractor__ is a non-integer, called a fractal, the attractor is a __strange attractor__ and can be identified with the properties of deterministic chaos.

2.3. Fractal Dimension

What is a fractal dimension? One of the oldest notions of dimension is that of a topological dimension D_T. For a point, $D_T = 0$; for a line, $D_T = 1$, and for a plane, $D_T = 2$. A first generalization is the __Hausdorff-dimension__ or __fractal dimension__ D_F. For simple sets, for example a limit cycle or a torus, the fractal dimension D_F is an integer and is equal to the topological dimension D_T. For an n-dimensional phase space, let $N(R)$ be the number of n-dimensional balls (or cubes) of radius R required to cover an attractor. Then the fractal dimension D_F is defined as

$$D_F = \lim_{R \to 0} \frac{\log N(R)}{|\log R|} \quad .$$

Let p_i be the probability that an arbitrary point (of an attractor) falls into cube i with radius R and let N(R) be the number of non-empty cubes. The generalized dimensions D_q of order q are given by

$$D_q = \lim_{R \to 0} \frac{I_q(R)}{\log(1/R)}$$

(I_q: Renyi-information of order q).

For q = 0 we find $D_0 = D_F$. D_1 is called the information dimension and D_2 is called the correlation dimension. It is the case that

$$D_0 > D_1 > D_2 > \dots$$

In practice, the correlation dimension D_2 is the generalized dimension easiest to estimate from attractors generated by experimental data [13], because

$$I_2 = - \log \sum_{i=1}^{N(R)} p_i^2 = - \log C(R)$$

where C(R) is a measure of the probability that two arbitrary points x, y will be separated by distance R. C(R) is called the correlation integral and can easily be computed:

$$C(R) = \lim_{N \to \infty} \frac{1}{N^2} \sum_{x \neq y} \theta (R - \| x-y \|)$$

where θ is the Heavyside function.

It then follows that

$$D_2 = \lim_{R \to 0} \frac{\log C(R)}{\log R} \quad \text{or}$$

$$C(R) \sim R^{D_2} .$$

The main point is that C(R) behaves as a power of R for small R. This means that it is possible to find a measure for the dimensionality of an attractor by evaluating C(R) and plotting log C(R) versus log R.

2.4. Experimental Procedure and Results

In order to analyze the dimensionality of field potentials, five cats with chronically implanted electrodes were studied. The chronic electrodes were implanted in the acoustical cortex (GEA), hippocampus (HI), and reticular formation (RF). In total, 15 experimental trials during slow-wave sleep (SWS) activity were evaluated. The intracranial EEG signals were digitized by a 12-bit AD converter and stored in the memory of an HP 1000-F computer. The sampling frequency was f_s = 100 Hz for all trials.

299

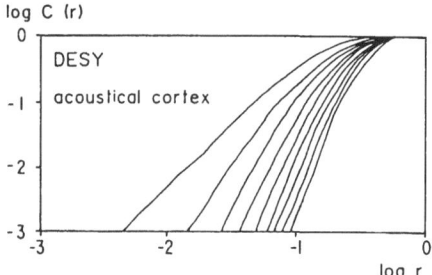

Fig. 1: Logarithm of cor-
relation integral C(r)
versus logarithm of r

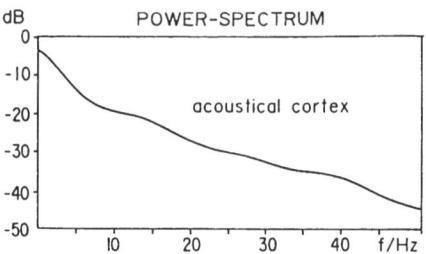

Fig. 2: Power-spectrum of the
EEG from the acoustical cortex
of the cat (slow-wave sleep)

Dimensions of the EEG signals were evaluated over a time period of about 20 sec (N = 2048) and 40 sec (N = 4096). Details of the software have been described elsewhere [14,11]. The phase space was constructed by using the time-delayed coordinates proposed by Takens. Theoretically, the evaluation of the dimension of the attractors should be independent of the arbitrary but fixed time increment τ. In practice, this independence is not generally valid. Investigations of low-pass filtered noise (non-deterministic signals) have shown that dD_2/dn depends on the time increment τ. It is evident that in the case of nondeterministic signals, $dD_2/dn \neq 0$ is observed for every embedding dimension.

Figure 1 shows the plot of log C(r) versus log r for the cats GEA.
One can detect that the slope of the curves converges towards a saturation value of D_2 = 4.93, which is the correlation dimension of the acoustical cortex. The power-spectrum of the EEG (Fig. 2) seems to be noise-like (1/f-noise). But in contrary to a real noise, the phase-space description and the evaluation of the correlation dimension lead to convergence which one cannot regard as a real noise (Fig. 3).

The mean value for 15 trials and 5 cats is given in Table 1

Table 1

D_{GEA}	= 5.06 ± 0.31
D_{RF}	= 4.58 ± 0.38
D_{HI}	= 4.37 ± 0.36

The main point of our investigations is that we can demonstrate significant differences betweeen cortical and subcortical structures of the cat brain and that
the dimensions of the various regions are relatively stable (standard deviation

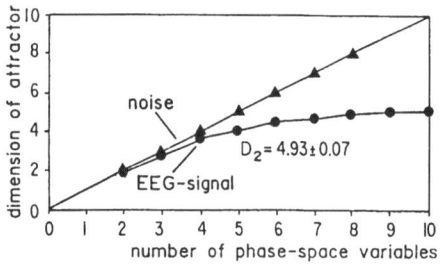

Fig. 3: Dimension D_2 versus
the number of phase-space
variables of the EEG (acoustical cortex) and a noise
signal

range smaller than 10%). Moreover, in 90% of the trials we find the maximal dimension in the acoustical cortex. That means, the data confirm the important relation

$$D_{GEA} > D_{RF} > D_{HI}.$$

2.5. Summary of the Nonlinear Analysis

We can summarize the results of our investigations in the following manner:

i) Obviously the power-spectral analysis of the slow-wave-sleep activity leads to the assumption that the EEG is randomlike (noise). In contrast, the phase-space analysis and the evaluation of the correlation dimension brings out that the EEG might not be confused with a noisy process. A noise signal does not have a finite dimension, whereas chaotic systems show finite dimensionality. Therefore, the SWS-EEG is a deterministic signal.

ii) The attractors of the investigated structures show significant differences during the slow-wave sleep, where a state of hypersynchrony is observed in all of the various brain structures. This is a kind of differentiated behavior that cannot be observed by the analysis of power spectra.

iii) The fact that we have computed dimensions between 4 and 6 gives rise to the assumption that the EEG might be generated by a nonlinear dynamical system with a maximum of 4 or 6 independent variables (differential equations). Assuming that the neural population of the brain may be regarded as a large number of coupled oscillators, the attractor will be n-dimensional if the oscillations are left uncoupled. But if they are coupled, the dimension will be reduced. Our results can be interpreted in the manner that a small number of coupled parameters should be able to generate the electroencephalogram.

3. Coherent States of the Brain during Cognitive Processes

3.1. Remarks on Cognitive Processes

Since the first measurement of the event-related potential (ERP) "P300" by SUTTON et al. [15] several paradigms have been used in order to correlate cognitive tasks and behavior with slow waves of the brain. However, there are only few reports which analyze prestimulus EEG activity during these tasks. Our earlier research which demonstrated the correlation between EEG and evoked potentials led us to start experiments which also indicated a strong relation between pre-stimulus EEG and the P300-wave [16-18]. We observed that during the application of various event-related paradigms the prestimulus EEG tends to attain a phase-ordered pattern prior to expected stimulation. Our preliminary experiments have now been extended to measure the event-related EEG prior to a cognitive task with a new paradigm. In the present report we will describe the results demonstrating the existence of regular, phase-ordered prestimulus EEG-rhythms, which tend to show a repeatable pattern formation preceding successful cognitive tasks.

3.2. Methods and Results

The experiments were carried out with sixteen volunteer healthy subjects, mostly students in the age of 19-21 years. The EEG has been recorded in vertex, parietal and occipital locations against the reference of the ear lobe (Cz, P3 and O1 in the 10-20 System). The EEG signals were amplified by using a Schwarzer EEG-Machine. The subjects sat in a sound-proof and echo-free room which was dimly illuminated. For stimulus preparation, evaluation of selective averaging procedure and digital filtering, a Hewlett Packard 1000F computer was used. The filtering of EEG and of event-related potentials was carried out by using a digital filtering method described elsewhere [3]. The digital filters used had no phase shift. As auditory stimulation 2000 Hz 80 dB tones of 800 msec of duration were applied with regular intervals of 2600 msec. Every third or fourth tone was omitted. The subjects were asked to predict and <u>mark mentally</u> the time of occurrence of the omitted signals. One second of the EEG prior to omitted stimulation was also recorded with the ERP.

When subjects learned and followed with success the rhythmicity which is contained in the paradigm they were usually able to increase their attention. Then, rhythmic pre-stimulus EEG-patterns could be observed. Most of the subjects reported that at the beginning of an experimental session with repetitive signals they had the difficulty to predict the time of occurrence of the stimulus omission. Usually, in the second part of the experiment they were able to predict the time of the omitted signal. Accordingly, in our signal analysis we applied a selective averaging by grouping approximately the first ten pre-stimulus sweeps at the beginning of the experiment and the last ten sweeps. Figure 4 illustrates comparatively the averages of the first and the last ten pre-stimulus EEG-epochs (digitally filtered between 1 - 25 Hz) which were measured at the vertex of a subject who reported that at the beginning of the experiment he felt himself unsecure and diffuse. Towards the end of the experimental session he could be more concentrated, thus performing his task much better. The average of the ten sweeps at the end of the experiment depicted a regular rhythmic behaviour with large amplitudes. The first ten sweeps tend also to the same rhythmicity, but the average is not very regular with low amplitudes.

The rhythms, that are in principle similar to those illustrated in Fig. 4, were observed in all the subjects. The alignment and phase reordering were not the same in all the subjects. The exact time of regularity and phase reordering showed fluctuations in a time period 500 to 0 msec prior to the event.

Figure 5 shows approximately 10 prestimulus EEG epochs of another subject (C) at the end (A) and at the beginning (B) of an experiment. It is easy to see that at the end of the experiment repeatable patterns were observed whereas at the beginning no coherent state was observed. The single EEG epochs are not in an ordered state.

Figure 6 shows results of an experiment with another subject. In this case the superposition of the last 9 sweeps from vertex and parietal leads is illustrated. Single sweeps were here digitally filtered in the frequency range between 7 and 13 Hz. The regularity of the single EEG-rhythms are in this case such that it is easy to observe that the shape of the EEG prior to stimulation reached a <u>template pattern</u>. Although the reactions of various subjects to the sequence of applied stimuli and omitted signals had in general a common character, slight fluc-

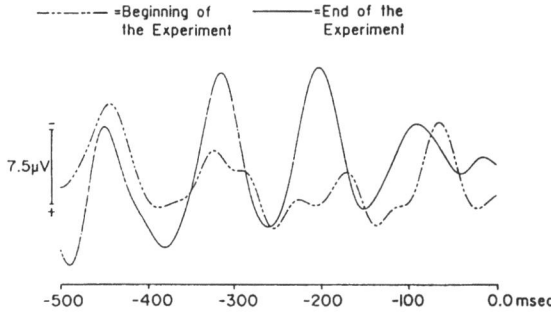

------ =Beginning of
the Experiment

———=End of the
Experiment

7.5μV

-500   -400   -300   -200   -100   0.0 msec

Fig. 4: Averages of the first and last ten prestimulus EEG epochs of the experiment, filtered in the 1-25 Hz frequency band

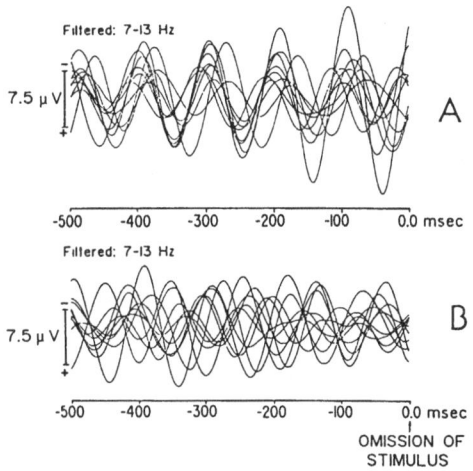

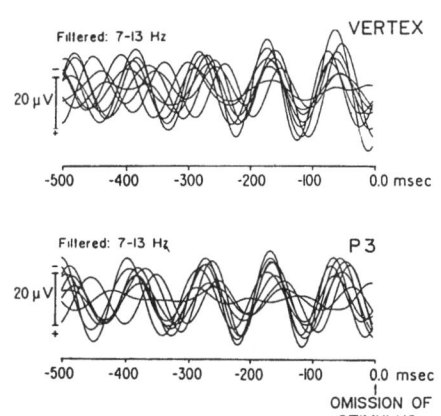

Fig. 5: (A) Approximately 10 prestimulus EEG-sweeps at the end of the experiment.
(B) Approximately 10 prestimulus EEG-sweeps at the beginning of the experiment

Fig. 6: The last 9 sweeps from vertex (Cz) and parietal (P3) leads filtered in the 7-13 Hz frequency band

tuations in the understanding of the paradigm and interpretation and performance differences were observed especially among the students. This was reflected also in prestimulus-EEG.

Since it is not possible to describe all these behaviors and rhythmic changes of event-related rhythms, we will describe a task-related correlation to EEG which seemed to reflect a subtile behavior during the experiments. Most of the subjects (N : 7) explained after the experiment that it was for them usually easier to mark mentally when the fourth stimulus was omitted. They explained the event in the following manner: "As I heard the third repetitive tone, I knew that now the fourth signal should be omitted. Then I tried to mark mentally the fourth omitted signal. I have done a better task to mark it mentally than in the cases

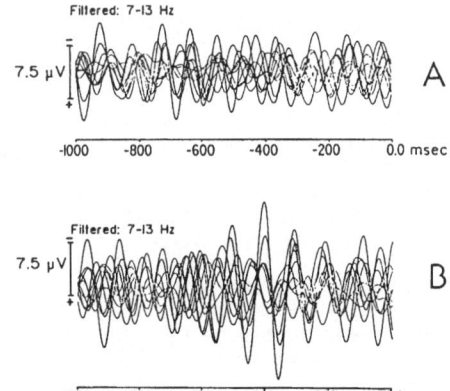

Fig. 7: Superposition of 10 EEG trials filtered in the 7-13 Hz frequency band for two groups of data (subject M):
First group (A): EEG prior to the third omitted signal.
Second group (B): EEG prior to the fourth omitted signal

303

of the missed third signal". According to the statements of the subjects we then performed a selective averaging. Averaging was made with two different categories of EEG signals: (1) EEG prior to the third omitted signal; (2) EEG prior to the fourth omitted signal. Figure 7 illustrates the superposition of filtered EEG trials for both groups of data for the subject (M).

The signals prior to the fourth tone are ordered having large regular amplitudes, whereas the EEG signals prior to the third tone showed a disordered behavior with smaller amplitudes. Only one of the subjects (subject C) had the contrary performance. She said that she tried to perform the task for the third signal and that she usually had good success. She was then usually no more concentrated or attentive to mark mentally the fourth tone. The EEG curve recorded from the vertex of subject (C) gave results opposite to those of subject (M) which are illustrated in Fig. 7. In other words, the EEG patterns showed coherent repeatable oscillations only prior to third omitted stimulation. This result is not illustrated in the present study.

3.3. Discussion of Coherent States during Cognitive Processes

It is largely recognized that the endogenous ERP components are related to cognitive processing of stimulus information or the organization of behavior, rather than evoked by the presentation of the stimulus. The event-related rhythms which have been presented in this study reflect the effort performed by the brain in the expectation and prediction of an event. They are purely endogenous, since they are emitted in relation to a mental task, not following a physical event or preceding a physical motion, since the subjects avoided every finger or glossokinetic artefacts. On the other hand, the cognitive task consisted in mental marking of an omitted signal and not a physical stimulation. Our results highly differ from reports describing cognitive EEG changes in frequency and amplitudes, since the experiments indicate that the EEG of a subject can reach patterns with a constant template during a constant mental task. This pattern has a defined phase-reordering and alignment during the execution of the mental task which starts approximately 500 msec prior to the event. Accordingly, we want to emphasize that the EEG might play a highly active defined role in processes of cognition, being involved especially in generation of percepts and short-term memory.

During this analysis we focussed our attention to EEG rhythmicity in a frequency range around 7-13 Hz during the described cognitive task. In our earlier studies we mentioned that activities of slower (1-7 Hz) and higher (40 Hz) frequency with phase-reordering could also be observed during cognitive tasks. We also assumed that synchronization of the electrical activity in delta (1-3 Hz), theta (3-8 Hz), and alpha (8-13 Hz) frequency ranges seems to occur during operative stages of the brain in which the brain processes information coming from sensory and cognitive signals [16]. We also want to emphasize here the results of FREEMAN [19] who showed that the spatial pattern of the 40 Hz-EEG appears to be related to the stimulation (odour) which an animal (rabbit) expects to receive. In recent studies, FREEMAN and SKARDA [20] showed that, by training rabbits to discriminate odours, a new spatial pattern appears with each odour, manifesting a learned regular pattern in the olfactory cortex. The "motor potentials" preceding movements, described firstly by KORNHUBER and DEECKE [21] (called also readiness potentials), attracted tremendous interest since these studies showed that changes in slow EEG-activity could be analyzed as an indicator of future motions to be performed by the cerebral cortex. In this report we want to emphasize that the analysis of the EEG prior to cognitive tasks may open new important aspects in the understanding of cognitive tasks and dynamic memory since, due to speedy and capacitive computers, several new applications will be possible by use of various analytical methods of single EEG-EP trials which are now described by several authors (see [22]).

4. Two Different Approaches Lead to One Conclusion

In part 2 of this study we have presented a nonlinear approach in order to demonstrate that the EEG in various structures of the cat brain are due to deterministic processes. BABLOYANTZ [9] already formulated that in human subjects the Slow-Wave-Sleep-EEG is also a deterministic signal. However, in part 3 of the present study we used a linear approach by digital filtering and were able to show that by using a sensory-cognitive paradigm the EEG can go over to coherent and ordered states which have repeatable nature. What do we intend by trying to bring together a nonlinear and a linear analysis? The question of this study is: try to find coherent and repeatable states of the brain's electrical activity. In a series of experiments which are now in press [23] we have been able to show that the transition from a disordered state to an ordered state also happened in the frequency range of 40 Hz and 4 Hz. When the probability of occurrence of a target signal is increased, the 10 Hz, 40 Hz, and 4 Hz EEG goes over to coherent phase-ordered states without the application of physical stimulation. These coherent states depict almost phase-ordered patterns. The manifestation of a strange attractor has an activity which appears to be random. However, the activity of a strange attractor is deterministic and reproducible if the input and initial conditions can be replicated.

By expecting repetitive sensory signals the subject seems to generate cognitive inputs due to repetitive mental effort. If the subject cannot predict mentally the occurrence of the expected target signal there is no averaged synchronization of the EEG in the 8-13 Hz or 40 Hz frequency channels. The 10 Hz and 40 Hz EEG go from disordered states to ordered coherent states (similar to evoked potentials described by BAŞAR [3]).

With the approach using the algorithm of Grassberger and Procaccia we have a theoretical prediction concerning the fact that one might be able to find reproducible patterns of the strange attractor EEG. By applying the methodology of part 3 we are, in fact, able to demonstrate that several coherent reproducible patterns can be found in the EEG prior to a defined target signal. These two approaches which, from the theoretical view point, are very different, can bring the brain scientist to the conclusion that the brain, in fact, might produce similar EEG-patterns during defined experimental conditions, the next days and weeks later. There are tremendous amounts of possibilities to use these coherent states of the brain manifested with EEG in order to elucidate several sensory-cognitive mechanisms.

Rhythmic behavior is one of the most ubiquitous properties of biological systems. One of the best examples of periodic behavior in cellular metabolisms are glycolytic oscillations [24]. However, temporal self-organization in biochemical systems is by no means limited to simple behavior [25]. Conceptual models which are based on observations of glycolytic oscillations [24], analysis of dissipative structures [26], and of coherent behavior in laser [27] are, without any doubt, very important landmarks which will also help to elucidate brain dynamics.

References

1. H. Berger: Nova Acta Leopoldina 38, 6 (1938)
2. W.J. Freeman: Mass Action in the Nervous System (Academic Press, New York 1975)
3. E. Başar: EEG-Brain Dynamics. Relation between EEG and Brain Evoked Potentials (Elsevier, Amsterdam 1980)
4. E. Başar: Amer. J. Physiol. 245 (4), R510 (1983)
5. E. Başar: In Synergetics of the Brain, ed. by E. Başar, H. Flohr, H. Haken, A.J. Mandell (Springer, Berlin, Heidelberg 1983) p. 183
6. E.Başar, J. Röschke: In Synergetics of the Brain, ed. by E.Başar, H. Flohr, H. Haken, A.J. Mandell (Springer, Berlin, Heidelberg 1983) p. 199
7. P. Grassberger, I. Procaccia: Physica 9D, 183 (1983)

8. A. Babloyantz, C. Nicolis: Phys. Lett. A111, 152 (1985)
9. A. Babloyantz: In Complex Systems Operational Approaches in Neurobiology, Physics and Computers, ed. by H. Haken (Springer, Berlin, Heidelberg 1985) p. 116
10. J. Röschke, E. Başar: Pflügers Archiv 405 (2), R45 (1985)
11. J. Röschke, E. Başar: In Dynamics of Sensory and Cognitive Processing by the Brain, ed. by E. Başar (Springer, Berlin, Heidelberg 1988) p. 203
12. F. Takens: Lect. Notes in Math. 898 (1981)
13. P. Grassberger, I. Procaccia: Physica 13D, 34 (1984)
14. J. Röschke: Eine Analyse der nichtlinearen EEG-Dynamik (Dissertation, Universität Göttingen, 1986)
15. S. Sutton, M. Braren, E.R. John, J. Zubin: Science 150, 1187 (1965)
16. E. Başar, C. Başar-Eroglu, B. Rosen, A. Schütt: Int. J. Neurosci. 24, 1 (1984)
17. E. Başar, H.G. Stampfer: Int. J. Neurosci. 26, 161 (1985)
18. H.G. Stampfer, E. Başar: Int. J. Neurosci. 26, 181 (1985)
19. W.J. Freeman: In Synergetics of the Brain, ed. by E. Başar, H. Flohr, H. Haken, A.J. Mandell (Springer, Berlin, Heidelberg 1983) p. 102
20. W.J. Freeman, C.A. Skarda: Brain Research Reviews 10, 147 (1985)
21. H.H. Kornhuber, L. Deecke: Pflügers Archiv 284, 1 (1965)
22. E. Başar: In Dynamics of Sensory and Cognitive Processing of the Brain, ed. by E. Başar (Springer, Berlin, Heidelberg 1988) p. 30
23. E. Başar, C. Başar-Eroglu, J. Röschke, A. Schütt: Int. J. Neurosci. (1988), in press
24. B. Hess, A. Boiteux: Annu. Rev. Biochem. 40, 237 (1971)
25. A. Goldbeter, O. Decroly: Amer. J. Physiol. 245, R478 (1983)
26. G. Nicolis, I. Prigogine: Self-Organization in Nonequilibrium Systems (Wiley, New York 1977)
27. H. Haken: Synergetics. An Introduction (Springer, Berlin, Heidelberg 1977)

The Creutzfeld-Jakob Disease in the Hierarchy of Chaotic Attractors

A. Babloyantz and A. Destexhe

Faculté des Sciences, Université Libre de Bruxelles,
Campus Plaine, C.P. 231, Bd. du Triomphe, B-1050 Bruxelles, Belgium

1. Introduction

Until recently model construction was the main tool for understanding the long time dynamical behavior of complex systems [1]. If the experimental measurement of a property of the system exhibited some regularity of behavior, such as for example periodic oscillations, this behavior was often formulated in terms of a set of coupled nonlinear differential equations. The solutions to these equations were compared with the experimental data. However, most often models fit the data only qualitatively and do not account for the variabilities seen in the actual measured time-dependent variables. Today we realize that these variabilities cannot be ignored as they may indicate the presence of dynamics radically different from the assumed models.

In other instances, the behavior of a system may appear quasi-random, thus being difficult to cast in terms of deterministic differential equations. However, the recent advances in the theory of nonlinear dynamics tell us that such random behavior may stem from deterministic systems described with only few degrees of freedom [2,3].

With the introduction of the notion of deterministic chaos, and the evaluation of dimensions of these chaotic attractors, today we are in a position to extend the theory of nonlinear systems into areas not explored before. We are also able to assess the dimensions from experimental data [4]. These dimensions quantify the system's dynamics and must be compared with the values obtained from theoretical models, thus imposing constraints to model construction.

Several definitions are provided for the characterization of attractor dimension, and various algorithms are proposed for their evaluation. The most practical and the most widely used algorithm was proposed by Grassberger & Procaccia [4]. They introduce the correlation dimension D_2 which may easily be computed from time series.

However, in order to implement such an algorithm, there is a need to resurrect all the pertinent variables describing the dynamics from the experimental measurements. In most cases only a single parameter is followed in time (time variables) or the temporal evolution of the system is measured simultaneously in several sites of the system (space variables).

Phase space portraits could be constructed from experimental data following four different procedures. With the help of Takens' theorem [5], one shows that m variables can be reconstructed from a single time series by introduction of (m-1) time lags τ. Also one can use the Eckmann-Ruelle conjecture [3] in which a measurement at each site of the system is regarded as a new variable. Singular systems analysis [6] is another promising way to construct phase spaces using either a single time series or measurements from various sites of the system.

In the second section of this paper, we use all four techniques for the construction of the phase portrait from experimental data. The time series were obtained from the electroencephalographic (EEG) recordings of a patient suffering from a neurological disorder known as "Creutzfeld-Jakob disease". In Sect. 3 Takens' theorem and the Eckmann-Ruelle conjecture are used to evaluate the correlation dimension of the attractors. This dimension is compared with the previous studies of beta waves, alpha waves, sleep stages two and four, REM sleep and the "petit mal" type of epileptic seizures. Section 4 is devoted to some considerations about the reliability of algorithms.

2. Phase Portraits of the Creutzfeld-Jakob Disease

In this section, we shall illustrate the various techniques for the construction of the phase space, which will be illustrated with the help of the EEG recorded during the terminal state of the Creutzfeld-Jakob disease in which it is believed that a virus attacks and gradually destroys the nerve cells. After a first stage of dementia, the patient enters a terminal coma with myoclonus. The EEG is then very coherent (like the "Burst Suppression Pattern" seen after administration of barbiturates) and regular patterns of stable slow waves reminiscent of the "petit mal" type of epileptic seizures appear. However, the phenomenon is of a much longer duration and shows a remarkable stationarity. Twenty minutes of EEG are shown in Fig.1 taken from a single lead and twelve of such leads were recorded simultaneously.

The most popular phase space construction results from a theorem shown by Takens [5]. This theorem states that for a dynamical system of n variables $X_1...X_n$, the space spanned by the m variables $\{ X_1(t) , X_1(t+\tau) , X_1(t+2\tau) , ... , X_1(t+(m-1)\tau) \}$ is at least topologically equivalent to the original phase space and therefore represents many of its dynamical properties. Moreover, m is restricted by the relation $m \geq 2d+1$, where d is the lowest integer greater than the dimension of the attractor. In this section we refer to such procedures as Takens'construction.

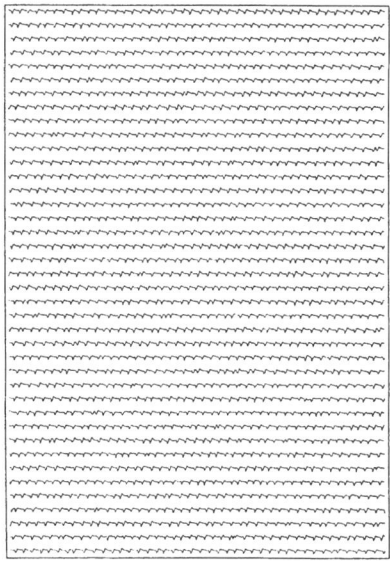

Figure 1. Electroencephalogram recording of the Creutzfeld-Jakob disease. A total of twenty minutes of EEG shows the remarkable stability of this slow wave pattern (30 s/line, 12 bit sampling at 250 Hz, 120 Hz filter).

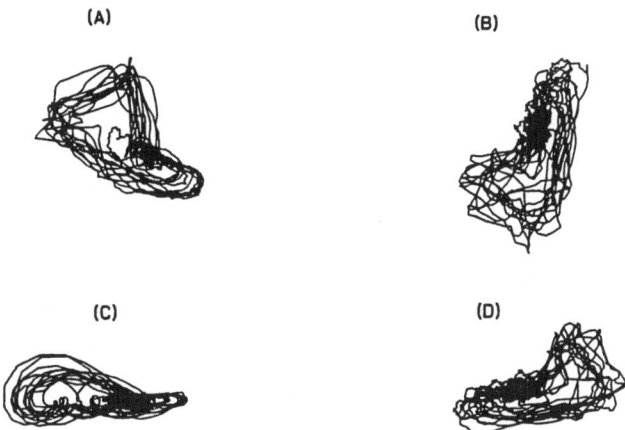

Figure 2. Phase portraits of the Creutzfeld-Jakob attractor. A similar structure is seen from four different three-dimensional constructions: (A) Takens' construction from one lead, (B) Eckmann-Ruelle's construction from simultaneous leads. (C) and (D) are respectively the (A) and (B) portraits constructed using the singular eigenvectors.

Recently it was conjectured that the space spanned by several simultaneous measurements of an observable in various sites of the system { $X_1(r_1,t)$, $X_1(r_2,t)$, $X_1(r_3,t)$, ... , $X_1(r_m,t)$ } may also yield a phase portrait which is topologically equivalent to the portrait obtained from the variables X_1... X_n of the system [3]. Such constructions may be obtained very easily from electroencephalograms or electrocardiograms as actually simultaneous measurements up to 256 leads may be recorded. In this Eckmann-Ruelle construction, the lag τ disappears from the construction and is replaced by another subjective quantity, namely, the inter-electrode distance, which is bounded by a minimum length imposed by the physical reality.

A more recent technique for phase space construction was introduced by Broomhead & King [6]. It is based on singular value decomposition which may be a noise-reducing procedure. In essence, in this technique one diagonalises the covariance matrix constructed from the phase space vectors of the previous constructions and one obtains orthogonal eigenvectors that may be used to reconstruct a phase portrait and evaluate its correlation dimension.

Figure 2 shows the phase portrait of the Creutzfeld-Jakob pathology constructed using all four techniques outlined above. It is interesting to note that the constructions of Takens, Eckmann-Ruelle and Eckmann-Ruelle with singular decomposition yield very similar three-dimensional phase portraits. In contrast, the portrait obtained with the help of the singular eigenvectors and from a single lead time series (Fig.2c) looks different. It is flatter and smoother, as noise, and also some of the finer twists of chaotic trajectories, has been eliminated from the portrait.

3. Dimensional Analysis

Once the variables spanning the phase space become available, the Grassberger & Procaccia algorithm can be used for the evaluation of the correlation dimension D_2. We have computed D_2 for all four phase portraits of Fig.2. Such a comparative study is in the process of completion for beta waves, alpha waves, sleep stages two and four, REM sleep, and "petit mal" epilepsy (Destexhe, Sepulchre & Babloyantz, to be published - these time series are shown in Fig.3).

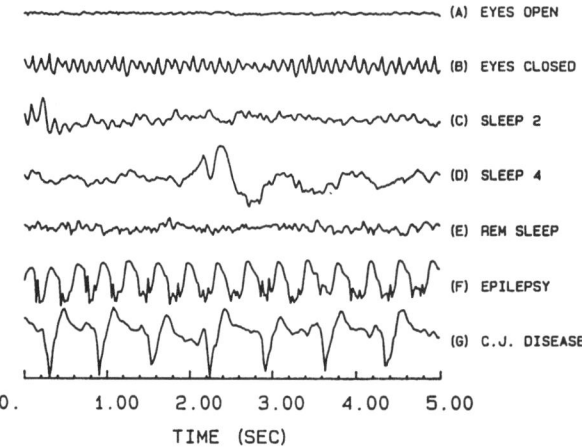

Figure 3. Short stretches of some of the most typical episodes of human EEG activity (identically scaled, 12 bit sampling).

In the case of the Creutzfeld-Jakob attractor, all four approaches give remarkably comparable results. We report here only the dimensions computed from the usual Takens' construction so as to compare them with the D_2 values already reported for other stages of brain activity. If the correlation dimension is computed from a single lead (Takens' construction), we find $D_2 = 3.8 \pm 0.1$. In the case of the Eckmann-Ruelle construction, the phase space was obtained from 12 EEG leads covering the entire scalp and a value of $D_2 = 3.8 \pm 0.2$ was seen. The agreement between these two values is remarkable.

Although at first sight the signal appears to be extremely coherent with obvious periodicities, we find a surprisingly high value for the correlation dimension as compared with the low value of $D_2 = 2.05 \pm 0.09$ observed for the "petit mal" type of epileptic seizure [7]. This finding may stem from the fact that the total phenomenon of the epileptic seizure was of 5 s duration whereas the Creutzfeld-Jakob disease was analyzed over 8 minutes (from a total phenomenon lasting several hours). The analysis of a 1 min recording gave the same result as 8 min whereas a 8 s time series gave an underestimated dimension of 3.1 ± 0.2.

In this respect, the Creutzfeld-Jakob disease may be compared with the normal cardiac activity. At first sight, the electrocardiographic signal appears to be periodic, however, in a recent paper [8] Babloyantz & Destexhe have shown that the normal cardiac activity is governed by deterministic chaos characterized by $D_2 = 4.4 \pm 0.4$.

The values of the correlation dimension of Creutzfeld-Jakob disease taken together with the D_2 computed previously for other stages of brain activity give a coherent image of brain dynamics. These values can be represented as a function of the width of the power spectrum (see Fig.4 and Fig.5). This width is estimated by taking the two extreme frequencies of 75% of the spectral energy.

Figure 4 shows that the broad spectra correspond to high dimensional attractors such as eyes open and REM sleep. For these two states we find, by using the Grassberger & Procaccia algorithm [4], extremely high values of D_2. These values were $D_2 = 9.7 \pm 0.7$ for beta rhythm [9] (eyes open in Fig.3a), $D_2 = 8.2 \pm 0.4$ for REM sleep [10] (Fig.3e). A word of caution is needed

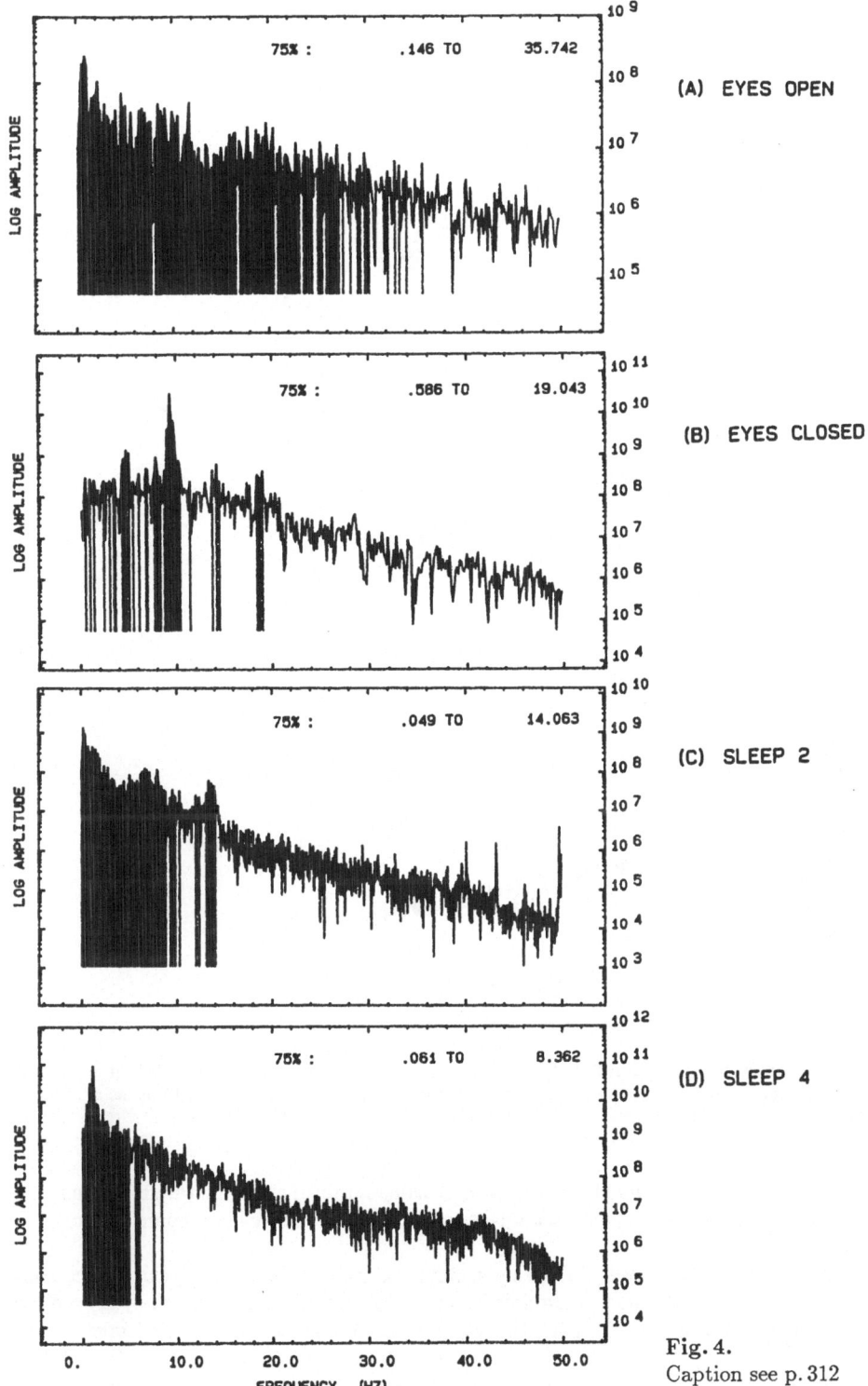

Fig. 4.
Caption see p. 312

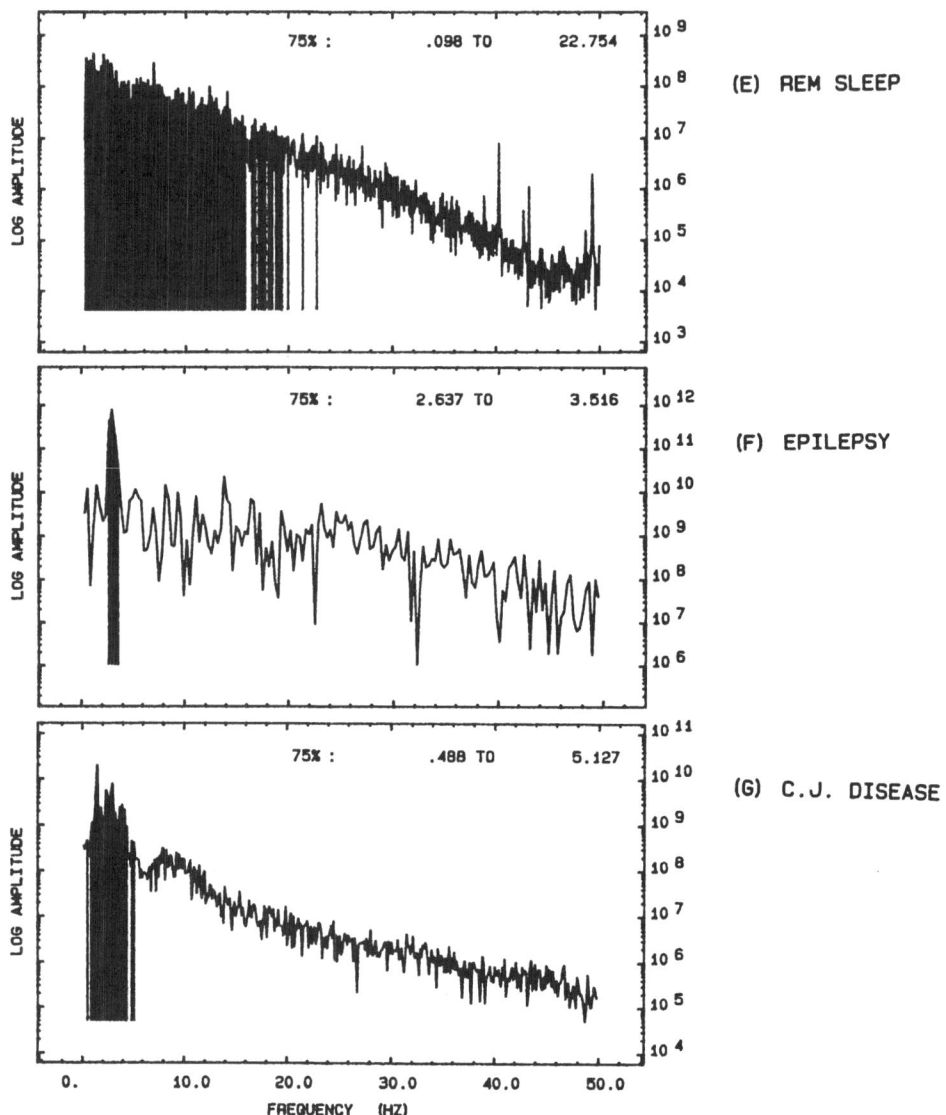

Figure 4. Power spectra of the typical EEG episodes seen in Fig.3. The main peaks forming 75% of the spectral energy are underlined with black solid vertical lines. Only for low-dimensional systems as (F) is the most of the spectral array contained in relatively few peaks.

here as we are not sure that the algorithm gives reasonable results for such high-dimensional systems. In any case, we can say that beta waves and REM sleep behave like coloured noise.

In an awake subject with eyes closed, alpha waves set in (Fig.3b). The dynamics becomes more coherent and switches into a deterministic chaotic activity [9], characterized by $D_2 = 6.1 \pm 0.5$ confirmed by other laboratories [11,12]. As expected, the spectral width also decreases. As the sleep cycle sets in, the brain activity enters a chaotic state of correlation dimension $D_2 = 5.01 \pm 0.03$ for the stage two [13] (Fig.3c) and $D_2 = 4.4 \pm 0.4$ for the stage four or deep sleep [10,13]

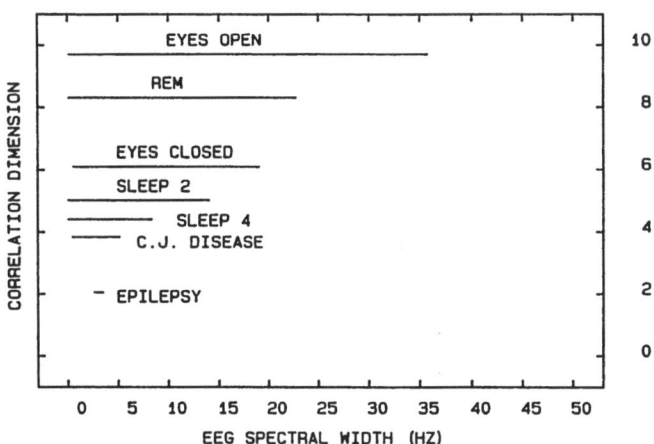

Figure 5. Dimension - power spectrum plot. The extreme frequencies from 75% of the spectral energy are drawn for each EEG episode. Active states are broad banded and high dimensional, while pathologies are of low dimension and their spectrum is restricted to a thin band of frequencies.

(Fig.3d). Similar values were obtained from intra-cranial EEG recordings from cats and rabbits [10,15]. As correlation dimension decreases, so does the width of the spectral band. The sleep stage four is the most coherent stage of the normal brain activity.

In various pathologies, the coherence of the brain dynamics increases further as the correlation dimension decreases. Finally in the "petit mal" type of epileptic seizure (Fig.3f), the most coherent state is reached. Here a near unison is seen in the neuronal activity and the correlation dimension drops to a value of $D_2 = 2.05 \pm 0.09$ [7]. Such a low-dimensional chaos is seen in three-variable differential equations such as the Rössler attractor [16]. As expected, the width of the power spectrum drops dramatically in this case.

Therefore we see from Fig.5 a consistent relationship between the correlation dimension and the main energy of the power spectrum. The correlation dimension appears as an increasing function of the spectral width.

The EEG is the sum over a very large number of neuronal membrane potentials and its magnitude is therefore a good measure of the synchrony between neurons. Low amplitude EEG reflects relatively desynchronized states whereas high amplitude waves are an indication of synchrony between neural masses. In Fig.6 the correlation dimension of various stages of the brain activity is plotted against a measure of the amplitude of the EEG. To obtain this measure, the signal is computed at one second intervals. The average value of these local amplitudes is represented by points in Fig.6. The highest and lowest amplitudes are included in the horizontal bar. During normal states of brain activity, the dimension of the chaotic attractors decreases as the amplitude of the waves increases. The reverse situation is seen in pathological conditions. The Creutzfeld-Jakob disease is characterized by a higher amplitude and a higher dimension than the epileptic seizure.

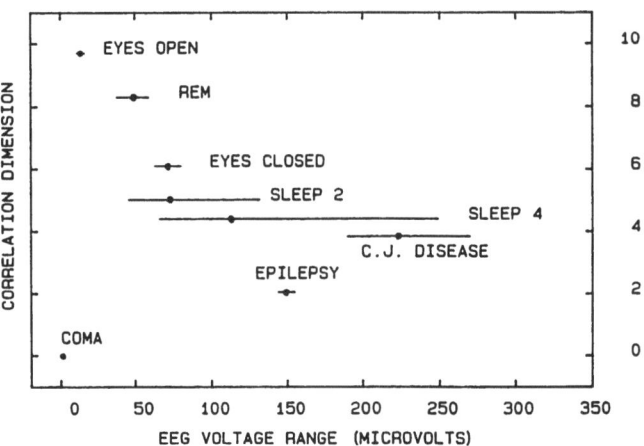

Figure 6. Dimension - amplitude plot. The variations of EEG voltage observed in successive stretches of one second are represented for each type of EEG behavior. The synchronization between neurons which occur in pathologies is reflected by high amplitudes and low dimensions.

4. Reliability of Algorithms

For several years, various algorithms for the phase space construction and dimension evaluation from experimental time series have been used to assess the existence of deterministic chaotic dynamics in problems ranging from hydrodynamics [17] to complex biological systems such as electroencephalograms [7,9-14] or electrocardiograms [8]. After a first wave of enthusiasm as the range of applications increased and conditions of applicability of the techniques relaxed, skepticism and criticisms arose. At the same time new techniques were proposed which aimed at reducing the few subjective aspects of the methods [6].

Most algorithms are tested on and are shown to give satisfactory results for simple maps or well-defined systems described by differential equations involving few variables, such as the Rössler attractor [16]. Such systems are free of experimental or intrinsic physiological noise. Therefore the success of the algorithm for such simple systems does not guarantee their applicability to real complex physiological processes involving many variables and endowed with experimental as well as physiological noise. In addition, in the nonlinear world, care must be taken before extrapolating the results from one dynamics to the next. An algorithm could fail for the characterization of one attractor and give good results for another system.

The subjective aspects of the methods are the following:

(a) The sampling frequency is a crucial factor for reliable analyses. Biological attractors may have very fine twists which are barely distinguishable from background noise. Too fine sampling will incorporate undesirable noise, and too wide intervals between samples may wash out some irregularities of the trajectories [9].

(b) Another crucial element is the choice of the time lag τ. In principle, for an infinite set of data points, all time lags must lead to the same results. However, with finite experimental data sets, a proper range for the values of τ must be found [18]. The Eckmann-Ruelle phase space construction avoids the problem of the lag τ. However, this quantity appears implicitly in the choice of the inter-electrode distance, which introduces a subjective element. One does not avoid this subjectivity of choice in singular vector technique. There also a window must be chosen according to criteria which are empirical.

(c) In biological systems when the stationarity of the data sets is of finite duration, the choice of the length of the data is crucial. We have explained elsewhere how to choose the recording time, the lag τ and the sampling frequency in the case of the EEG [9]. Our experience in this field shows that if care is taken to eliminate the subjective factors in the implementation of the algorithms, one can get satisfactory results for biological complex systems where no other simple way of access to the system's dynamics exists.

5. Discussion

In several papers [7,9,10,13] we have reported the existence of deterministic chaos in various stages of the normal as well as pathological brain activity. Our previous results together with the new values of the Creutzfeld-Jakob disease, as summarized in Figs. 5 and 6, constitute a coherent representation of the cerebral activity.

In this paper we have shown how from the recording of the brain waves we can have access to the dynamics of the system and quantify various stages of the cerebral activity. From Figs.5 and 6, we can infer that the brain activity constantly switches between various chaotic attractors, losing or gaining coherence as it is needed by the biological functions of the organism.

The fact that brain attractors obey deterministic chaotic activity is not surprising as these objects, although obeying   deterministic dynamics, are nevertheless unpredictable and are sensitive to the initial conditions. Therefore they are endowed with great information-processing capability [19], a most conspicuous fact of the cerebral activity.

We believe that with all the problems raised regarding the application of the algorithms, the dimension analysis is a powerful tool for the study of complex biological phenomena which cannot be handled otherwise. It is especially valuable when it is used as a comparative study, as in the case in this paper, which follows the evolution of the coherence of the brain waves in various stages of the cerebral activity. The various attractors appear to follow a well-defined hierarchy where some key properties, such as the EEG amplitude, the spectral width and the correlation dimension, seem to be intimately correlated. It is expected that this hierarchy could be reproduced by models describing neural networks (work in progress).

We thank J.A.Sepulchre for his computer help and for interesting discussions. A.D. is a fellow from the Institut pour l'Encouragement de la Recherche Scientifique dans l'Industrie et l'Agriculture.

References

1. A. Babloyantz: in **Molecules, Dynamics and Life** (Wiley, New York,1986)
2. P.Bergé, Y.Pomeau & C.Vidal: in **L'Ordre dans le Chaos** (Hermann, Paris, 1984)
3. J.P.Eckmann & D.Ruelle: Rev. Mod. Phys. **57**, 617 (1985)
4. P.Grassberger & I.Procaccia: Physica 9D, 189 (1983)
5. F.Takens: in **Dynamical Systems and Turbulence** Eds. D.A.Rand & L.S.Young, Lecture   notes in mathematics **898**, 366 (Springer, Berlin 1981)
6. a- E.R.Pike, J.G.McWhirter, M.Betero & C.de Mol: IEEE Proc. **131**, 660 (1984)
    b- D.S.Broomhead & B.King: Physica 20D, 217 (1986)
7. A.Babloyantz & A.Destexhe: Proc. Natl. Acad. Sci. USA **83**, 3513 (1986)

8. A.Babloyantz & A.Destexhe: Biological Cybernetics **58**, 203 (1988)
9. A. Babloyantz & A. Destexhe: in **Temporal disorder in human oscillatory** systems Eds. L. Rensing, U. an der Heiden and M.C. Mackey, Springer Series in Synergetics **36**,48 (1987)
10. A.Babloyantz & A.Destexhe: in **Proceedings of the first IEEE International Conference on Neural Networks,** Eds M. Caudill and C Butler **Vol IV,** 31 (1987)
11. S.P.Layne, G.Mayer-Kress & J.Holzfuss: in **Dimensions and Entropies in Chaotic Systems** Ed. G.Mayer-Kress (Springer,Berlin 1986)
12. I.Dvorak & J.Siska: Phys. Jett **118A,** 63 (1986)
13. A.Babloyantz, C.Nicolis & M.Salazar: Phys. Lett. **111A,** 152 (1985)
14. J.Röschke & E.Basar: in **Dynamics of Sensory and Cognitive Processing by the Brain,** Ed. E.Basar, Springer Series in Brain Dynamics, **Vol 1,** 203 (1988)
15. W.J.Freeman: Brain Res. Reviews **11,** 259 (1986)
16. D.E.Rössler: Ann. N.Y. Acad. Sci. **316,** 376 (1979)
17. A.Brandstater & H.L.Swinney: Phys. Rev. **35A,** 2207 (1987)
18. A.M.Fraser & H.L.Swinney: Phys. Rev. **33A,** 1134 (1986)
19. J.S.Nicolis: in **Hierarchical Systems** (Springer, Berlin 1985)

Part VI

Ecological, Epidemiological and
Economical Organization

Ecosystems Under Varying Ambient Conditions

J. Rössler[1], *M. Kiwi*[2], *and M. Markus*[3]

[1]Departamento de Física, Facultad de Ciencias,
 Universidad de Chile, Casilla 653, Santiago, Chile
[2]Facultad de Física, Pontificia Universidad Católica,
 Casilla 6177, Santiago 22, Chile
[3]Max-Planck-Institut für Ernährungsphysiologie,
 Rheinlanddamm 201, D-4600 Dortmund 1, Fed. Rep. of Germany

Abstract

Ecosystems governed by random dynamic non-linear equations are studied in detail. The process is precisely characterized, both for the non-markovian and the markovian case, including short range order. A new feature, the appearance of early chaos, is introduced and analyzed. The Lyapunov exponent is defined and evaluated. Its precise meaning is discussed and its significance to characterize the dynamics of a random succession of events is investigated and clarified.

1. Introduction

A variety of different natural phenomena are described, at least qualitatively, by simple random recursive relations. For example, biological populations with no generational overlap (see [1] and references cited therein), populations with generational overlap [2,3], and also one-dimensional alloys [4], or quasi-one-dimensional ribbons [5] are described by equations of the type

$$X_{t+1} = f_t (X_t). \qquad (1)$$

Here, the variable X_t describes the state of the system for a specific value of the discrete variable t (time or site, depending on the example), while its dynamics is summarized in the application $f_t(X)$, which also depends on the discrete t value.

In this contribution we circumscribe ourselves to the particular case of the logistic map, that is

$$f_t(X) = C_t X(1-X), \qquad (2)$$

where we assume C_t = A,B to be a dicotomic parameter (we do so in order to include the minimal number of variables in our analysis). In the case of ecosystems, ambient conditions are described by C_t. We study the external perturbation of the system by allowing C_t to change, as a function of t, with varying degrees of randomness including even the case of cyclic variation.

The specific choice of the logistic map is not very restrictive, since two maps which are "topologically conjugated" exhibit the same features (see [6] and references therein) as far as the stable or chaotic nature of trajectories on these maps is concerned. In particular, the logistic map is topologically conjugated with any application g : [a,b]→[a,b], such that g(a) = g(b) = a, and with only one parabolic maximum in the [a,b] interval. Of special interest to us is that the logistic and the RICKER map (see [1]) are topologically conjugated. It is also worth mentioning that the work of CHANG et al. [6] is a special case of ours, where C_t is of period 2, C_{2t}=A, C_{2t+1}=B and 2t→t.

The main objective of the present contribution is to generalize and extend the results of our previous work [1], with special emphasis on the description and understanding of <u>early chaos</u> in the iterative process defined by Eqs. (1) and (2). We denote by early chaos the appearance of positive Lyapunov exponents in the iterative process (2) for values A and B smaller than r_o = 3.569945672...; r_o is the critical value of r above which the ordinary logistic map, with C_t=r=constant, shows chaotic behavior [7].

A hint of early chaos can already be found in the work of CRUTCHFIELD et al. [7]. They added noise to the logistic map to find that it hastened the appearance of chaos. Here we allow for large fluctuations in the value of C_t which leads us to the observation of new, qualitatively different phenomena, including the appearance of chaos when both A and B are significantly smaller than the critical value r_o.

The other main subject we discuss in this paper is the extent, or realm of applicability, of the concept of chaos in a process governed by a random law. In this context we will compare time averages (obtained with fixed initial conditions), with averages taken over the initial conditions (seeds) at a specified fixed time. It will be shown that both averages are far from being equal; thus, an "ergodic theorem" for these processes cannot be established. This leads to the clarification of the meaning of chaotic or non-chaotic, for a dynamical process which is carried out with random perturbation C_t.

This contribution is organized as follows: after this Introduction we discuss, in section 2 , the evolution of systems described by random dynamic laws. In section 3 our main concern is the appearance of early chaos. In section 4 the meaning of chaos generated by a random perturbation is discussed and clarified.

2. Systems governed by a Random Dynamic Relation

2.1. Characterization of the Random Perturbation

The dynamic process of Eq. (2) is fully determined by the perturbation sequence S = { C_1, C_2,... C_t,... C_n } and the initial condition (seed) X_o=$X_{t=0}$. The history of the system $\mathcal{X}$ = {X_1, X_2,...X_t,...X_N} is generated uniquely, from [S,X_o], by the repeated use of Eq. (2). Obviously we assume the final time N to be very large, actually N→∞.

To characterize the degree of randomness, or conversely of short range order, of the sequence S we introduce the probabilities

$$P_{\alpha|\beta}^{(j)} = N_{\alpha\beta}^{(j)}/N_{\alpha}.$$

Here α=A,B, N_{α} is the number of times C_t= α in the sequence S and $N_{\alpha\beta}^{(j)}$ is the number of pairs (C_t, C_{t+j}) = (α,β) in S. We limit our attention to systems which can be fully characterized by just one pair probability $P_{\alpha|\beta}^{(j)}$, in addition to P_{α} = N_{α}/N.

The pair probabilities can be expressed in terms of a single parameter γ_j, known as the Cowley parameter (see [8]), using

$$P_{\alpha|\beta}^{(j)} = P_\beta + \gamma_j \, (\delta_{\alpha\beta} - P_\beta). \tag{3}$$

If there is only direct correlation between the values of C_t and C_{t+1} (Markov process), then the induced correlation for the j-th nearest-neighbor obeys $\gamma_j = (\gamma_1)^j$.

In total absence of correlation (random sequence) $\gamma_j = 0$, for all j. If $\gamma_1 > 0$ the number of pairs (A,A) or (B,B) is larger than in the random case; the limit $\gamma_1 = 1$ indicates full segregation: {...AAAABBBB...}. $\gamma_1 < 0$ indicates a tendency towards nearest-neighbor AB pairs; $\gamma = -1$ corresponds to the sequences {ABABAB.....} or {BABABA...}.

2.2. Description of the Dynamical System

Rigorously, the history $\mathcal{X} = \{X_1, \ldots X_N\}$ depends both on the seed X_0 and the sequence S. However, just as in many other cases in science, detailed knowledge is not required to extract pertinent information (for example, full knowledge of a many-body quantum mechanical wave function is not necessary to obtain the thermodynamics of a physical system). In the $N \to \infty$ limit it is possible to characterize the history of our system independently of X_0 and S, as long as the latter is a random process and that we know the probabilities P_α and $P_{\alpha\beta}^{(j)}$.

An illustrative example of the preceding statement is given by the distribution functions [4]

$$D_\alpha(X)dX = \lim_{N \to \infty} \frac{1}{N} \{ \text{ \# of } X_j \epsilon [X, X+dX] \text{ with } X_j = f_\alpha(X_{j-1}) \} , \tag{4a}$$

$$D(X) = D_A(X) + D_B(X), \tag{4b}$$

which for random processes are stationary, and independent of X_0 and S, in the $N \to \infty$ limit. In fact, following COHEN [3], and for the special case of a Markov process, we have obtained the following functional relation for the $D_\alpha(X)$:

$$D_\alpha(X) = [\alpha(\alpha - 4X)]^{-\frac{1}{2}} \sum_{\sigma = \pm 1} \sum_\beta P_{\beta|\alpha} \, D_\beta \left(\frac{1}{2} + \sigma \sqrt{\frac{1}{4} - \frac{X}{\alpha}} \right) , \quad \alpha = A, B \quad . \tag{5}$$

The argument of D_β on the right hand side above represents the two branches of the inverse function of application (2).

The functional equation (5) accounts for the structural complexities [4] of the distribution function D(X); in particular, for the singularities which originate on the repeated application of f_A, f_B on X = 0.5. For example, $f_A(0.5) = A/4$, $f_A f_B f_A(0.5)$, $f_A f_B f_B f_B(0.5)$, ... are singularities of $D_A(X)$. In Fig. 1 we display an illustrative plot of D(X) for $\gamma_1 = 0$, $P_A = P_B = 0.5$. In general the distribution D(X) is a fractal, as we have verified by evaluating the generalized dimensions of GRASSBERGER [9], for some specific values of A and B.

In the context of our work the meaning of the Lyapunov exponent might seem dubious at first sight, because of the imbricated evolution of the sequence S = {C_t} and the dynamic system {X_t}. Thus, in this section, we will define the Lyapunov exponent λ as an indication of the sensitiveness of the system to a change in initial conditions, for a fixed sequence S. Analytically

$$e^{N\lambda} \equiv \left| \frac{\partial}{\partial X_0} X_N(S, X_0) \right| , \quad N \to \infty. \tag{6a}$$

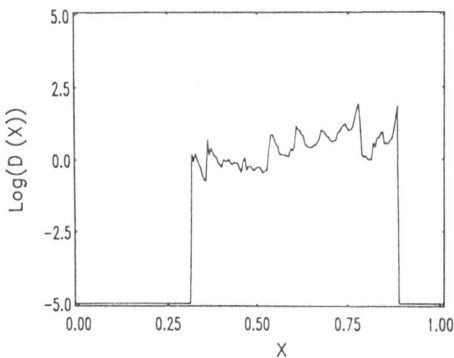

Fig. 1: Distribution function D(X), obtained through a time average for a random parameter sequence with $\gamma_1=0$, A=3.54, B=3.1225

Using the chain rule of differentiation and combining with Eq, (4a), we obtain for a Markov process

$$\lambda = \sum_{\alpha,\beta=A,B} \int_0^1 dX \; P_{\alpha|\beta} \; D_\alpha(X) \ln \left| \frac{\partial f_\beta}{\partial X} \right| \qquad . \qquad (6b)$$

When we consider two replicas of the system, with different seeds X_o and $\tilde{X}_o$, and we subject both of them to the same time chain of random processes, it is natural to expect that $\lim_{t\to\infty}(\tilde{X}_t-X_t)\to0$ when $\lambda<0$. Conversely, the opposite behavior would manifest itself for $\lambda>0$. We will return to this discussion and some of its implications in section 4. For the time being we display in Figs. 2-4 maps of the Lyapunov exponent on the AB-plane, for $2\leqslant A,B\leqslant4$. The chaotic regions ($\lambda>0$) are painted black. In the non-chaotic zones the magnitude of λ is represented by varying tonalities of grey, which range from nearly white for $\lambda=0$, to very dark for $\lambda\to-\infty$. In Figs. 2 and 3 we have chosen $P_A=P_B=0.5$. In Fig. 2 the Cowley parameter $\gamma_1=0$ (random sequence), while in Fig. 3a $\gamma_1=-1$ (binary superstructure). In the latter case only two sequences are possible: $S=(ABABAB...AB)$ or $\tilde{S}=(BABABA...BA)$, and S is related to the quartic map studied by CHANG et al. [6] since it corresponds to the repeated application of $f_A(f_B(X))=f_{AB}(X)$.

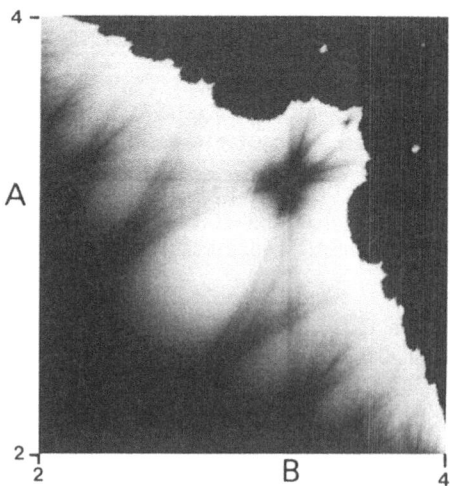

Fig. 2: Lyapunov exponent (given by different shadings of grey) from the logistic equation, as a function of A and B, for a random parameter sequence with $\gamma_1=0$

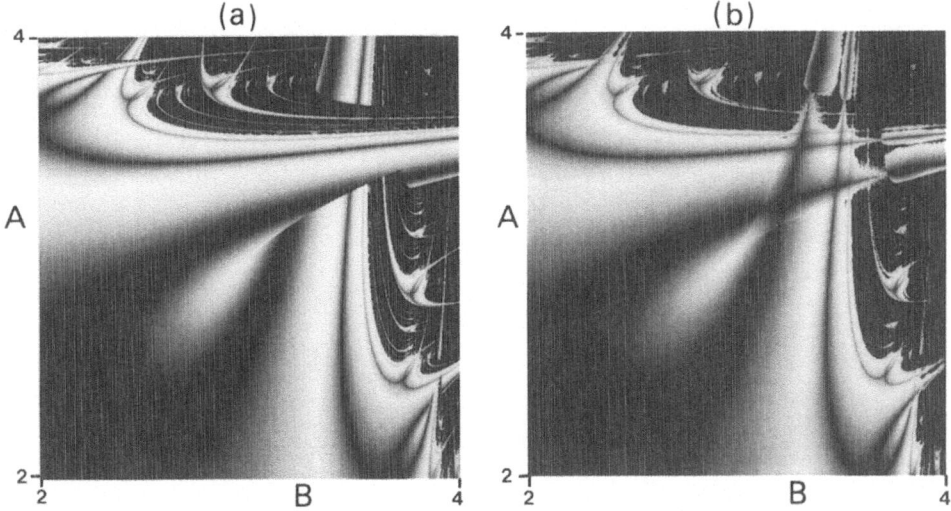

Fig. 3: Lyapunov exponent (given by different grey shadings) from the logistic equation, as a function of A and B (a) Periodic parameter sequence S=(BABABA...) (b) Random parameter sequence with γ_1=-0.98

The asymmetry of Fig. 3a under the exchange A↔B is immediately seen, with ni-tid discontinuities associated with the transition from one branch of λ to an-other. The origin of this asymmetry can be traced to the fact that the function $f_{AB}(X)$ has several basins of attraction in some regions of the AB-plane. For exam-ple, if $f_{AB}(X)$=X has three solutions, with X^* being the intermediate (unstable) one, then, if $1-X^*<f_{AB}(0.5)$ and $X^*<f_{AB}(A/4)=f_A(f_B(f_A(0.5)))$, at least two attraction basins R_1 and R_2 do exist: $R_1=[1-X^*,X^*]$ and $R_2=[X^*,A/4]$ with $f_{AB}(R_j) \subseteq R_j$, j=1.2. It is also verified that the attraction basins of $f_{BA}(X)$, denoted by T_1 and T_2, are related to the previous R_1 and R_2 by $T_1=f_B(R_2)$ and $T_2=f_B(R_1)$.

In order to obtain a more complete understanding of the relation between the existence of more than one attractor for $f_{AB}(X)$ and the asymmetry of Fig. 3a, we mention that the value of the Lyapunov exponent λ for one particular point (A,B) may depend on the attractor employed to evaluate it. For example, the seed X_o=0.5 belongs to the attraction basin R_1; since $f_B(R_1)=T_2$ the Lyapunov exponent of point (B,A), evaluated in T_2, coincides with the one of (A,B) evaluated in R_1. But, if we keep our seed X_o=0.5, then λ at point (B,A) is evaluated in the T_1 basin. Thus, it is not necessary that we obtain the same value calculated in (A,B) starting from the same seed. As we move away from the causal situation, i.e. for γ_1>-1, the multivaluedness disappears. This is illustrated in Fig. 3b, where γ_1=-0.98. In Fig. 4 we display another causal case: the period 8 cycle defined by S=($A^7BA^7BA^7B$...). Here, γ_8=+1, P_B=1/8. While the multivaluedness of λ is quite ap-parent for this non-markovian process, its most striking feature is the onset of "early chaos" for very small values of A and B, as compared with r_o=3.56994567...; for example, for A=3.36 and B=2.25. We will return to this point in section 3.

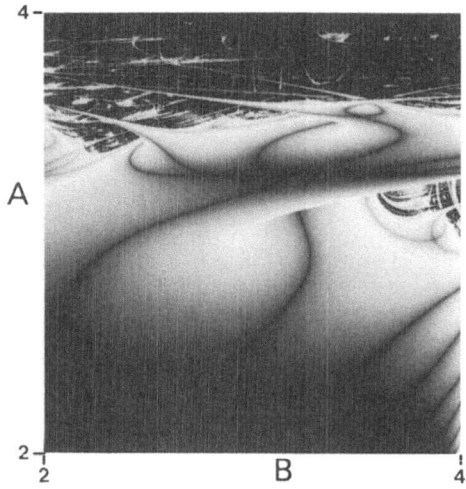

Fig. 4: Lyapunov exponent (given by different shadings of grey) from the logistic equation, as a function of A and B, for the periodic parameter sequence $S=(A^7BA^7BA^7B...)$

2.3. Superstable Curves

There is still another salient feature of the above-mentioned figures which deserves attention: the narrow dark grooves in the lighter regions, especially in Figs. 3 and 4. They can be readily related to superstability, which is what occurs when $\lambda \to -\infty$. We will not enter into details here, since these "backbones" are discussed at length by CHANG et al. [6]. We just mention that our figures constitute an excellent illustration of the different types of intersections (both coupled and decoupled dynamically) of the superstable trajectories described in [6]. Of particular interest is the intersection of many lines found in Fig. 4, which we found to be typical for long periodic cycles. An extensive analysis of such cycles will be published elsewhere.

Superstable lines are also present in Fig. 2, albeit much more diffuse than for the non-Markovian results of Figs. 3a and 4. Their origin can be traced to cyclic sequences, within the random process, of sufficient length. But, they are much less likely than short cycles and thus their tenuous appearance.

3. Early Chaos

It was already mentioned that chaos can manifest itself for values of A and B both well below the critical value $r_o=3.569945672...$ For the period 8 cycle defined by $S=(A^7BA^7B...)$ the situation is well illustrated with Fig. 4. However, similar behavior also occurs in Markovian random process, as illustrated in Fig. 2. In fact, Fig. 2 shows two black "bays" with early chaos. In one of these chaotic bays one obtains chaos, for example, at A=3.54 and $2.96 \leqslant B \leqslant 3.20$. This chaotic region ($\lambda>0$) is also seen in Fig. 5 where we display the Lyapunov exponent λ versus B for fixed A=3.54. Note that the case of Fig. 1 lays in this bay. In the other bay with early chaos in Fig. 2, it does appear e.g. for A=3.56 and $3.46 \leqslant B \leqslant 3.51$.

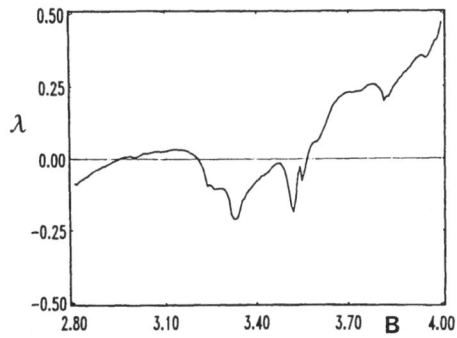

Fig. 5: Lyapunov exponent versus B, for fixed A=3.54 (random parameter sequence with γ_1=0)

3.1. Early Chaos: Random Perturbation Case

To understand the appearance of early chaos for A,B< r_o it is of paramount importance to realize that the functional composition implicit in Eq. (1) leads to a completely new dynamics. Consequently, there is no good reason to expect that the Lyapunov exponent λ of the composite process could be obtained as some kind of average of the exponents $\lambda(A)$ and $\lambda(B)$, associated with the applications f_A and f_B, respectively. (For example, $\lambda(A=3.1225)\cong-0.35$, $\lambda(B=3.54)=-0.0244$, but for the composite Markov process with p_A=0.5=p_B and γ_1=0, one obtains γ=+0.03).

When we re-examine the distribution function D(X) of Fig. 1, evaluated for A=3.52 and B=3.1225, we do observe that it has no relation whatsoever with the "pure" A or B results. The latter consist in four and two Dirac delta functions, respectively. Moreover, these fixed points do not even show up as strong peaks of D(X). In fact, D(X) is a distribution which although irregular, covers densely the interval $[f_B(A/4),A/4]$. Important weight is found in regions where $|\partial_x f_\alpha|>1$; in particular at the edge A/4=f_A(0.5), where $|\partial_x f_\alpha|$ reaches its maximum. According to Eq. (6b) these peaks provide important contributions towards a positive λ. Conversely, a peak in the neighborhood of X=0.5, where $|\partial_x f_\alpha|=0$, yields a negative contribution.

It should be stressed that a peak at <u>exactly</u> X=0.5 corresponds to a superstable situation, in which case the multiple images of X=0.5 are precisely X=0.5 itself. As mentioned above and in Ref. [6] these superstable points originate lines in the AB-plane, each of them being associated to a particular cycle. It is thus natural to expect that the chaotic behavior will be quenched in the vicinity of these lines, which is well illustrated by Fig. 2. In fact, the super-stable "backbones" associated with f_{AA}(0.5)=0.5, f_{AB}(0.5)=0.5 and f_{AAAA}(0.5)=0.5, ...generate deep needles of order which perforate the chaotic region. For example, for A=3.54 there is a λ<0 gap between 3.207<B<3.567, as seen more clearly in Fig. 5.

In summary, it is to be expected that chaotic behavior (λ>0) will occur for moderately large A and B values, due to the peaks of D(X) in regions with $|\partial_x f_\alpha|>1$; in particular, in relation with the peaks located at f_A(0.5)=A/4 and f_B(0.5)=B/4. However, the latter does not hold when a short periodic cycle is near the condition of superstability.

3.2. Early Chaos: Non-Markovian Sequences

Let us now consider repeated cycles of the $A^M B$ type, for which chaos is manifest for A and B values well below the critical $r_o = 3.56994...$ The causal character of these sequences will allow a more detailed and precise discussion than in the preceding random case.

For $3.0 < A < 3.4495...$ the iterative application $f_A(f_A(....f_A(X)))\equiv f_A^{(M)}(X)$ leads (see [7]) to the well-known stable bi-cycle $\{X_-(A), X_+(A)\}$, where

$$X_\pm (A) = [1+A\pm \sqrt{(1-A)^2-4} \,]/2A. \tag{7}$$

If λ is evaluated when the bi-cycle has reached stability one obtains negative values, due precisely to the stability of the process. But if instead of taking $M \to \infty$ we periodically intercalate an f_B application, i.e. $S=\{A^M B A^M B A^M B...A^M B\}$, where B is adjusted to satisfy

$$f_B(X_+(A)) \cong X^*(A) = 1-\frac{1}{A}. \tag{8}$$

Here, X^* is the (unstable, A>3) fixed point of f_A, i.e. $f_A(X^*)=X^*$. For M reasonably large and for a suitable seed X_o, the application f_B will find the system around $X_t \cong X_+(A)$. Thus, according to (8)

$$f_B(X_t) = X_{t+1} \cong X^*(A). \tag{9}$$

Next, between (t+2) and (t+M+1), f_A is repeatedly applied. The system goes through a transient, oscillating slightly around X^* at first, which yields positive contributions to λ, since X^* is unstable for f_A. However, little by little the system leaves the vicinity of X^* and at time (t+M+1) it already satisfies $X_{t+M+1} \cong X_+(A)$ (the last stages in the approach $X_t \to X_+(A)$ may occur quite rapidly if $A \cong 3.23606...$, the superstable value). To effectively reach the $X_+(A)$ branch, X_{t+1} has to enter this process with the right "phase": that is, for M even (odd) $X_{t+1} > X^*$ ($X_{t+1} < X^*$); these requirements impose restrictions on the seed X_o. An "adequate" choice of the A and B values allows the system to remain for a long time in the vicinity of the unstable point, yielding positive contributions to λ. In this way, we can ascribe the chaotic character of the $A^M B$ cycle to a situation of "eternal transients". The role of the f_B application is simply to reset the transient when the system is nearing stability.

A similar chaotic regime, related to $f_B(X_-(A)) \cong X^*$, does also exist. However, it implies much larger values of A and B.

The ordering and disordering processes in non-markovian systems are susceptible to experimental observation through measurements on physical systems, like electronic circuits subject to periodic pulses. The most likely candidate for experimental verification seems to be a Josephson junction, as implemented by OCTAVIO et al. [10].

4. The Meaning of Chaos for a Random Process

The definition of the Lyapunov exponent λ, given in section 2, is related to the sensitivity of state X_t to a change in the seed X_o, when a _fixed_ sequence S is considered. However, an ecologist for example, does not have a priori knowledge of the sequence S which is generated by a changing environment. So, his ability

a question of purely academic interest with no practical implica-
e words chaotic or non-chaotic would not make sense, when applied
/ironments.

ere is an alternative interpretation of Lyapunov exponent, suitable
il test: assume that several replicas of the system under study are
1 different initial states $\{X_{0,\nu}|\nu=1,2\ldots L\}$, where ν is the replica
them evolving in time according to the same sequence S. One could
example, the evolution of a large number of similar independent
ibject to the <u>same variation</u> of ambient conditions, as represented

chaotic process it is plausible to expect that all systems with
e initial conditions will converge to the same final state. Con-
chaotic process, with $\lambda>0$, one would expect that systems starting
iitial conditions will reach very different final states.

ı put these ideas on a more quantitative basis we have carried out
ilations of the time evolution of L systems, with the same sequence
g from L different seeds: $X_{0,\nu} \rightarrow X_{t,\nu}$. In the end, we evaluated the
$_{,\alpha}(X)$ associated with the set $\{X_{t,\nu}\}$, where $\alpha=C_{t-1}=A,B$. This $D_{t,\alpha}(X)$
asted with $D_\alpha(X)$ of Eq. (4a), obtained with fixed initial condi-
ie average.

:matic discussion of the above-mentioned calculations it is conve-
fy the random processes into two categories: a) For some values of
he support of $D_\alpha(X)$ is compact (i.e. a support without gaps); and
alues of A and B the support has 2, 3 or more disconnected bands,
, with $\Gamma_i \cap \Gamma_j = \phi$ if $i \neq j$. Also, they do alternate in time, i.e. if
$_2 \epsilon \Gamma_2, \ldots X_{t+s} \epsilon \Gamma_s$. This situation arises when A = B and if both A and
correspond to cycles of period s in the pure logistic map; in
ey correspond to widening the s fixed points of the application
example we found that two bands do exist when the following in-
satisfied:

$$B_{cr}(A) = \frac{27}{[9A+2A^2\{(1-\frac{3}{A})^{3/2}-1\}]} < B < A < 1+\sqrt{6}=3.449. \qquad (10)$$

the same as above with $A \rightleftarrows B$.

some results obtained through the numerical calculations:

ibution D(X) has a single band, then the new distribution $D_t(X)$,
ing over initial conditions, is <u>independent</u> of the choice of the
s holds practically always, as long as the seeds are not manipu-
udiced way.

gle band regime, if $\lambda<0$ and once transients have died out, the
(X) is

$$D_t(X) = \delta[X-X(S_t)], \qquad (11)$$

$\ldots C_{t-1}$). If instead $\lambda>0$ the distribution has a certain width.

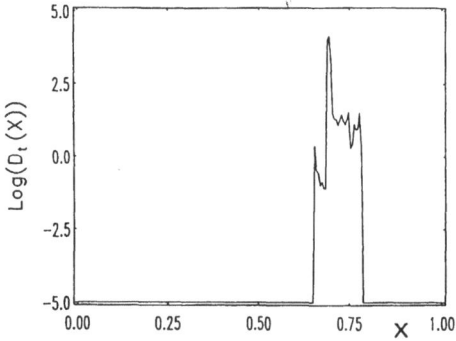

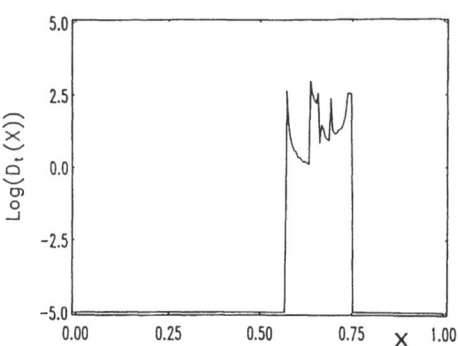

Fig. 6a: Distribution function $D_t(X)$, obtained through a seed average, at time t=3000. The last application was f_B and the values of A=3.54, B=3.1225 and γ_1=0 are the same as in Fig. 1

Fig. 6b: Same as Fig. 6a, but at time t=3500 and last application was f_B

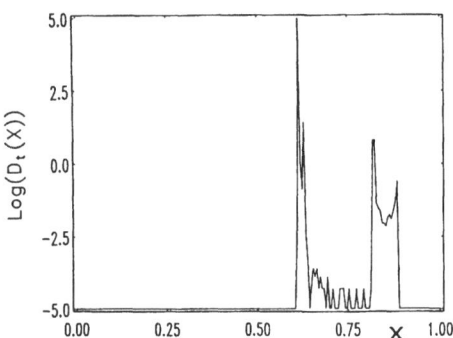

Fig. 6c: Same as Fig. 6a, but at time t=4000 and last application was f_A

iii) We also have compared, in the chaotic single band regime, the distributions $D_\alpha(X)$ and $D_{t,\alpha}(X)$, to find that they do not coincide with each other. Indeed, $D_{t,\alpha}(X)$ does depend on t; in addition we notice that it is much more concentrated around a few peaks than $D_\alpha(X)$; this is particularly notorious for early chaos.

In Figs. 6a-6c we display the logarithm of the distribution $\log[D_t(X)]$ for A=3.54 and B=3.1225, a set quite typical for early chaos. It is important to keep in mind that this is a semi-log plot, so that the actual peaks appear very severely reduced. Even so it is quite apparent that the distribution, which looks somewhat diffuse for t=3000 in Fig. 6a, becomes even wider after another 500 time units in Fig. 6b, to sharpen quite drastically around a few peaks after a total of 4000 time units, in Fig. 6c. (Recall it is a semi-log plot).

At this point a comparison of Figs. 6a-6c with Fig. 1, which was plotted for the same AB-pair of values, is in order. Clearly Fig. 1, which depicts a time average, is much more diffuse than any of the seed averages.

We have also quantified the "pathological" character of the distribution $D_t(X)$, evaluating some of the generalized dimensions defined by GRASSBERGER [9].

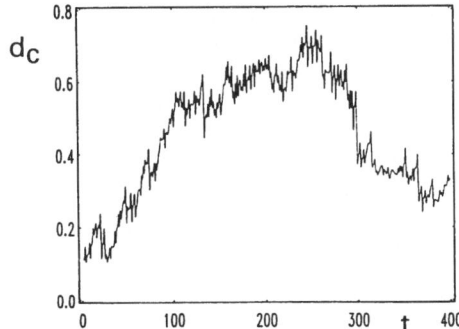
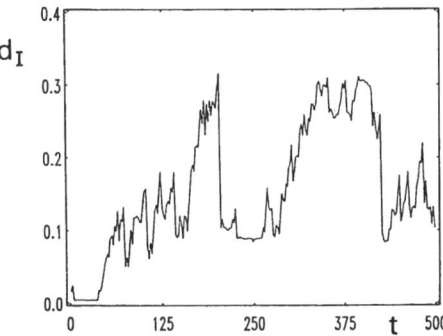

Fig. 7: Correlation dimension for $\gamma_1=0$, $\epsilon=10^{-6}$, A=3.54, B=2.97

Fig. 8: Information dimension for the same conditions as Fig. 7

For example, for the pair A=3.54 and B=2.97 which is at the edge of early chaos, we calculated the time dependence of the correlation dimension d_c of the distribution $D_t(X)$. This time dependence is shown in Fig. 7. d_c fluctuates between 0.1 and 0.8, while the d_c of D(X), as defined by Eq. (4), is nearly equal to 1. To give an intuitive feeling to these values of d_c, we mention that in our computations of $D_t(X)$ carried out with 2000 seeds we found that the number of pairs satisfying $|X_{t\mu}-X_{t\nu}|<8\times10^{-7}$ fluctuates between the orders of 10^2 and 10^6. Instead, for the distribution D(X) we only obtained around 10 pairs satisfying the same requirement.

In Fig. 8 we display the time variation of the information dimension d_I as defined in Ref. [9], which shows notorious variations in the degree of order of the distribution $D_t(X)$. When the limit A→B is taken the distributions $D_{t,\alpha}(X)$ and $D_\alpha(X)$, become more similar to each other, as well as the associated generalized dimensions. On the other hand, when strongly chaotic situations are investigated, the generalized dimensions related to $D_t(X)$ do increase.

iv) When $D_\alpha(X)$ is split into several bands $\Gamma_1,\Gamma_2...\Gamma_s$, the distribution $D_{t,\alpha}(X)$ obviously also appears split. For $D_{t,\alpha}(X)$ to be independent of the choice of seeds, it is necessary to normalize the distribution of each band Γ_j with the number of seeds associated to the attraction basin of Γ_j, at time t. That is, with the number of $X_{0,\nu}$ which flowed into Γ_j, at time t. With this normalization all what was said for the single band situation also applies here.

In summary, if the sequence $S=\{C_t\}$ is unknown (e.g. changing environmental conditions), but there is a set of system replicas all subject to the same succession of random events, then we can state:

a) If the dynamic process is non-chaotic the different replicas converge to a few final states. There are as many final states as bands of $D_\alpha(X)$. While we are unable to predict the precise position of these states within the different bands, we can evaluate the relative distance between replicas belonging to the same band. This distance vanishes as t→∞.

b) If the dynamic process is chaotic the final relative distance between different replicas cannot be obtained. However, if $0\leqslant\lambda<<1$ (weak chaos) all replicas belonging to the same band will be nearly in the same state. As chaos becomes stronger, and the entropy of the distribution $D_t(X)$ increases, our <u>a priori knowledge</u> about the relative distance between states does decrease.

Thus, knowledge about the attraction basin of each of the bands of D(X) plus the value of λ gives us some information on the relative distance of the final states, even when we ignore <u>a priori</u> the random sequence S which describes the changing environment.

Acknowledgements

We thank Mr. José Rogan and Mr. Jürgen Marsch for help with the computation and illustrations and Mrs. Ch. Riemer for efficient typing of the manuscript.

Jaime Rössler acknowledges the support of DIB and Fondecyt, Miguel Kiwi the support of DIUC and Fondecyt.

Mario Markus thanks Professor Benno Hess for his support and guidance over the years, and dedicates this contribution to him on occasion of his 65th birthday.

References

1. M. Markus, B. Hess, J. Rössler, M. Kiwi: In <u>Chaos in Biological Systems</u>, ed. H. Degn, A.V. Holden and L.F. Olsen (Plenum Publ. Co., N.Y. 1987) pp. 267-277
2. S.D. Tulyapurkar, S.H. Orzack: Theor. Pop. Biol. <u>18</u>, 314 (1980)
   S.D. Tulyapurkar: Theor. Pop. Biol. <u>21</u>, 114 (1982)
3. J. Cohen: Advan. Appl. Prob. <u>9</u>, 18 (1977)
4. J. Rössler, G. Martinez, M. Kiwi: Solid State Comm. <u>61</u>, 395 (1987)
5. Y. Liu, K.A. Chao: Phys. Rev. <u>B33</u>, 1010 (1986)
6. S.J. Chang, M. Wortis, J.A. Wright: Phys. Rev. <u>A24</u>, 2669 (1981)
7. J.P. Crutchfield, J.D. Farmer, B. Huberman: Phys. Repts. <u>92</u>, 45 (1982)
8. J. Rössler, E. Lazo: J. Phys. <u>C14</u>, 3499 (1981)
   J.M. Cowley: Phys. Rev. <u>77</u>, 669 (1950)
9. P. Grassberger: Phys. Lett. <u>97A</u>, 227 (1983)
10. M. Octavio: Phys. Rev. <u>B29</u>, 1231 (1984)
    M. Octavio, C.R. Nasser: Phys. Rev. <u>B30</u>, 1586 (1984)

Periodic and Chaotic Dynamics
in Childhood Infections

W.M. Schaffer, L.F. Olsen, G.L. Truty, S.L. Fulmer, and D.J. Graser

Department of Ecology and Evolutionary Biology,
The University of Arizona, Tucson, AZ 85721, USA

1. Introduction

Most biologists, when confronted with a time series, assume that the dynamical possibilities are quite limited. Either a system sits still, presumably at equilibrium, or it may oscillate with a fixed period. Anything else, they will probably tell you, is evidence of noise - observational error or chance perturbations from without. By these criteria, most biological systems, especially at the population level, are extremely noisy. Hence, it is no accident that mathematical biology places a heavy emphasis on stochastic models as well as on statistical techniques designed to extract the "deterministic" component of the signal.

Such attitudes are reflected in a recent study by ANDERSON et al. [1] of childhood epidemics. Prior to the computation of smoothed power spectra, the data were successively log-transformed, mean-corrected, detrended, and tapered. In essence, the authors went to great lengths to make the data better fit a simple dynamical model, namely, the output of a series of coupled linear oscillators. Lest the reader mistake our intent, we hasten to add that such manipulations are standard components of traditional time series analysis.

Our own approach [2-4] to this problem has been somewhat different. On the one hand, we recognize the value of looking for periodicities in the data and have done so ourselves (see below). At the same time, we also believe it important to consider a broader range of possible dynamics. This includes various kinds of chaos, in which the irregularities are inherent, rather than the result of chance vagaries. In addition, we believe that any claim to an understanding of the phenomena requires that one be able to write down biologically reasonable equations which, for suitable parameter values, generate behavior reminiscent of what one sees in nature.

2. SEIR Models

One way of studying the dynamics of childhood diseases is embodied in so-called SEIR models [5]. Here, the host population is divided into categories: Susceptible, Exposed, but not yet infectious, Infectious and Recovered and immune. Generally speaking, individuals enter the susceptible category at birth. Following exposure to the pathogen, they move progressively through the other categories, i.e., in terms of the schematics in Figure 1, from left to right. In diseases such as measles and chickenpox, immunity is permanent. When an individual enters the recovered state, he or she stays there. In other diseases, immunity is transient, and there will be transitions from R back to S. Finally, note that all categories can be exited by dying.

The SEIR scheme is very general and can be implemented in various ways. The simplest approach assumes that the population size is constant and neglects the complicating effects of age-structure and spatial distribution. Then one has the following differential equations:

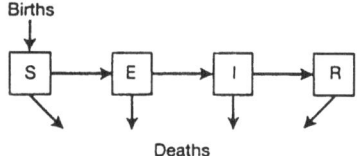

Births

Deaths

Fig. 1: Flow diagram for the SEIR model. Individuals enter the susceptible class, S, at birth, and thereafter the exposed, E, infective, I, and recovered, R, classes by contracting the disease, becoming infectious and recovering. In diseases such as measles, immunity is permanent. Hence, there is no feedback from R to S

$$dS(t)/dt = m [1-S(t)] - b S(t) I(t)$$

$$dE(t)/dt = b S(t) I(t) - (m+a) E(t) \qquad (1)$$

$$dI(t)/dt = a E(t) - (m+g) I(t) \quad .$$

Of the four parameters, m, a and g, can be measured directly. Specifically,

$1/m$ is the average life expectancy;

$1/a$ is the average latency period;

$1/g$ is the average infectious period.

For childhood diseases in first world countries, $1/m$ is on the order of 10^2; $1/a \sim 10^{-1}$, and $1/g \sim 10^{-2}$ [6]. In contrast, the contact rate, b, (average number of susceptibles contacted yearly per infective) must be estimated from the other parameters and the average age of infection [7]. Typically, b ranges from 10^2 to 10^3.

As written above, the SEIR equations have trivial dynamics. Either the disease dies out, or it persists at some constant level. Adding seasonal variation in contact rates changes things considerably. Note that this modification is not simply a bit of mathematical whimsy. Diseases such as measles and chickenpox strike principally at school children, and schools, whatever else they may be, are incubators of infection. Indeed, detailed studies [8] of measles' incidence in England suggest that one can see not only the effect of summer vacation, but also of Christmas recess.

In this spirit, several authors [9-11] have investigated the effects of replacing constant contact rates with the function

$$b(t) = b_0 (1 + b_1 \cos 2\pi t) \quad . \qquad (2)$$

Here, b_1 indexes the degree of seasonality. On increasing this parameter, one observes what may be called the main period-doubling route to chaos. Here, the initial state is a simple periodic orbit corresponding to a yearly cycle in prevalence.

Coexisting with the main set of period-doubling are other bifurcation sequences. Typically, these begin with an n-cycle, i.e. the time-one or Poincaré map consists of n points, and lead to an n-piece chaotic attractor. In general, the amplitude of the fluctuations associated with these subsidiary attractors is much larger than those on the main route. Some bifurcation diagrams (for a more realistic contact function) are given by KOT et al. [12].

Figures 2 and 3 give examples of the dynamics along the main period-doubling sequence. In particular, we contrast a chaotic orbit with a simple limit cycle in

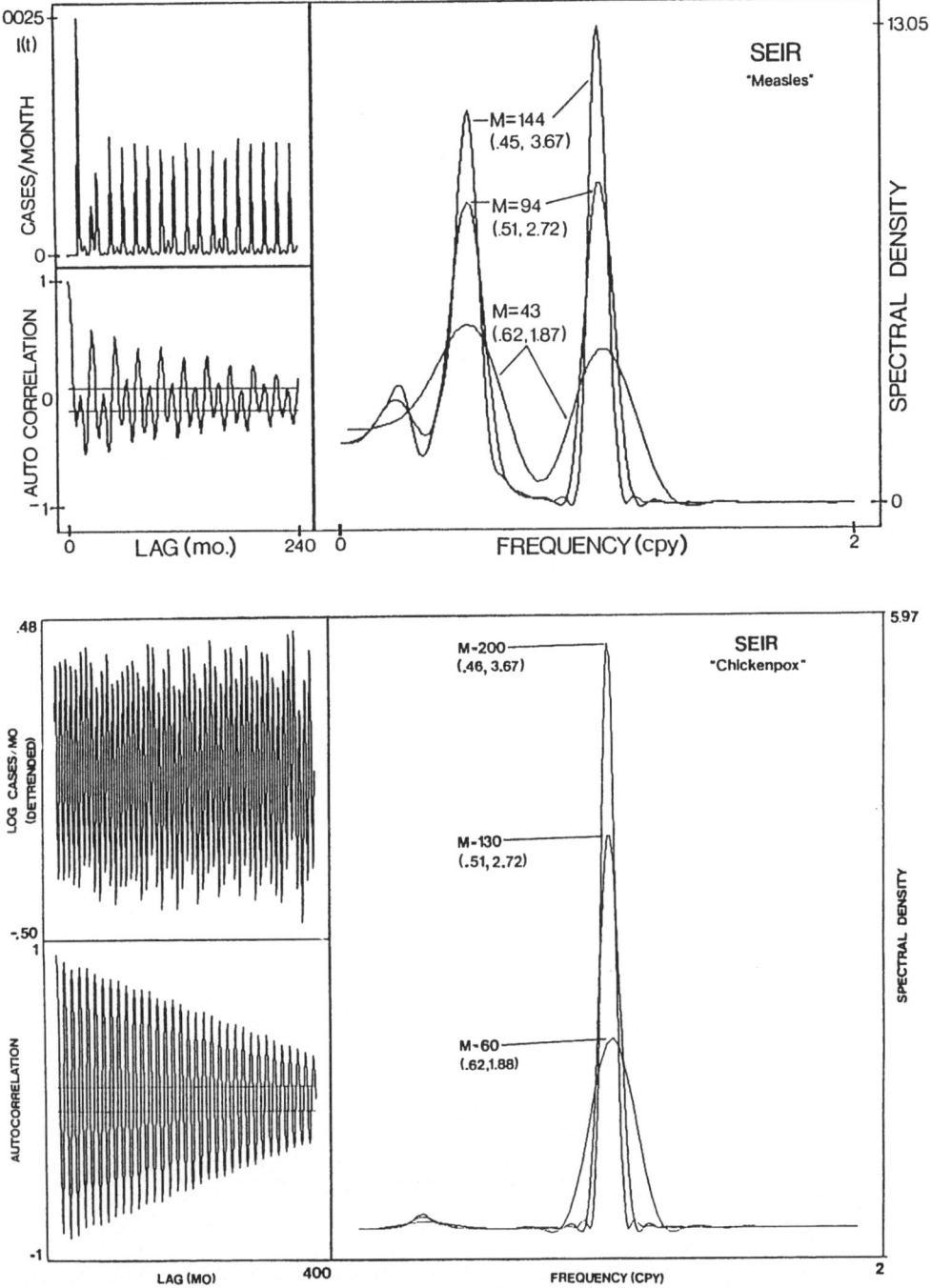

Fig. 2: Spectral analysis of SEIR trajectories along the main period-doubling sequence. Top: A chaotic trajectory. m = 0.02; a = 35.84; g = 100; b_0 = 1800; b_1 = 0.28; Bottom: A periodic orbit with noise. m = 0.02; a = 36.0; g = 34.3; b_0^1 = 308.7; b_1 = 0.28

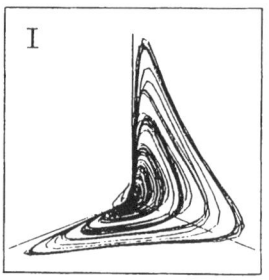

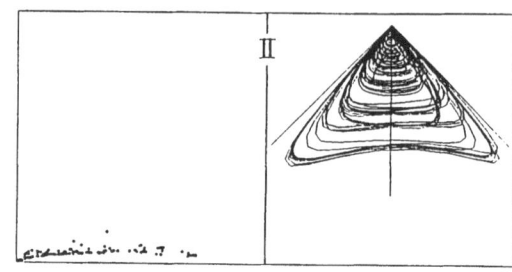

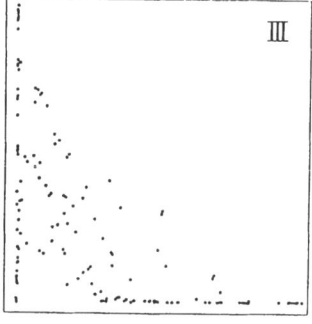

Fig. 3a: Nonlinear analysis of a chaotic SEIR trajectory. I: Orbit reconstructed from the number of infectives I(t), and embedded in three dimensions. II: Orbit viewed from above and sliced (right); Poincaré section (left). III: "One-dimensional" map computed from the Poincaré section

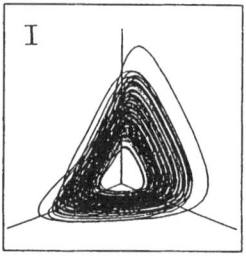

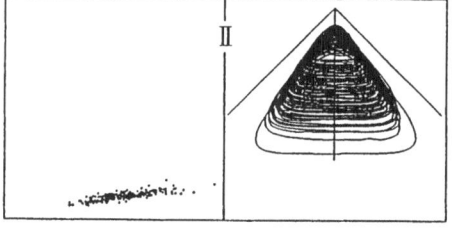

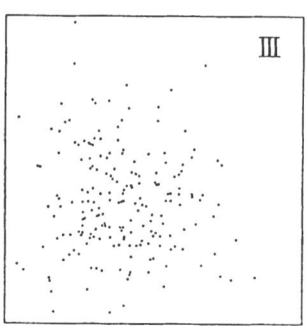

Fig. 3b: Nonlinear analysis of a periodic SEIR trajectory. Two per cent uniform noise has been added to the simulation

the presence of noise. The figures are labeled "Measles" and "Chickenpox" because the parameters used to generate the data correspond roughly to what one expects for these diseases.

Figure 2 shows standard spectral analyses: transformed and detrended time series (upper left), autocorrelation function (lower left) and smoothed power spectra (main part of picture).

In Figure 3, we present nonlinear analyses. Here, the data are embedded in three dimensions (I) as prescribed by TAKENS [13]. The resulting orbit is then sliced (II, right) thereby yielding a Poincaré section (II, left). From the latter, we attempt to construct a one-dimensional map. Points on the section are projected to a regression. Each point is assigned a number, X_i, indicating its distance along the curve, and the X's then plotted in temporal sequence, that is, X_{i+1} vs. X_i, for all i (III).

Summarizing these computations, we remark that the "chickenpox" simulation gives all the hallmarks of a noisy limit cycle, which it is. The only spectral feature is the yearly cycle; the reconstructed orbit suggests a noisy closed orbit; and attempts to extract a one-dimensional map yield an essentially random scatter of points. By contrast, the "measles" data are far more erratic. In addition to the yearly peak, which may be more or less pronounced [14], the power spectrum has a prominent peak in the low frequency range. Moreover, the reconstructed trajectory has obvious structure, being somewhat reminiscent of, but more complicated than the simple RÖSSLER [15] attractor. The complications are reflected by the Lyapunov dimension, which is about 2.5 [2], and the form of the "one-dimensional" map. For short time series, the latter construction suggests a noisy one-humped curve, but for longer runs, it is apparent that its structure is actually more complex.

3. Analysis of Childhood Infections

In all, we have now studied 17 data sets for four childhood diseases - chickenpox, measles, mumps and rubella - in a variety of first world cities. Figures 4 and 5 give some examples, Here, monthly case reports are analyzed as in Figures 2 and 3. What can we conclude from all these pictures? The following remarks appear reasonable:

a) Time Series Analysis. All of the diseases fluctuate in incidence. For most, there is a seasonal component which is reflected by a prominent spectral peak at 1 cycle per year (cpy). In some cases, this is the only noteworthy feature. In others, there are other peaks corresponding to long term fluctuations superposed on the yearly cycle. Sometimes, these low frequency oscillations contribute little to the total variance. In other cases, they rival the yearly cycle. Finally, there are a few instances in which the long term fluctuations dominate.

Grouping the data by disease, it seems that chickenpox always gives a noisy limit cycle with a period of one year. The other diseases are more variable. Generally speaking, the secondary peaks decline in frequency as one goes from measles through mumps to rubella. That is, the long-term cycles in measles have periods of 2-3 years. In mumps, the corresponding figure is 3-4 years, and in rubella, about 5-7 years. Table 1 summarizes the results of spectral analysis for all diseases and cities.

In terms of the SEIR equations, we can say that whereas the chickenpox data are certainly consistent, the measles data generally have too much power associated with the yearly cycle. That is, while the chaotic SEIR equations can exhibit both low and higher (1 cpy) frequency peaks, it is uncommon for the latter to dominate. Whether this reflects a fundamental flaw in the model, or simply indicates the need for a more realistic forcing function remains to be determined.

b) Nonlinear Analysis. Initially [3], it was suggested that the New York and late (1928-63) Baltimore measles data were consistent with the sort of chaos associated with essentially two-dimensional attractors such as the RÖSSLER attractor [15]. In fact, this interpretation is probably in error [2]. A more reasonable assertion is that measles dynamics correspond more closely to those of the SEIR model in the chaotic region. OLSEN's [4] studies of childhood diseases in Copenhagen, together with the data presented here, would seem to substantiate

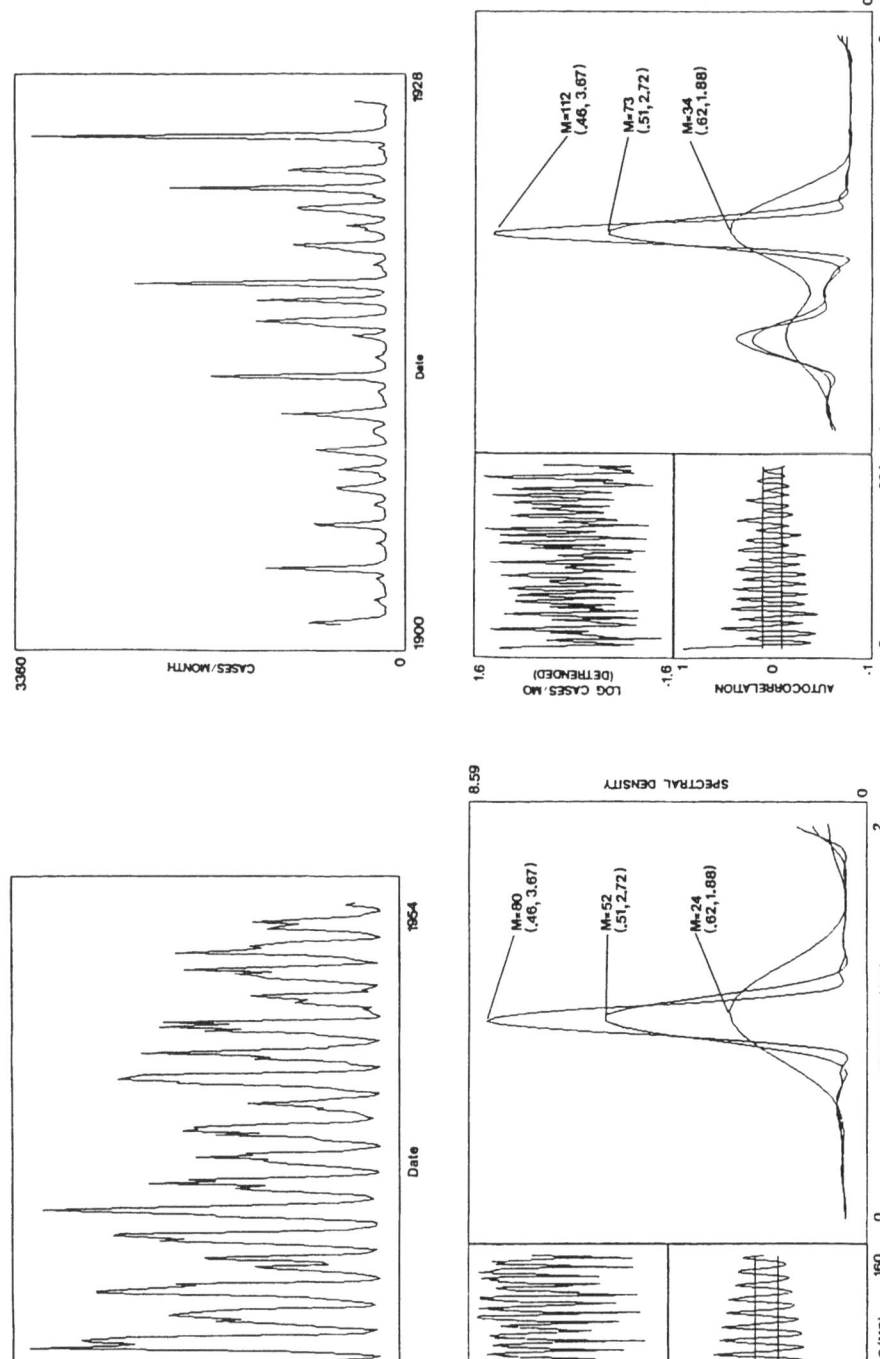

Fig. 4a: Time series and spectral analysis of chickenpox epidemics in St. Louis, 1934-1953

Fig. 4b: Time series and spectral analysis of measles epidemics in the city of Baltimore, 1900-1927

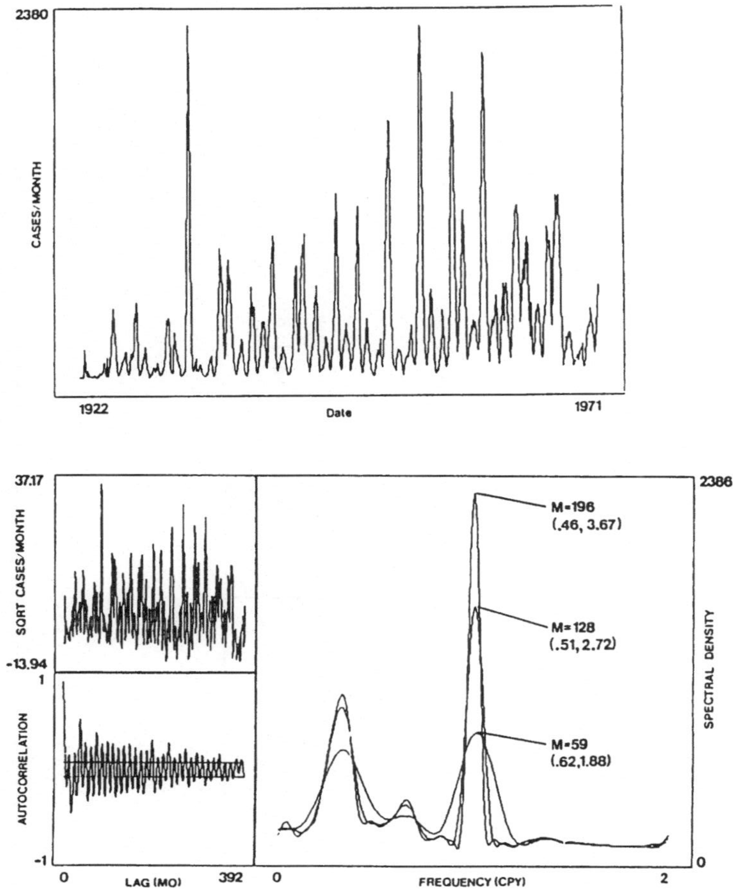

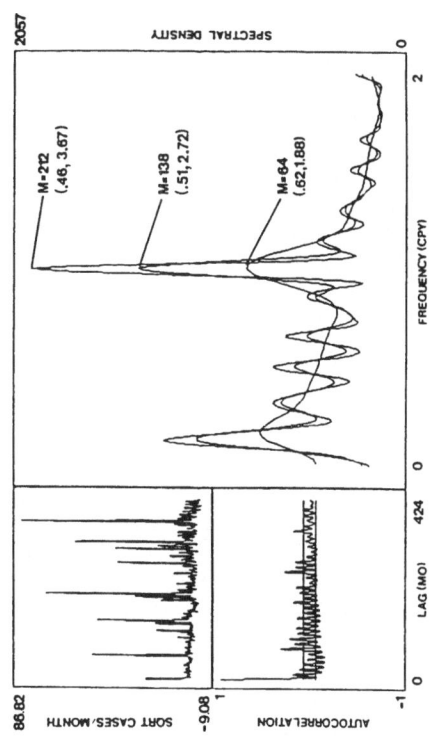

Fig. 4c: Time series and spectral analysis of mumps
epidemics in Milwaukee, 1922-1970

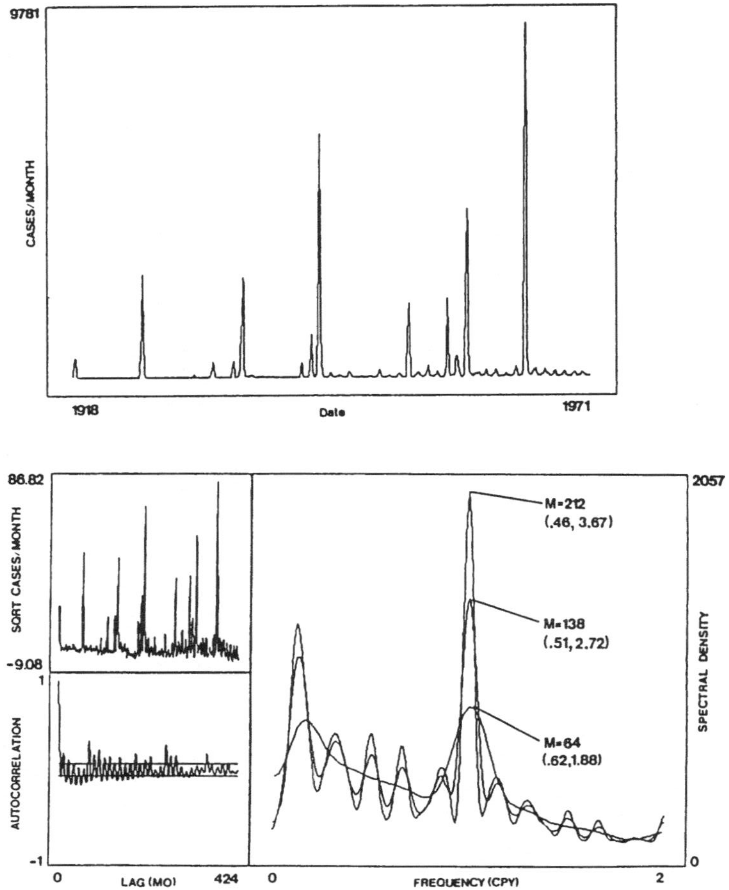

Fig. 4d: Time series and spectral analysis of rubella
epidemics in Milwaukee, 1918-1970

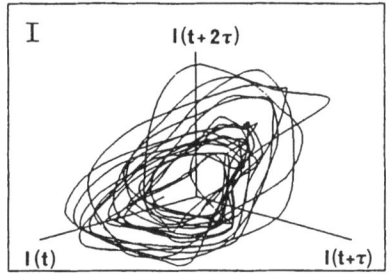

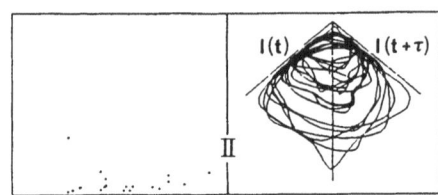

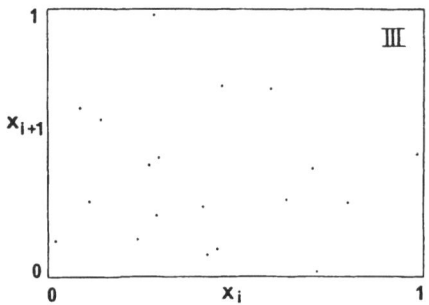

Fig. 5a: Nonlinear analysis of chickenpox epidemics in St. Louis, 1934-1953. Monthly case reports were subjected to three-point smoothing and interpolated with cubic splines. Sequence of diagrams as in Fig. 3

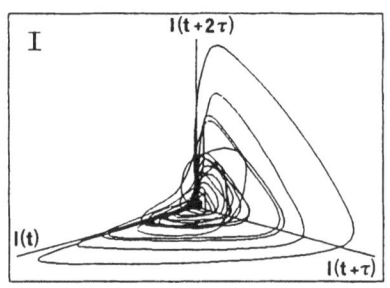

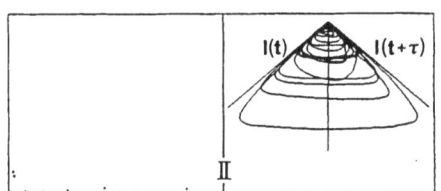

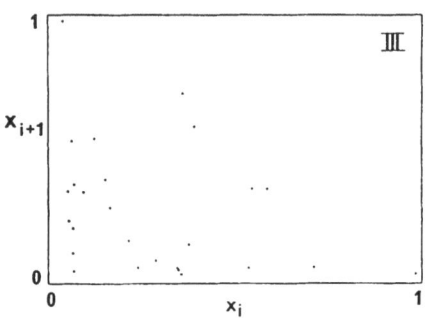

Fig. 5b: Nonlinear analysis of measles epidemics in Baltimore, 1900-1927

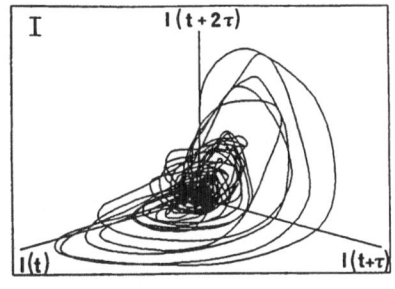

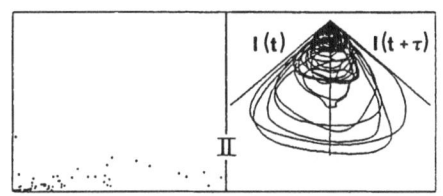

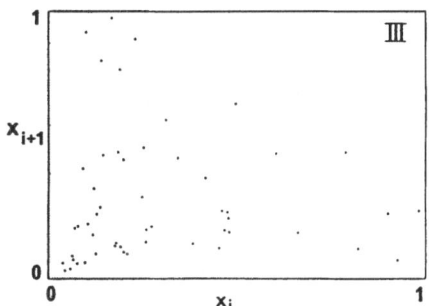

Fig. 5c: Nonlinear analysis of mumps epidemics in Milwaukee, 1922-1970

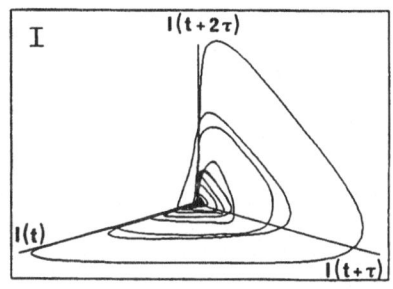

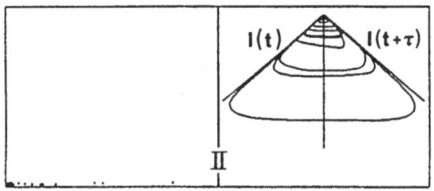

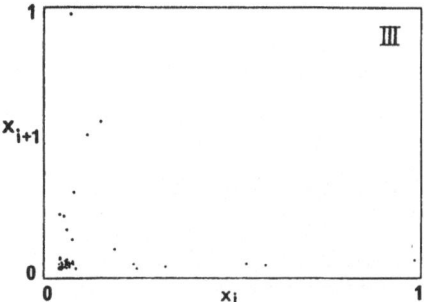

Fig. 5d: Nonlinear analysis of rubella epidemics in Milwaukee, 1918-1970

Table 1: Spectral Analysis of Childhood Diseases

CITY	DATES	Spectral peaks (cpy)		
CHICKENPOX				
Copenhagen	1938 - 1967	-	-	1.00*
Milwaukee	1916 - 1965	-	-	0.98*
New York	1928 - 1963	-	-	0.99*
St. Louis	1934 - 1953	-	-	0.98*
MEASLES				
Aberdeen	1883 - 1902	-	0.46*	0.89
Baltimore	1900 - 1927	-	0.45	0.99*
Baltimore County	1928 - 1963	-	0.40*	0.99
Copenhagen	1927 - 1966	-	0.40*	0.99
Milwaukee	1916 - 1965	-	0.40	1.00*
New York	1928 - 1963	-	0.41	1.01*
St. Louis	1934 - 1954	-	0.34	1.01*
MUMPS				
Copenhagen	1927 - 1966	-	0.25*	1.00
Milwaukee	1922 - 1970	-	0.32	1.00*
New York	1928 - 1963	-	0.37	0.99*
RUBELLA				
Copenhagen	1938 - 1967	-	0.21*	1.00*
Milwaukee	1918 - 1970	0.14	-	1.00*
St. Louis	1934 - 1953	0.14	0.49	1.00*

* Asterisks denote major spectral features.

this view. In fact, if one looks at all of the maps (Fig. 6) that have been extracted from measles time series, the correspondence with the model is really quite striking. Further evidence (correlation dimension, Lyapunov exponents), favoring this interpretation will be presented below.

Similarly, actual epidemics of chickenpox generate orbits and maps suggestive of the SEIR model in the limit cycle region - provided the simulation is carried out in the presence of small amounts of noise. Thus two of the diseases exhibit dynamics which are consistent both among cities and with the available theory.

Mumps and rubella, on the other hand, are more problematic. In the first case, the data are extremely variable. In New York, mumps epidemics resemble the chickenpox pattern; in Copenhagen, measles; and in Milwaukee, something intermediate. Interestingly, the SEIR model (with modified forcing function) when run for parameters appropriate for mumps, has at least three coexisting attractors (one chaotic, two periodic) [12].

Finally, rubella seems to be quite different entirely. In particular, note the clustering of map points near the origin.

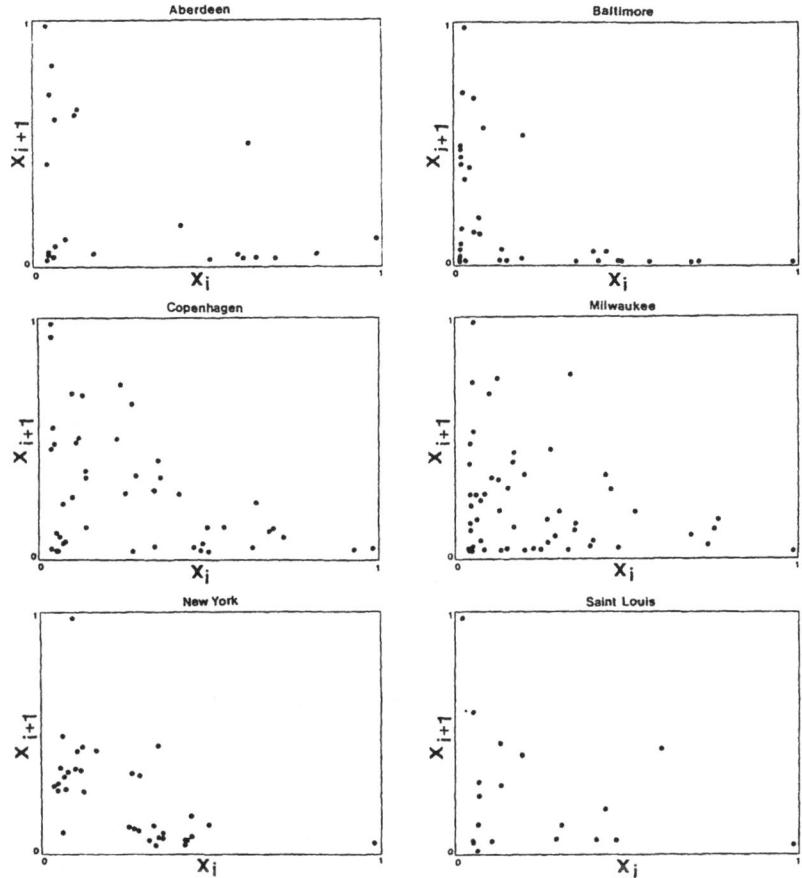

Fig. 6: "One-dimensional" maps extracted from epidemics of measles. Compare with
Figure 3a

4. Further Analysis

Because of sensitivity to initial conditions, the sequence of orbital excursions
of a chaotic time series may not be informative. Accordingly, one looks for statistical properties reflecting the overall organization of the flow. Quantities
that have attracted interest in this regard include Lyapunov exponents, metric
entropy and various fractal dimensions. In particular, a positive Lyapunov exponent is often taken as evidence of chaos, while an attractor's dimension is
thought in some sense to index its complexity. During the past several years,
there has been considerable progress in devising algorithms [16-18] for estimating these quantities from experimental data. Unfortunately, most of these methods
still require substantial amounts of data.

4.1. Correlation Dimension

Figure 7 illustrates the problems that one can encounter. Here we attempt to estimate the fractal dimension of measles epidemics in Aberdeen and Baltimore using
the algorithm developed by GRASSBERGER and PROCACCIA [16]. For each data set, we

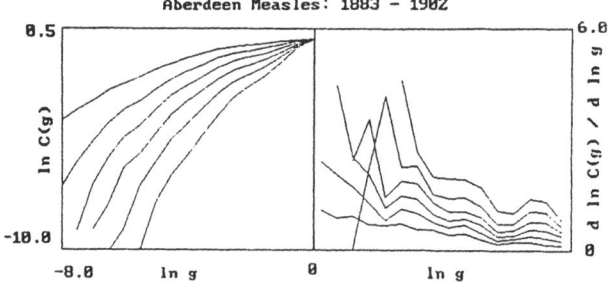

Aberdeen Measles: 1883 - 1902

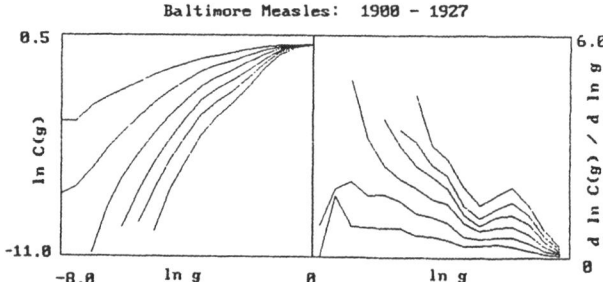

Baltimore Measles: 1900 - 1927

Fig. 7: Attempting to estimate fractal dimensions for childhood epidemics with the Grassberger-Procaccia method. Left: Correlation integrals for successively higher embedding dimensions (from 1 to 6) plotted for various length scales. Right: Slopes of the correlation integral

plot the logarithm of the correlation integral, ln C(g), against a series of length scales, and the slopes, (d ln C(g)/d ln g), vs. ln g (right). Theory suggests that for small length scales, and with infinite amounts of data,

$$\ln C(g) = n \ln g . \tag{3}$$

The important mathematical result is that $n \leq d_I$, where d_I is the information dimension.

When applying this procedure to experimental data, one successively embeds the time series in higher dimensions - and hopes that the exponent converges. We emphasize the word 'hopes', because it is generally true that the width of the scaling region, i.e., the range of g values for which   (3) applies, shrinks at higher embeddings. This problem is generally exacerbated when the data are highly non-uniform. That is, "spiky" data sets, such as we see both in the SEIR model and diseases such as measles and rubella, are more difficult to resolve. Thus, failure of a time series to yield a well-defined correlation dimension can mean either that the quantity does not exist or that there is not enough data to resolve it. Sadly, these are the alternatives suggested by Figure 7.

What to do? Several authors [19,20] have applied smoothing procedures, such as singular value decomposition, directly to the data. An alternative approach is suggested by traditional time series analysis. Here, one divides the data set into samples, computes the periodogram for each, and then averages over the different frequencies. Such procedures are motivated by the fact that the periodogram is not a consistent estimator, i.e., the variance does not tend to zero for large samples. In estimating correlation dimensions, we therefore proceed as follows:

a) For each data set, we compute time-one, i.e., first return, maps by lagging the data at yearly intervals. This is not only easy to do, it also makes mathematical sense since the systems are subject to periodic forcing. Since the data are collected monthly, we can compute 12 such maps for each time series. In principle, each of the maps should have the same dimension, Lyapunov exponents, etc. In practice, because of the non-uniform nature of the dynamics, we will get different estimates. It is this nonuniformity that we wish to smooth.

b) Having computed the maps, we apply the Grassberger-Procaccia method to each. For each length scale, we then average the slopes. To these we then add one to recover an estimate for the original time series.

The results of applying the modified method to the Aberdeen and Baltimore measles data are shown in Figure 8. We emphasize that for these data, one does not obtain a well-defined scaling region. (Nor is scaling obtained in simulations when the length of the data set is so short.) On the other hand, the slopes do not continue to rise as we increase the embedding dimension. Instead, there is rough, but clearly apparent, saturation with the maximum values for the time series approximating 2.4-2.6. Table 2 summarizes our results for all of the diseases and cities. In this regard, one may say the following:

a) The method does not create order out of nothing. That is, when applied to a set of random numbers, the data are space-filling or close to it.

b) The purportedly chaotic time series, measles, rubella and Copenhagen mumps yield dimension estimates of 2.0 to 3.1. Most of the values are closer to 2.5, which, as noted above, is the Lyapunov dimension for the SEIR model in the chaotic region [2].

c) For the purportedly noisy limit cycles, chickenpox, New York mumps and Milwaukee mumps, the estimates tend to be higher, generally in excess of three.

d) The difference, if it is real, accords with previous results [21] on the effects of adding noise to one-dimensional maps and simple flows. That is, limit cycles in the presence of noise yield higher estimated dimensions than nearby (in parameter space) chaotic orbits.

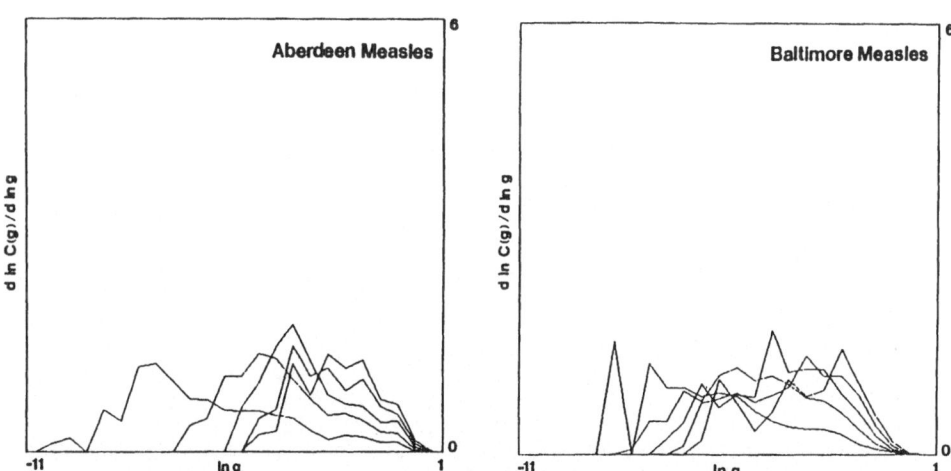

Fig. 8: Modified Grassberger-Procaccia technique. For each length scale, the slope of the correlation integral is averaged over the twelve time-one maps. Although there is no clearly defined scaling region, the maximum slopes saturate with increasing embedding dimension

Table 2: Dimensional Analysis of Poincaré Maps*

CITY	EMBEDDING DIMENSION				MEAN**
	2	3	4	5	
	CHICKENPOX				
Copenhagen	1.89	2.18	2.31	2.35	2.28
Milwaukee	1.93	1.94	2.33	2.47	2.25
New York	2.21	2.03	2.01	2.96	2.33
St. Louis	1.37	1.88	1.92	1.49	1.76
	MEASLES				
Aberdeen	1.34	1.60	1.71	1.37	1.56
Baltimore (1900-27)	1.17	1.19	1.28	1.54	1.34
Baltimore (1928-63)	1.10	1.20	1.27	1.78	1.42
Copenhagen	1.72	1.96	1.93	2.32	2.07
Milwaukee	1.60	1.71	1.30	1.69	1.57
New York	1.23	1.55	1.75	1.76	1.68
St. Louis	1.08	1.39	1.14	1.14	1.22
	MUMPS				
Copenhagen	2.29	1.87	2.14	1.88	1.96
Milwaukee	1.98	1.92	2.59	2.07	2.19
New York	1.99	2.23	2.45	2.81	2.50
	RUBELLA				
Copenhagen	1.81	2.20	1.86	1.82	1.96
Milwaukee	1.88	2.01	1.68	1.50	1.73
St. Louis	2.21	1.45	1.62	1.15	1.41
	NOISE				
Uniform	2.41	2.63	3.51	4.03	-
Gaussian	2.39	3.04	3.52	4.86	-

* To obtain the dimension estimates (i.e., for the original time series) discussed in the text, add 1.0 to the numbers in this table.

** Average of embedding dimensions 3-5.

4.2. Lyapunov Exponents

An alternative way of characterizing the behavior of dynamical systems is to compute their Lyapunov exponents [22]. Essentially, these are generalized eigenvalues, evaluated over the entire attractor. When the equations of motion are known, estimating the exponents is straightforward [17,23,24]. For experimental data, it is generally only possible to measure the largest exponent. Here [17], one estimates rates of orbital divergence from the reconstructed time series. Like the Grassberger-Procaccia algorithm for estimating the correlation dimension, current methods for estimating the largest Lyapunov exponent choke on nonuniformity. For example, when studying the Rössler attractor, it is easy to obtain reasonable estimates of the largest exponent (about 0.13 bits/sec) from $x(t)$, which varies

Table 3. Lyapunov Exponents for SEIR Equations[*]

b_1	Equations	Wolf[*]	Coyote[**]
		"MEASLES"	
.27	0.12	0.10 ± .01	.11 ± .003
.28	0.46	-.70 ± .10	.42 ± .01
.29	0.49	-.77 ± .09	.53 ± .02
		"CHICKENPOX"	
.28	-	0.43 ± .11	.19 ± .09

[*] Parameters for Wolf Method: Evolution time = 1 y; Embedding dimension = 3; Time lag = 0.25 y; Minimum length scale = [.01,.06]; Maximum length scale = [.05,.10].

[**] Parameters for Coyote·Method: Evolution time = 1; Embedding dimension = 2; time lag = 1; Length scales as in Wolf method. Data log transformed.

smoothly, and, virtually impossible to do so from z(t), which is more highly non-uniform [25].

Table 3 summarizes the results of trying to compute the positive Lyapunov exponent for the SEIR model for several parameter values. Here, we compare WOLF's [17] method with what we propose to call the Coyote (a modified wolf that lives in Arizona) procedure. Coyote estimates the exponent for each of the time-one maps and averages for each of the parameters required by the algorithm. To further diminish the effects of nonuniformity, the data were first log transformed and then rescaled on [0,1]. For reference, we also give values of the positive exponent computed directly from the equations.

The results of these computations are pretty unambiguous. For b_1 (the magnitude of seasonal forcing) = 0.27, the "measles" attractor is a two-piece chaotic orbit which is still relatively uniform. Both Wolf and Coyote do a good job. For higher values of b_1, the time series is spikier. Here, Wolf gives negative values, while Coyote yields estimates that are much closer to those obtained directly from the equations. We conclude that for non-uniform data Coyote is the way to go. (The corresponding approach for autonomous systems would be to average over multiple Poincaré sections.)

Table 4 summarizes the estimates so obtained for the data. For chickenpox, the computed exponents range from 0.1 to 0.3; for measles, from 0.3 to 0.7; for mumps, 0.2 to 0.6, and for rubella, 0.2 to 0.3. Significantly, we believe, the estimates for chickenpox and measles are very close to what one obtains from the SEIR equations as given in Table 3.

5. Conclusion

Implicit in the notion of chaos is the hope that processes and phenomena historically beyond the ken of "hard" science may become amenable to more exact description and understanding [26,27]. In fact, most of the well-documented examples of chaotic behavior have been in highly controlled situations [27]. In this regard, the childhood disease data are somewhat unique. Although much remains to be done, it does not seem unwarranted to suggest that what we have here is not only an

Table 4. Lyapunov Exponents for Diseases*

CITY	DISEASE			
	CHICKENPOX	MEASLES	MUMPS	RUBELLA
Aberdeen	-	.29 ± .05	-	-
Baltimore (early)	-	.40 ± .06	-	-
Baltimore (late)	-	.56 ± .04	-	-
Copenhagen	.32 ± .02	.60 ± .03	.55 ± .05	.25 ± .06
Milwaukee	.12 ± .02	.70 ± .03	.41 ± .05	.26 ± .05**
New York	.19 ± .03	.45 ± .02	.23 ± .03	-
St. Louis	.32 ± .03**	.44 ± .02	-	-***

* Parameters as follows: Evolution time = 1 y; Embedding dimension = 2; Minimum length scale = .035; Maximum length scale = [0.5, .10].

** Embedded in three dimensions. *** Insufficient data.

example of chaos in the "wild", but also a case in which we can account for the chaos in terms of models that are biologically sensible. As optimists, we look forward to determining just what modifications to the equations will bring their properties more into line with observations. As realists, we acknowledge the possibility that additional data may subvert the generalities that appear to be emerging.

Acknowledgements

This work was supported by grants from the National Institutes of Health (WMS) and the Danish Natural Science Research Council (LFO).

References

1. R.M. Anderson, B.T. Grenfell, R.M. May: J. Hyg. Camb. 93, 587 (1984)
2. W.M. Schaffer: IMA J. Math. Appl. Med. Biol. 2, 221 (1985)
3. W.M. Schaffer, M. Kot: J. theor. Biol. 112, 403 (1985)
4. L.F. Olsen: In Chaos in Biological Systems, ed. by H. Degn, A.V. Holden, L.F. Olsen, NATO ASI Ser., Vol. 138 (Plenum, New York 1987) p. 249;
   L.F. Olsen, W.M. Schaffer, G.L. Truty: Theor. Pop. Biol., in press
5. R.M. May, R.M. Anderson: Nature 280, 459 (1979)
6. R.M. Anderson: In Population Dynamics of Infectious Diseases. Theory and Applications, ed. by R.M. Anderson (Chapman and Hall, New York 1982) p. 1
7. K. Dietz: Lect. Notes Biomath. 11, 1 (1976)
8. P.E.M. Fine, J.A. Clarkson: Int. J. Epidem. 11, 5 (1982)
9. I.B. Schwartz, H.L. Smith: J. Math. Biol. 18, 233 (1983)
10. H. Smith: J. Math. Anal. Appl. 64, 467 (1978)
11. J.L. Aron, I.B. Schwartz: J. theor. Biol. 110, 665 (1984)
12. M. Kot, W.M. Schaffer, G.L. Truty, D.J. Graser, L.F. Olsen: Ecol. Mod. (1988), in press
13. F. Takens: In Dynamical Systems and Turbulence, Warwick, ed. by D.A. Rand, L.S. Young (Springer, New York 1981), p. 366
14. W.M. Schaffer: In Chaos in Biological Systems, ed. by H. Degn, A.V. Holden, L.F. Olsen, NATO ASI Ser., Vol. 138 (Plenum, New York 1987) p. 233

15. O.E. Rössler: Phys. Lett. <u>57A</u>, 397 (1976)
16. P. Grassberger, I. Procaccia: Physica <u>9D</u>, 189 (1983)
17. A. Wolf, J.B. Swift, H.L. Swinney, J.A. Vastano: Physica <u>16D</u>, 285 (1985)
18. A.M. Albano, A.I. Mees, G.C. de Guzman, P.E. Rapp: In <u>Chaos in Biological Systems</u>, ed. by H. Degn, A.V. Holden, L.F. Olsen, NATO ASI Ser., Vol. 138 (Plenum, New York 1987) p. 207
19. D.S. Broomhead, G.P. King: Physica <u>20D</u>, 217 (1986)
20. A.I. Mees, P.E. Rapp, L.S. Jennings: preprint
21. W.M. Schaffer, S.E. Ellner, M. Kot: J. Math. Biol. <u>24</u>, 479 (1986)
22. J.-P. Eckmann, D. Ruelle: Rev. Mod. Phys. <u>57</u>, 617 (1985)
23. I. Shimada, T. Nagashima: Prog. Theor. Phys. <u>61</u>, 1606 (1979)
24. G. Bennetin, L. Galgani, A. Giorgilli, J.-M. Strelcyn: Meccanica <u>15</u>, 9 (1980)
25. W.M. Schaffer, G.L. Truty: <u>Dynamical Software: II. User's Manual and Introduction to Chaotic Systems</u> (Dynamical Systems, Inc., Tucson 1987)
26. C. Nicolis, G. Nicolis: Nature <u>311</u>, 529 (1984)
27. P. Grassberger: Nature <u>323</u>, 609 (1986)

Analysis of Life Processes and
Economic Systems by Methods of Thermodynamics

H.G. Busse and B. Havsteen

Biochemisches Institut im Fachbereich Medizin, Universität Kiel,
Olshausenstraße 40, D-2300 Kiel, Fed. Rep. of Germany

1. Introduction

Systems are ubiquitous in our life. We encounter them repeatedly in the realm of technology (computers, television, cars, engines, etc.) and biology (interaction between living cells, animals, plants, societies, etc.). All these systems possess the common traits that they have a structure and participate in dynamic processes. A typical example of such a system is a radio. Here, the circuit diagram, in an abstract form, illustrates the structure, whereas the voltages and the currents display the associated dynamics. A more complete description of such properties of systems is given in the discipline: systems theory.

Recently, a group of scientists have investigated the relationships between industry, natural environment, population growth, and other factors in a world model [1]. The studies, which were aided by computers, were supported by diagrams of the structure of the model [2]. Similar mathematical models can be applied to industrial plants. The property common to all such systems is, that human beings are somehow involved in the processes in these systems. Hence, by virtue of the decisions of the human species, the system contains nondeterministic variables. Moreover, the models so far employed suffer from the disadvantage of failing to yield a detailed description of all relevant properties, since the number of variables and mutual relationships exceeds the limit which the human brain can process. Nevertheless, such models may give a description of overall (global) quantities. The situation is the same in the field of thermodynamics. In the latter case it is, e.g., impossible to describe the detailed motion of individual molecules in a cubic centimeter of air because their number is too large. Nevertheless, laws could be found which describe important properties of the system. These laws do not pertain to the individual sub-systems (the molecules) but rather to properties of the system as an ensemble.

The laws of thermodynamics belong to this category. They define relationships between system parameters, such as energy, entropy, and temperature.

The purpose of this paper is to generalize the thermodynamic concept and to apply it to some systems which so far have not been described in that fashion. This approach will be exemplified by an analysis of an economic system, the keeping of a primitive store and by the significance of economic forces in spatial pattern formation.

2. Electrical Circuit Diagrams as Thermodynamic Networks

Firstly, we shall consider the thermodynamic concept in the analysis of electrical and thermodynamic network diagrams, since the theory in this area is highly developed. Then we shall extend the concept to economic systems.

The graphical documentation of the structure of electronic systems in circuit diagrams is well known. This graphical method of representation very much facilitates the analysis and construction of complex systems. Moreover, the diagrams are a comprehensive documentation of the essential parts of the systems. Network theory is an advanced theory that deals with circuits such as the ones encountered in electronic systems. In particular, two universal laws are related to the structure of the systems:

1. Kirchhoff's current law. This is a conservation law for local charges which states that the sum of the arriving currents at an intersection of conductors is equal to the sum of the departing ones.

2. Kirchhoff's voltage law. This is a law of uniqueness which defines a single-valued function on the circuit, the voltage. It states that the sum of all voltages in a closed loop is zero.

These laws are illustrated in Fig. 1a for a system consisting of components having only two terminals (ports). The dynamics may become mathematically intractable if the elements of the network are nonlinear and have more than two ports, e.g. transistors. However, conservation and uniqueness are still satisfied and methods have been developed which also permit the incorporation of such complex elements into network theory.

This useful tool of electronics has been extended to the realm of thermodynamic networks of irreversible processes [3], to which electrical circuits belong and also coupled chemical reactions, mechanical engines and physical transport processes. In the latter theory, the electrical potential is generalized to the chemical potential and the electrical current to the thermodynamic flow, e.g. a reaction velocity. To demonstrate the analogy, the thermodynamic network of a chemical reaction system is shown in Fig. 1b. The details of the procedure for construction of such networks are given in the literature [4].

The new feature inherent to thermodynamic systems is that they are governed by two additional laws:

1. Energy conservation: Energy can be transformed into any other type of energy but does not disappear.

2. The system possesses an additional potential function, the entropy, which in a closed, i.e. limited system, cannot decrease with increasing time.

Usually, in thermodynamics certain potential functions, such as energy and entropy, are transformed into other potential functions, e.g. the Gibbs' potential and the enthalpy, which more adequately describe a process. The usual thermodynamics does not prescribe space and time dimensions. The thermodynamics of irreversible processes was developed by introducing these coordinates. The func-

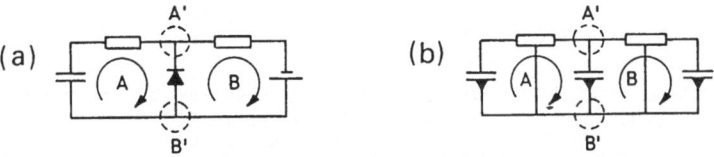

Fig. 1: (a) A and B are meshes in the electrical circuit diagram consisting of capacitors, resistors and a diode. The sum of the potential differences across the elements of a mesh is zero. The same applies to the currents in the nodes A' and B'. Both rules (Kirchhoff's laws) are closely related to the structure of the diagram. (b) Thermodynamic circuit of coupled chemical reactions represented in a similar way as in (a), the electrical circuit. In this more general case, Kirchhoff's laws also apply, e.g. to the meshes A and B and to the nodes A' and B'. The equivalent potentials are the chemical potentials and the currents are the rates of reaction

tions energy and entropy become spatial density distributions of the system and a new function, the entropy production appears, which contains the irreversible processes. Hence, the systems treated by this theory possess irreversible processes and are therefore in general non-Hamiltonian. In thermodynamics of irreversible processes the law of an ever increasing entropy (see law 2 above) is expressed by the entropy production as the sum of the products of the thermodynamic driving forces and their respective flows:

$$\tau = \sum_i (I_i \cdot U_i), \tag{1}$$

where τ is the rate of entropy production, U_i the thermodynamic driving force and I_i the flow. The meaning of τ one may verify in an electric circuit, where τ is related to the rate of heat production by the resistive elements. Each of these generate a heat equal to the product of the voltage across it and the current which runs through it. Since I_i and U_i are defined for an individual branch (an element) of the network, the structure of the latter expresses itself in the terms of the entropy production. Capacitive and inductive elements do not appear in τ since they do not dissipate energy (produce heat), but only store it. The entropy law of thermodynamics requires that $\tau \geq 0$. This ineqality reduces the freedom of dynamic processes in the network since it defines the direction in which the sytem develops. Thus, the value of $I \cdot U$ for a resistor (dissipative element) must be positive, i.e. the current must flow from a higher to a lower potential, and it is implied that addition of heat to the resistor cannot reverse the direction of the current which flows through it.

3. Networks of Economic Processes

Next, we shall construct the network of an economic process, the keeping of a store, and discuss the network in the sense of thermodynamic networks. A simple inventory model [5] demonstrates the main features of this process. It may serve as a very simple example. In short, the model may be described as follows.

A manufacturer must supply a customer with a product at a constant rate for a total time interval of T. The expenses of the manufacture comprise:

1. The cost of storage of the product (variable expenses)

$$E_S = S \int n(t) \cdot dt,$$

where S is the rate of cost of storage of a unit of product, $n(t)$ is the instantaneous (i.e. at time t) number of units in stock.

2. The cost to initiate a production series (fixed expenses per cycle)

$$E_P = P \cdot m,$$

where P is the investment required to start a production series and m is the number of series initiated during the total time T of the contract.

Then the total expenses are

$$E = E_S + E_P = S \int n dt + P \cdot m \qquad .$$

If a total of N units of product are ordered during the time interval T, then the number of units M produced within a series is $M = N/m$. Since on the average the store contains M/2 units, the mean total expenses are

$$E = S \frac{M}{2} T + P \frac{N}{M} \qquad . \tag{2}$$

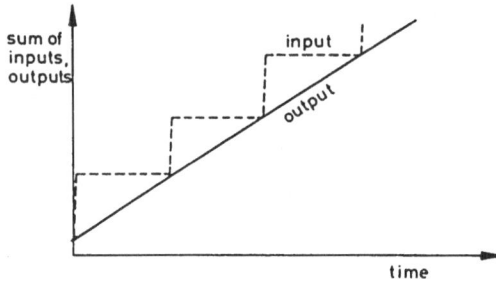

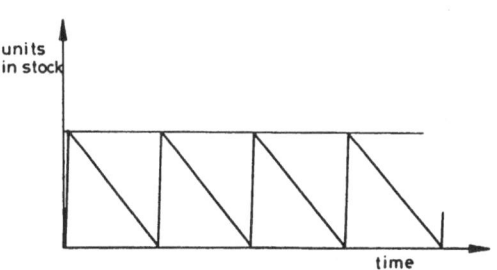

Fig. 2: Time course of the units of products in stock of a simple model, describing the keeping of a store (see text)

Equation (2) may be recast to display a form similar to the one of (1) by computing the rate of expenses:

$$e = \frac{E}{T} = \left(\frac{S}{2} \cdot \tau\right)\left(\frac{M}{\tau}\right) + \left(\frac{P}{M}\right)\left(\frac{N}{T}\right), \tag{3}$$

where τ is the total time needed to fill the store with the units of one production series. Equation (3) demonstrates that the rate is a sum of arithmetric products (see (1)). The transport of a unit of product to or from the store is denoted as the flow I_{in}, I_{out}, respectively:

$$I_{out} = \frac{N}{T} = \text{the number of units shipped to the customer per time unit}$$

$$I_{in} = \frac{M}{\tau} = \text{the input to the store of units from the production cycle per time unit.}$$

Hence, (3) acquires a form similar to that of (1) (see Fig. 2):

$$e = \left(\frac{S\,\tau}{2}\right) I_{in} + \left(\frac{P}{M}\right) I_{out} \tag{3'}$$

in which terms may be interpreted in thermodynamic notation. Here, the potentials or driving forces are the costs per unit of product (usually in terms of their price) and the flows are the transactions per time unit.

A graph of the network of this process can be drawn on the basis of (3') in the same way as it was done for the thermodynamic systems. Each term of (3') corresponds to a dissipative element (see Fig. 3). The non-dissipative processes do not appear in (3') and, hence, can only be deduced from the time course of the system. The element, which has to be added to complete the circuit, must represent the storing capability of the inventory. Thus, in this model, it may be simulated by a quantity-storing element, a capacitor; because of its similarity with an electronic capacitor, an analogous symbol is used (Fig. 3).

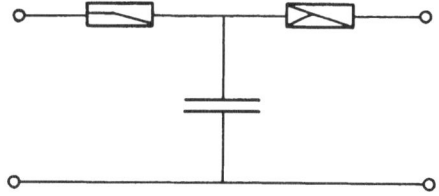

Fig. 3: A circuit diagram of the model in Fig. 2, the structure of which is derived from the function of the total cost, that has similar properties as the entropy of thermodynamics. The dissipative elements are represented as diodes, since the flow of transactions through them is unidirectional. The store itself acts as a capacitor, which may be charged or discharged by products

A simple model of an economic system has here been presented to illustrate the principles needed to analyze such systems. More complex models can be treated similarly, if the cost-producing processes are additive and if the terms can be partitioned into force and flow factors. A partition could easily be obtained for the more complicated inventory systems. Then, a network representation could be derived by well-established methods [3]. Graphs of this kind can be rather large and complex but may serve as a useful basis for the discussion of dynamic events produced in the structure.

4. The Significance of the Analogy

Operations research has proven to be a useful tool in the organization of business enterprises and military operations. It is also an important aid in the analysis of the dynamics of these systems. The development of graphical methods for the construction and investigation of models would greatly expand the scope of and reduce the process time in operations research. Such graphs can be obtained by adaptation of the procedures known for electrical and thermodynamic circuits to operations research, since the analogy between the fields is so extensive that several laws and methods should be generally applicable. This may be exemplified with Tellegen's theory which can be applied to any network in thermodynamics as well as in electronics and undoubtedly also in operations research.

The limits to growth have been vividly discussed in wide circles. A similar problem in the field of thermodynamics attracted some years ago much attention. That was the concept of an ever growing entropy function. Its significance is still in the focus of interest. The growth properties of quantities in operations research pose virtually the same problem which appears to be more general. Thus, the relation between the natural sciences and operations research is in many respects evident.

Although here only a model of a simple, economic process has been presented to illustrate the principles, a similar technique may be applicable to large networks, e.g. representing a cooperation or a society. Much effort has already been given to emphasize the "thermodynamic" features of a society, and the gross national product (GNP) has been suggested as a useful potential function [6]. Moreover, attempts have been made to introduce the entropy law of thermodynamics into the analysis of structures and dynamics of societies.

5. Economic Implications of Spatial Organization

The significance of the thermodynamic treatment of economic processes is even more impressive, if some phenomena, recently studied, are compared. In the theory of the thermodynamics of irreversible processes, the following phenomenon is ob-

352

served: If the physical processes in a system depart far enough from the thermo-dynamic equilibrium [7] to exceed a certain threshold, the system may suddenly change properties. An example is the heat exchanged across a thin liquid layer, a process which usually occurs by heat conduction. If, however, the temperature gradient across the layer exceeds a critical value, the situation changes. The heat in the liquid layer is then exchanged by convection and the previously homo-geneous layer breaks up into small (mostly) hexagonal cells in which the liquid circulates along closed paths. Thereby, heat is absorbed on the hot side of the layer and transported to the cold interface (see Fig. 4). The process is known as the Bénard phenomenon [8]. Similar phenomena are observed in initially homoge-neous solutions containing chemical reactions which proceed far from their equi-libria. Here, hexagonal cells of reaction are also formed [9]. Similarly, algae cells which usually live separately may under special conditions form such cells [10]. Since thermodynamics relates these phenomena to nonlinearities in the laws govering the processes of the system, such events may be expected also to occur in economic systems. This is a consequence of their ability to expand in space and of the nonlinearity of the interactions between components.

The common features may be illustrated with a simple economic model. Let us assume that a product is manufactured in the quantities n^+ and ordered in the quantities n^-. Let us further assume that it always is possible to balance these two quantities of the system. The two balance equations are in differential form:

$$\frac{\partial}{\partial t} (n^+) + \mathrm{div} (J^+) = q^+$$

$$\frac{\partial}{\partial t} (n^-) + \mathrm{div} (J^-) = q^-$$

with t = time, x,y = space coordinates and

$$\mathrm{div} (J) = \frac{\partial}{\partial x} (J) + \frac{\partial}{\partial y} (J).$$

q^+ and q^- are the production rates of n^+ and n^-; J^+, J^- are the transport rates of n^+ and n^-, respectively.

The form of the equations is the same as that of other conservation laws [11]. The equations express that the quantities n^+ and n^- are conserved if interacting forces are absent (q^+, q^- = 0). The thermodynamic theory requires the existence of constitutive relations which state the relation between the thermodynamic forces and the quantities of the components. The analogy between the price in economics and the chemical potential in physical systems is one of the essential results which follows from the inventory model considered. The constitutive equation in the economic model is the relation between the price and the quantity n of the product, i.e. the curves illustrating the dependence of the price upon the quantity n^+ offered and quantity n^- demanded, $p^+ = p^+(n^+)$ and $p^- = p^-(n^-)$, respectively. (The analogous ideal chemical potential is defined as $\mu = \mu_0 + RT \ln(n)$, where μ_0, R, T are parameters and n the quantity of a chemical entity.) If the quantities n^+, n^- are received or shipped, the cost of transportation must be added to the price. Therefore, the price may also be a function of the space coordinate x, i.e. $p^+ = p^+(n^+,x)$ and $p^- = p^-(n^-,x)$.

Next, the relation which describes the flows of the product in space must be formulated. The laws of thermodynamics require that in the absence of external forces a gradient in the chemical potential causes the components to flow. There-fore, a law similar to the diffusion law of thermodynamics should apply to the economic case:

$$J^\pm = a^\pm \cdot \frac{\partial p^\pm}{\partial x} \quad (a^\pm = \text{coefficients}).$$

353

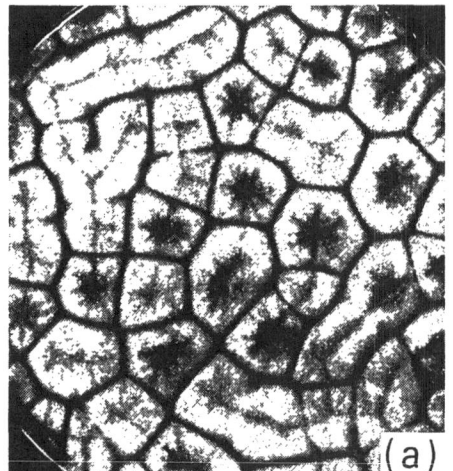

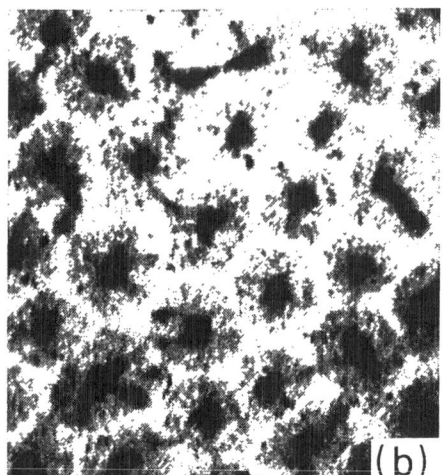

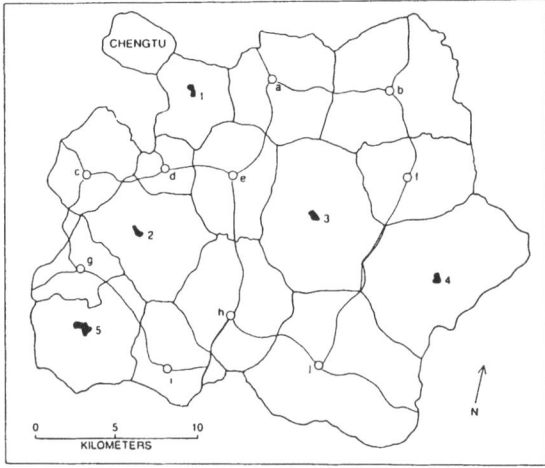

Fig. 4: Polygonal (mostly hexagonal) cell pattern formation in different systems:
(a) In a thin liquid layer (Bénard phenomenon) which is heated uniformly from beneath. Within each cell circular convection transports heat from the hot lower to the cold upper interface (details in [8], picture from S.C. Müller).
(b) In a 3 mm thin layer of a culture of the algae *Euglena gracilis*, the pattern is formed from a uniform culture if the cell density and the oxygen-carbon dioxide ratio are within defined limits (see [10]).
(c) Arrangement of market towns southeast of Chengtu in China in 1949. Each town has six neighbors which may be rearranged by abstraction to a hexagonal lattice (see [13]). Towns labelled with letters are smaller than the numbered ones. Lines connecting the towns are major roads of trade. The line surrounding a town limits its estimated market area.
Common to all these patterns is that they are not formed a priori, but arise as soon as definitive threshold conditions are reached

Since the price a priori is not fixed to any particular point in space, its spatial relation resides in the spatial distribution of the quantity n, that is $p^{\pm} = p^{\pm}(n^{\pm}(x))$.

The equation above may then be recast as

$$J^{\pm} = D^{\pm} \frac{\partial n^{\pm}}{\partial x} \quad (\text{with } D^{\pm} = a^{\pm} \frac{\partial p^{\pm}}{\partial n^{\pm}}).$$

If the ideal case, in which D^{+} and D^{-} are constants within the range of interest, may be assumed, the balance equations acquire the form

$$\frac{\partial n^{+}}{\partial t} + D^{+} \frac{\partial^2 n^{+}}{\partial x^2} = q^{+} (n^{+}, n^{-}, p^{+}, p^{-})$$

$$\frac{\partial n^{-}}{\partial t} + D^{-} \frac{\partial^2 n^{-}}{\partial x^2} = q^{-} (n^{-}, n^{+}, p^{-}, p^{+}) \quad .$$

In this form, the set of equations is very similar to the set of equations which describe the processes in chemically reacting systems [12] capable of displaying spatial structures (as e.g. Fig. 4). Hence, similar spatial structures should also be found in regional economics, under conditions such as those found in analogous chemical systems. Spatial distributions induced by the exchange of products were actually observed in the early stages of pattern formation in rural markets in China, Guatemala and Germany [13]. The evolution of towns seems to be closely related to the commercial interaction laws. Also in this case, the theory predicts the development of hexagonal patterns. The structures which were observed were in fact not far from these predictions.

The assumption that such pattern may arise from the set of differential equations derived implies that the production rates q^{+} and q^{-} contain nonlinear terms. Most production functions will be nonlinear, since the exchange of products and the production itself are determined by the price. The price, however, is nearly always a nonlinear function of the quantity of available product. The specific functions q^{+} and q^{-} are strongly dependent upon the nature of the model investigated and will not be discussed here. The analogy to chemical processes is not very helpful because chemical production rates are based on the statistical encounter of the parts of a molecule to form the product molecule. Since in economics goods are not manufactured by chance, only in rare cases can an analogue to a chemical model be established.

Both the relation between the price and the quantity n and the law describing the flow of products under the pressure of a price gradient are constitutive relations in the sense of the theory of systems. There is, however, one distinctive difference between these relations. The former is reversible and the latter is irreversible. The irreversibility appears here in the transportation of the product which requires that a service is rendered. This expense cannot be regained by retransporting the product to its point of origin. In thermodynamics, the irreversibility of processes is expressed by the entropy production which must be positive or zero. Since this entropy production function was the starting point for the construction of the economic analogy to the thermodynamic theory, the discovery that irreversible processes also are found in economic networks could have been foreseen. The same is true of the observation in economics, where spatial pattern may appear under the pressure of a steady cost rate, since in thermodynamics structured pattern may arise under the pressure of a steady entropy production. Finally, a third similarity should be mentioned. In thermodynamics, the spatial structures are usually observed in close connection with oscillatory phenomena which often are involved in the process of pattern formation. The market patterns, which were referred to above, are similarly related to periodic processes, in this case to the time sequence of the rural markets. In the thermodynamics of irreversible processes, closed dissipative systems progress toward equilibria. The existence of an equilibrium is expressed by the Onsager relations. Since an oscillating system does not proceed toward its equilibrium, it may only remain for a finite time on the oscillatory path (the limit cycle). The system has to be open to oscillate for an infinite time. The same applies to pattern-producing systems. Since the economic systems here are treated similarly, the concept is probably also applicable to them.

6. Structures Generated by Bonds

One of the remarkable characteristics of the theory of thermodynamic systems appears in chemical reaction systems. It is the formation of bonds between atoms to give molecules. In abstract terms the individuals of a system may form clusters which stick together by bonds. In economics the contracts between participants may be considered as the bond-forming force which binds together the participants in that contract in a specific manner. In this view, a number of individuals bound together by contracts form an institution [14,15] (the probable analogy to a chemical molecule). Transactions which break and reconstitute the bonds may be used to construct dynamics-like reactions in chemical systems. The interesting features are the costs associated with transitions and the capital (or energy) stored in a bond.

This picture of the processes in an economy probably could be applied to derive models on which the analogy to thermodynamic systems could be treated in more detail. In particular, the relation of the uncertainty which is minimized in contract and the entropy which is maximized in thermodynamic systems might be studied by such models. However, this approach would be more adequately treated by the statistical theory of thermodynamics and not by the phenomenological theory discussed in this paper.

7. Summary

Network diagrams have proven to be useful in the representation and analysis of electronic circuits. Recently, their application has been extended to the thermodynamic treatment of a wide variety of systems. It appears that this technique also may be used on more complex and highly organized systems, e.g. networks in the field of economics. This suggestion is here exemplified by the analysis of a simple model system, the keeping of a store. The analogy between the cost function in economics and the thermodynamic entropy concept is used to construct the network diagram of the inventory model and to explain pattern formation in rural networks.

Acknowledgement

We greatly acknowledge the very stimulating discussions with Dr. K. W. Schatz and Dr. S. Danø.

References

1. D. Meadows: The Limits of Growth (Earth Island Ltd., London 1972)
2. J.W. Forester: World Dynamics (Wright Allen Press, Cambridge 1971)
3. G.F. Oster, A. Perelson, A. Katchalsky: Nature 234, 393 (1971)
4. B. Hess, E.M. Chance: FEBS Lett. 25, 119 (1972)
5. C.W. Churchman, R.L. Ackoff, E.L. Arnoff: Introduction to Operations Research (John Wiley, New York 1959)

6. R.E. Overbury: Nature <u>242</u>, 561 (1973)
7. G. Nicolis, J. Portnow: Chem. Rev. <u>73</u>, 365 (1973)
8. S. Chandrasekhar: <u>Hydrodynamic and Hydromagnetic Stability</u>
   (Clarendon Press, Oxford 1961)
9. A.M. Zhabotinskii, A.N. Zaikin: J. theor. Biol. <u>40</u>, 45 (1973)
10. K. Brinkmann: Z. Pflanzenphysiol. <u>59</u>, 364 (1968)
11. J. Meixner, H.G. Reik: In <u>Handbuch der Physik</u>, Vol. III/2, ed. by S. Flügge
    (Springer, Berlin 1959)
12. G.M. Frank: <u>Oscillatory Processes in Biological and Chemical Systems</u>, Vols.
    1,2 (Nauka, Moscow 1967)
13. S. Plattner: Sci. Amer. <u>232</u>, May, 66 (1975)
14. D.C. North: Zeitschrift gesamte Staatswissenschaft <u>140</u>, 7 (1984)
15. T.W. Hutchinson: Zeitschrift gesamte Staatswissenschaft <u>140</u>, 20 (1984)

Index of Contributors